An Introduction
to Probability
and Its Applications

Prentice-Hall Series in Statistics

Richard A. Johnson and Dean W. Wichern, Series Editors
University of Wisconsin

Johnson and Wichern, *Applied Multivariate Statistical Analysis*

Larsen and Marx, *An Introduction to Mathematical Statistics and Its Applications*

Larsen and Marx, *An Introduction to Probability and Its Applications*

Morrison, *Applied Linear Statistical Methods*

Remington and Schork, *Statistics with Applications to the Biological and Health Sciences, Second Edition*

An Introduction to Probability and Its Applications

Richard J. Larsen
Vanderbilt University

Morris L. Marx
University of Oklahoma

Prentice-Hall, Inc., Englewood Cliffs, New Jersey 07632

Library of Congress Cataloging in Publication Data

LARSEN, RICHARD J.
    An introduction to probability and its
applications.
    Bibliography: p.
    Includes index.
    1. Probabilities.   I. Marx, Morris L.   II. Title.
QA273.L3519 1985   519.2   84-17860
ISBN 0-13-493453-9

Editorial/production supervision
    and interior design: *Kathleen M. Lafferty*
Cover design: *Edsal Enterprises*
Manufacturing buyer: *John B. Hall*

Printed in the United States of America

10   9   8   7   6   5   4   3   2   1

0-13-493453-9   01

Prentice-Hall International, Inc., *London*
Prentice-Hall of Australia Pty. Limited, *Sydney*
Editora Prentice-Hall do Brasil, Ltda., *Rio de Janeiro*
Prentice-Hall Canada Inc., *Toronto*
Prentice-Hall Hispanoamericana, S.A., *Mexico*
Prentice-Hall of India Private Limited, *New Delhi*
Prentice-Hall of Japan, Inc., *Tokyo*
Prentice-Hall of Southeast Asia Pte. Ltd., *Singapore*
Whitehall Books Limited, *Wellington, New Zealand*

To our wives

Contents

3 Combinatorics 96

4 Random Variables 169

5 Expected Values 228

⑥ Special Distributions   *305*

⑦ Limit Theorems   *347*

Bibliography   *382*

Appendix   *387*

Answers to Selected Questions   *391*

Index   *402*

Preface

An Introduction to Probability and Its Applications presents an introductory mathematical treatment of probability. The development is rigorous but we rely heavily on examples and case studies to motivate and illuminate the mathematics. Historical information has been included to give students a feeling for the evolution of the subject.

We have hoped to write a book adaptable to the needs of a diverse audience. The first six chapters cover all the material traditionally found in a one-semester or two-quarter undergraduate course. The seventh chapter, on limit theorems, is considerably more difficult and requires a knowledge of real analysis or, at the very least, advanced calculus. All seven chapters constitute a beginning graduate-level course in non-measure-theoretic probability.

Motivation has been given high priority. We have tried to make the book readable for the student and "teachable" for the instructor. Extensive examples and case studies based on real data are the vehicles we use to demonstrate the relevance of probability. It was our deliberate choice not to narrow the focus of the book to any particular area of science, economics, or mathematics. We feel that probability is so strongly interdisciplinary that its impact on students is greatest when they are exposed to the full breadth of its applications. Mixed in, then, with the usual gamut of examples dealing with genetics, gambling, reliability, medicine, and business, students will find new applications to such disparate areas as cryptology, criminology, history, sports, and current events. We have also paid attention to the lighter side of the subject by including some not-so-serious examples. Rigor and relevance feed the mind but they must not be allowed to starve the soul. By indulging in a bit of whimsy here and there, we have tried to let some of the *fun* of probability shine through.

Our emphasis on applications notwithstanding, the topics we cover and the order in which they appear are quite traditional. Chapter 2 states Kolmogorov's axioms and develops the fundamental properties associated with probability functions. Combinatorics is taken up in Chapter 3. In deference to the renewed interest in discrete mathematics being sparked by the computer revolution, we discuss combinatorics at considerably greater length than is typical of other probability texts. Chapter 4 is devoted to random variables and contains all the standard univariate and bivariate results. Special cases of random variables—the normal, the Poisson, and so on—are profiled in Chapter 6. Expected values are treated in Chapter 5.

There is more than enough material in the first six chapters for an undergraduate course in probability. The excess allows for flexibility. Instructors have the option of orienting their courses in a variety of directions, depending on which chapters they elect to pursue in depth. These first six chapters can also be used as the first half of a two-semester course in mathematical statistics. A student, for example, can go from this text to Chapter 5 of our *An Introduction to Mathematical Statistics and Its Applications* (Prentice-Hall, 1981).

The addition of Chapter 7 makes the book long enough and sufficiently challenging to be appropriate for beginning graduate students. Limit theorems, by their very nature, are a sophisticated topic mathematically. To do them justice requires a familiarity with certain ideas from both real and complex analysis. In Chapter 7 we trace the historical and mathematical evolution of limit theorems, starting with the first law of large numbers published by James Bernoulli in 1713 and ending with a twentieth-century result, the Lindeberg-Lévy version of the central limit theorem.

This has been an enjoyable book to write. Probability is a fascinating topic; even if it had absolutely no application to the real world, its elegance and charm would be irresistible. We hope these pages capture a part of that charm and can share with the reader some of the enthusiasm and affection we feel for the subject.

Thanks are due to many people for their contributions to the development and completion of this project. We are indebted to the many researchers who graciously allowed us to use portions of their data for examples and case studies. And for their valuable criticisms and suggestions, we thank the reviewers: Paul N. DeLand, California State University at Fullerton; M. Lawrence Glasser, Clarkson College; Paul T. Holmes, Clemson University; John D. Neff, Georgia Institute of Technology; Elizabeth Papousek, Fisk University; and Olaf P. Stackelburg, Kent State University.

We would also like to express our sincere appreciation for the editorial support and encouragement given by Bob Sickles, who guided this endeavor from its inception, and to Kathleen Lafferty, who edited the final manuscript.

RJL
Nashville, Tennessee

MLM
Norman, Oklahoma

Introduction

We sail within a vast sphere, ever drifting
in uncertainty, driven from end to end.
Pascal

1.1 A BRIEF HISTORY

Tomb inscriptions and other archaeological evidence reveal that as early as 3500 B.C.
Egyptians were using astragali, the four-sided bones found in heels of dogs, sheep,
and horses, as primitive dice to determine moves in board games. Anyone playing
with dice, of course, quickly develops a certain amount of intuition about chance and
the behavior of random events. It follows, then, that *some* knowledge of probability—
however rudimentary—was every bit as old as human socialization itself. Still, any
kind of formal analysis of chance events, which is what this book is all about, is a
much more recent development.

During the late Middle Ages and early Renaissance, an awakening of interest in
the mathematics of gambling occurred. Among the handful of early scholars strug-
gling to formalize the notion of probability with an eye to applying it to games of
chance was the celebrated Galileo. But progress was slow; as a discipline, probability
proved difficult to get started.

A breakthrough finally came in 1654. In that year a correspondence between the
renowned French mathematicians, Blaise Pascal and Pierre de Fermat, began. A
countryman of the two, Antoine Gombaud, Chevalier de Meré, had brought two old
problems to Pascal's attention, one concerning odds in dice games, the other relating

to the equitable division of stakes in contests interrupted before completion. Pascal was intrigued by the problems and communicated his thoughts to Fermat, thus prompting an exchange of letters that many consider the genesis of mathematical probability.

A year later a young Dutchman, Christiaan Huygens (best known for his work in optics), visited Paris and was told of the work of Pascal and Fermat. Not able to meet with Pascal, who had withdrawn to religious and philosophical contemplation, Huygens elected to pursue the subject on his own. Two years later, in 1657, his efforts produced the first treatise on probability, *De Ratiociniis in Aleae Ludo* (*Calculations in Games of Chance*), published as an addendum to a mathematical work by his fellow Dutchman, Van Schooten. As a catalyst, Huygens' work had tremendous impact. Many scholars not previously interested in probability found Huygens' ideas a real challenge. They began to generalize his problems and pose new ones.

The next major event in the early history of probability was the posthumous publication of Jacob Bernoulli's *Ars Conjectandi* (*The Art of Conjecture*) in 1713. Bernoulli was a member of the famous Swiss family that produced eight distinguished scientists; his own stature in the mathematical community was considerable. *Ars* included Huygens' *De Ratiociniis* as its opening chapter. It also contained the first published theorem on the limiting behavior of probability in repetitions of a simple chance experiment. Variations on that idea are still topics of research today.

Any discussion of the evolution of probability would be incomplete without something being said about its sister discipline, *statistics*. Today, statistics has two meanings: It refers to collections of data and to a set of mathematical procedures (based on probability) for drawing inferences. Originally, it meant only the former. Although lists of data are as old as the art of writing, only when Europe shed its feudal character and opted for industrialization did demographic and other kinds of data become really important. The first "statistical" treatment of biological and social phenomena was due to John Graunt in 1662, when he published *Natural and Political Observations Made upon the Bills of Mortality,* a book based on birth and death records gathered in plague-strickened England. Shortly thereafter, another Englishman, Sir William Petty, developed what became known as Political Arithmetic, "the art of reasoning by figures upon things relating to government."

Actuarial problems afforded scholars an early opportunity to combine probabilistic reasoning with statistical data. In 1671 de Witt ushered in the modern theory of pensions with the publication of *Waerdye van Lyf-Renten.* By 1700 the work of Edmund Halley on mortality tables and life expectancy was beginning to convince the business community that many of these ideas were eminently practical.

By the time *Ars Conjectandi* was published, the disciplines of probability and statistics were both well recognized but still, for the most part, separate. Over the next century and a half, though, their interrelationship became abundantly clear. Carl Friedrich Gauss, a towering figure in the history of science and mathematics, invented a variety of statistical techniques for the purpose of better analyzing astronomy data. Most notably, he pioneered the now-familiar bell-shaped distribution, a mathematical model sometimes referred to as the *Gaussian probability law.*

Eighteenth-century scholarship in probability culminated in the work of Pierre-Simon Laplace. (Even if Laplace had not been a major figure in the history of mathematics, he would have been notable for having survived—indeed, prospered—

through the turbulent era that included the reign of Louis XVI, the French Revolution, Napoleon's Empire, and the restoration of the Bourbons.) In 1812, his influential *Theorie Analytique des Probabilities* summarized what was known about probability up to that time, unified basic research in the field, and discussed a number of new applications. By the end of the first quarter of the nineteenth century, probability was well established as a mathematical discipline having the potential to be of great benefit to the physical sciences.

About this same time, Adolphe Quetelet, a most remarkable Belgian academician, began to notice surprising statistical regularities in the human estate. As he so poetically commented,

> Thus we pass from one year to another with the sad perspective of seeing the same crimes reproduced in the same order and calling down the same punishments in the same proportions. Sad condition of humanity! The part of prisons, of irons and of the scaffold seem fixed for it as much as the revenue of the state. We might enumerate in advance how many individuals will stain their hands in the blood of their fellows, how many will be forgers, how many will be poisoners, almost as much as we can enumerate in advance the births and deaths that should occur.[1]

For a good part of the nineteenth century, Quetelet traveled widely throughout Europe, championing the use of probability models as a way of describing social and biological phenomena.

By the mid-1800s probability was recognized not only as a useful practical tool, but also as a source of challenging problems. Russian mathematicians, in particular, were working hard to find new and more-powerful limit theorems, generalizations of the work begun by Bernoulli and DeMoivre. Out of all this research came the growing realization that probability was related in a very fundamental way to ideas in mathematical analysis being developed by Borel, Lebesgue, Hausdorff, and several others at the turn of the century.

At the 1900 International Congress of Mathematicians held in Paris, David Hilbert presented a list of problems whose solutions he claimed would be important steps in the further development of mathematics. Not the least of these was his sixth problem, a call for the axiomization of probability. The Russians responded. In 1933, Andrei Kolmogorov published *Grundbegriffe der Wahrscheinlichkeitsrechnung* (*Foundations of the Theory of Probability*). Like Euclid's axioms for geometry, Kolmogorov's postulates were simple and intuitively plausible, yet sufficiently abstract that theorems could be proved independently of whatever meaning was given to the underlying concepts. But *Grundbegriffe* did more—Kolmogorov had phrased the major concepts of probability in the language of analysis, thus firmly embedding probability in the mathematical mainstream.

Today, probability theory, like its ancient source the astragalus, presents four faces. We see it as an intuitive concept, a formal mathematical discipline, the essential substructure of statistics, and an applied subject providing models for an extraordinary diversity of natural phenomena.

---

[1] From Frank H. Hankins, "Adolphe Quetelet as Statistician," in Studies in History, Economics, and Public Law, vol. xxxi, no. 4 (New York: Longman, Green and Co., 1908), p. 497.

Probability

2

Lest men suspect your tale untrue, keep probability in view.
John Gay

Coincidences, in general, are great stumbling-blocks
in the way of that class of thinkers who have
been educated to know nothing of the theory of probabilities.
Edgar Allan Poe

2.1 INTRODUCTION

In Chapter 1 we sketched a brief history of probability and discussed its early relationship with statistics. Our next objective is to develop a probability "calculus." To do that, we need to examine in detail the mathematics that evolved from those very first problems with which Pascal and Fermat struggled so gamely some 325 years ago.

Today, *our* probability problems are still much the same as theirs—at least those we will see in Chapter 2—but how they are set up and the context in which they are viewed have changed dramatically. Much of the current terminology, for example, comes straight from set theory, and the subject's logical foundations derive from a 20th-century axiom system. Perhaps even more significantly, the *scope* of probability has broadened enormously. What began as a set of rules for dealing with games of chance now finds meaningful applications in virtually every area of science.

Chapter 2 addresses all these points. It begins by defining notation (most of which should already be quite familiar) and resolving some fundamental questions: What does a probability represent? How is it defined? Does it always exist? What properties does it have? How is it manipulated mathematically? Chapter 2 concludes with a series of theorems and examples that show how these abstract mathematical principles, when properly extended, give rise to a subject of enormous practical significance.

This is an important chapter. Many of the topics covered later in the book are little more than generalizations, specializations, and reformulations of the material in Sections 2.2–2.9.

2.2 THE SAMPLE SPACE

We begin our development of the fundamentals of probability theory by defining four very simple, but important, notions—*experiment, sample outcome, sample space,* and *event.* The latter three, all carry-overs from classical set theory, give us a familiar mathematical structure within which to work; the former is new and provides the conceptual mechanism for casting real-world phenomena into probabilistic terms.

By an *experiment* we mean any procedure that (1) can be repeated, theoretically, an infinite number of times and (2) has a well-defined set of possible outcomes. Thus, rolling a pair of dice qualifies as an experiment; so do measuring a hypertensive's blood pressure and doing a spectrographic analysis to determine the carbon content of moon rocks. Each of the potential eventualities of an experiment is referred to as a *sample outcome, s,* and their totality is called the *sample space, S.* To signify the membership of s in S, we write $s \in S$. Any designated collection of sample outcomes, including individual outcomes, the entire sample space, and the null set, constitutes an *event.* The latter is said to *occur* if the outcome of the experiment is one of the constituent members of that event.

Example 2.2.1

Consider the experiment just mentioned of rolling two dice. Imagine one of the dice to be black; the other, white. Making up the sample space is the set of 36 possible outcomes shown in Figure 2.2.1. To simplify notation, we can represent each sample outcome by a vector (x, y), where x and y are the numbers showing on the black die and the white die, respectively (see Figure 2.2.2).

For a craps shooter, several events defined on the sample space of Figure 2.2.2 take on special significance. For example, the event E, "shooter rolls a 7," contains six different sample outcomes. Using standard set notation, we write

$$E = \{(x, y) : (x, y) = (1, 6), (2, 5), (3, 4), (4, 3), (5, 2), (6, 1)\}$$

If the nature of the outcomes in S is clear from their context, the initial generic designation—in this case, (x, y)—will be deleted. Thus, the event F, "shooter throws an easy 8," consists of four sample outcomes, $(5, 3)$, $(3, 5)$, $(6, 2)$, and $(2, 6)$, and we write, $F = \{(5, 3), (3, 5), (6, 2), (2, 6)\}$ (see Figure 2.2.2.)

Comment

Our first task in Section 2.3 will be to assign a *probability measure* to the various outcomes of an experiment. Here, if two two dice were fair—that is, perfectly balanced—we would expect each of the 36 outcomes to be equally likely, and the particular outcome, such as $(1, 3)$, should occur, *in the long run,* $\frac{1}{36}$ of the time. More formally, if G denotes the event "$(1, 3)$ is thrown," we say that

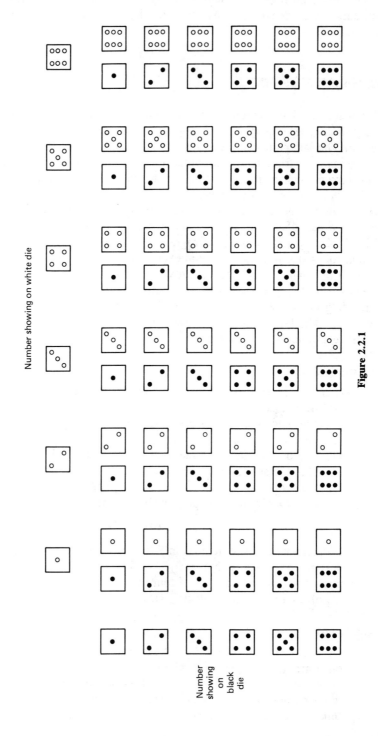

Number showing on white die

Number showing on black die

Figure 2.2.1

6

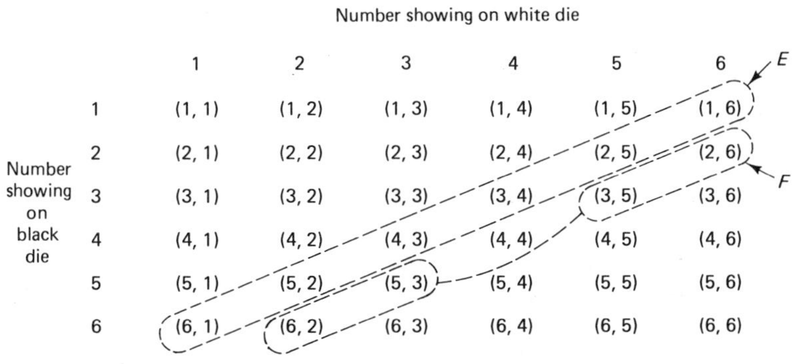

Number showing on white die

	1	2	3	4	5	6
1	(1, 1)	(1, 2)	(1, 3)	(1, 4)	(1, 5)	(1, 6)
2	(2, 1)	(2, 2)	(2, 3)	(2, 4)	(2, 5)	(2, 6)
3	(3, 1)	(3, 2)	(3, 3)	(3, 4)	(3, 5)	(3, 6)
4	(4, 1)	(4, 2)	(4, 3)	(4, 4)	(4, 5)	(4, 6)
5	(5, 1)	(5, 2)	(5, 3)	(5, 4)	(5, 5)	(5, 6)
6	(6, 1)	(6, 2)	(6, 3)	(6, 4)	(6, 5)	(6, 6)

Number showing on black die

Figure 2.2.2

the *probability of G is* $\frac{1}{36}$. (Intuitively, what probability should be associated with the event E, "shooter rolls a 7"?)

Example 2.2.2

Having introduced the first of what will be many dice examples, it seems only appropriate to give equal space to another favorite pasttime of the probabilist—the lottery. To keep the numbers workable, imagine a lottery consisting of just five tickets, two of which will be drawn and designated as the winners. (The tickets are to be drawn one at a time, without replacement.) If the two prizes to be awarded are the same, what is the sample space?

Let the tickets be numbered 1 through 5; then there are 10 possible pairs of numbers that can be drawn (see Figure 2.2.3). Note that the *order* of the tickets in each sample outcome is irrelevant, since the prizes are identical. That is, it makes no sense to distinguish between "first ticket drawn is a 3, second ticket drawn is a 5" and "first ticket drawn is a 5, second ticket drawn is a 3." (How many outcomes would the sample space contain if the rules of the lottery dictated that the first ticket drawn be given one prize and the second ticket drawn be given a lesser prize? What would S be if the prizes were different, but the drawing was done *with* replacement—that is, the first ticket was put back before the second was drawn?)

(1, 2)	(2, 4)
(1, 3)	(2, 5)
(1, 4)	(3, 4)
(1, 5)	(3, 5)
(2, 3)	(4, 5)

Figure 2.2.3 Lottery sample space.

Comment

Lotteries provide a convenient model for a variety of problems. Consider the experiment of dealing a five-card hand in the game of poker. We will see in Chapter 3 that the corresponding sample space—the set of all possible (unordered) five-card hands—numbers 2,598,960. Of those more than two and a half million hands, a total of 4, for example, make up the event, "player is dealt a royal flush" (10, jack, queen, king, and ace of the same suit).

Example 2.2.3

An ecologist has trapped a male bear, a female bear, and a cub. Interested in studying their daily foraging behavior, she intends to fit them with radio-transmitting collars so they can be kept under electronic surveillance after they are released. To allow each bear to be distinguished one from another, each collar transmits on a different frequency. Write out a sample space describing the ecologist's options in assigning the collars to the bears.

Here a typical outcome in S can be thought of as an ordered triple (x, y, z), where bear x is assigned the collar with the lowest frequency, bear y the middle frequency, and bear z the highest frequency. Let m, f, and c denote the male bear, the female bear, and the cub, respectively. Then S consists of *six* ordered triples:

$$(m, f, c) \qquad (m, c, f) \qquad (f, m, c)$$
$$(f, c, m) \qquad (c, m, f) \qquad (c, f, m)$$

(How many outcomes would S contain if *four* bears needed collars, all transmitting on different frequencies?)

In many situations, such as Examples 2.2.1–2.2.3, it is quite feasible—indeed, desirable—to define an experiment's sample space by simply enumerating it. But not always. There are times when making a complete listing of S is either unreasonable or impossible; then our only recourse is to characterize the sample space by defining, functionally, a typical s. The next two examples are cases in point.

Example 2.2.4

A probability-minded despot offers a convicted murderer a final chance to gain his release. The prisoner is given 20 chips, 10 white and 10 black. All 20 are to be placed into two urns according to any allocation scheme the prisoner wishes, with the one proviso being that each urn contain at least one chip. The executioner will then pick one of the two urns at random and, from that urn, pick one chip at random. If the chip selected is white, the prisoner will be set free; if it is black, he "buys the farm." Construct a sample space describing the prisoner's possible allocation options.

Note, first, that even though there are two urns to be filled, we need consider only one in characterizing our sample space because whichever chips are not in the first urn must necessarily be in the second. Accordingly, let (x, y) denote the allocation scheme whereby x white chips and y black chips are put into urn I. While all the admissible (x, y)'s *could* be enumerated in this case, such a list would be quite lengthy. Characterizing S by imposing a suitable set of constraints on x and y promises to be a much more tractable solution. Specifically, let

$$S = \{(x, y) : x = 0, 1, \ldots, 20; \ y = 0, 1, \ldots, 20; \ 1 \le x + y \le 19\}$$

(Intuitively, which allocation affords the prisoner the greatest chance of survival?)

Example 2.2.5

A computer programmer has written a subroutine for solving a general quadratic equation, $ax^2 + bx + c = 0$. What is the corresponding sample space for the coefficients a, b, and c, and how would the event E, "equation has two equal roots," be characterized? (Note: The "experiment" here would be a potential user of the subroutine selecting a particular set of values for a, b, and c.)

First, the sample space: Because presumably no combinations of finite a, b, and c are inadmissible, we characterize S by writing,

$$S = \{(a, b, c) : -\infty < a < \infty, \ -\infty < b < \infty, \ -\infty < c < \infty\}$$

Recall from algebra that a quadratic will have equal roots if and only if its discriminant, $b^2 - 4ac$, vanishes. Thus

$$E = \{a, b, c) : b^2 - 4ac = 0\}$$

(How would the event F, "roots are real," be described?)

Question 2.2.1 Engine blocks coming off an assembly line are numbered serially. During one particular work shift, the six blocks produced are numbered 17850 through 17855. An inspector selects two of the blocks at random to test for stress damage. Write out the sample space showing all possible pairs of blocks that might be inspected.

Question 2.2.2 An urn contains six chips numbered 1 through 6. Three are drawn out without replacement. What outcomes are in the event "second smallest chip is a 3"?

Question 2.2.3 Consider the quadratic equation,

$$x^2 + 2bx + c = 0$$

Let A be the event "equation has complex roots." Characterize event A in terms of a set of (b, c)-values.

Question 2.2.4 In the game of craps, the person rolling the dice (the *shooter*) wins outright if his or her first toss is a 7 or an 11. If his or her first toss is a 2, 3, or 12, he or she loses outright. If the first toss is something else—for instance, a 9—that number becomes the *point* and he or she keeps rolling the dice until he or she throws another 9, in which case he or she wins, or a 7, in which case he or she loses. Characterize the sample outcomes contained in the event "shooter wins with a point of 9." How many outcomes are in S?

Question 2.2.5 To determine an *odd person out*, $m \ (= 2k + 1)$ players each toss a coin. If one player's coin turns up differently than all the others, that person is declared the odd person out. Suppose $m = 3$ coins are tossed.
a. List the outcomes in the sample space.
b. Let E be the event that no one is declared an odd person out. Which outcomes are in E?
c. Can you see a pattern in your answer to part (a) that would suggest, without enumeration, the number of outcomes in S if m were equal to 7?

What we have investigated so far is the sample space appropriate for a *single* performance of an experiment. Thus, if we roll a pair of dice one time, the complete set of possible results consists of the 36 entries listed in Figure 2.2.1. Many experiments, though, are really the composite result of a series of subexperiments, where each subexperiment has its own sample space (which may or may not be the same from subexperiment to subexperiment). As a simple example, suppose we were to toss a pair of dice *twice*. Then the outcomes would be of the form (x, y, u, v), with x and y being the faces showing on the first pair of dice and u and v the faces showing on the second. Clearly, the number of such outcomes is 1296 ($36 \cdot 36$): For each of the 36 possible outcomes on the first throw, there are 36 possible outcomes on the second. In the language of set theory, what we are describing to be the sample space for the pair of throws is the *Cartesian cross product* of the sample spaces for the individual throws.

In general, if $S_1, S_2, \ldots, S_k$ are the sample spaces for the 1st, 2nd, $\ldots$, kth subexperiments, respectively, and if S is the sample space describing the *sequence* of k subexperiments, we write

$$S = S_1 \otimes S_2 \otimes \cdots \otimes S_k$$

where $S = \{(s_{i_1}, s_{i_2}, \ldots, s_{i_k})\}_{i_1, i_2, \ldots, i_k}$. Convince yourself that if each S_i contains n_i outcomes, the composite S will contain n outcomes, where $n = n_1 \cdot n_2 \cdots n_k$. (We will return to counting questions of this sort in Chapter 3.)

Example 2.2.6

The feathers of a frizzle chicken occur in three variations (or *phenotypes*)—extreme frizzle, mild frizzle, and normal (see Figure 2.2.4). It has been suggested that two alleles (F and f) control frizzle and that F and f interact according to a phenomenon known as *incomplete dominance* (82). What the latter predicts is that progeny whose genetic complement is (F, F) will be extreme frizzles and those whose genotype is (F, f) will be mild frizzles, while the (f, f) offspring will be normal. Suppose two chickens having mild frizzle are crossed. Describe the sample space that gives the possible genetic makeups for the offspring.

Here the experimental result of interest—the genetic complement of the progeny—is the composite of two subexperiments: (1) the allele produced by the first parent (x) and (2) the allele produced by the second parent (y). Denote

Normal

Mild frizzle

Extreme frizzle

Figure 2.2.4 Frizzle chickens.

the two associated sample spaces by S_1 and S_2. (Both S_1 and S_2 have two outcomes.) As Figure 2.2.5 shows, the sample space S for which we are looking contains four members $(2 \cdot 2)$ and is the Cartesian cross product of S_1 and S_2. (Outcomes (F, f) and (f, F), of course, are biologically equivalent.)

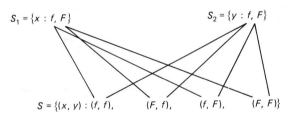

$S_1 = \{x : f, F\}$ $S_2 = \{y : f, F\}$

$S = \{(x, y) : (f, f), \quad (F, f), \quad (f, F), \quad (F, F)\}$

Figure 2.2.5

Example 2.2.7

A graduating engineer has signed up for three job interviews next week. She intends to categorize each as being either a success or a failure depending on whether or not it leads to a plant trip. Write out the appropriate sample space. What outcomes are in the event "second success occurs on third interview"? What outcomes are in the event "first success never occurs"?

Here each individual interview can be thought of as a subexperiment S_i, where the two possible outcomes are s (success) and f (failure). The sample space, S, describing her interview activities for the entire week is then the Cartesian product of three such subexperiments: $S = S_1 \otimes S_2 \otimes S_3$. Either by the formula given earlier $(n = n_1 \cdot n_2 \cdot n_3 = 2 \cdot 2 \cdot 2)$ or by a simple enumeration, we see that S consists of 8 outcomes:

$$S = \{(s, s, s), (s, s, f), (s, f, s), (f, s, s), (s, f, f), (f, s, f), (f, f, s), (f, f, f)\}$$

where (s, s, f), for example, denotes successes on the first two interviews and a failure on the third. If E is the event "second success occurs on third interview" and F the event "first success never occurs," then

$$E = \{(s, f, s), (f, s, s)\} \quad \text{and} \quad F = \{(f, f, f)\}$$

(What outcomes would be in the event "successes occur either on the first or the third interview or on both"?)

Question 2.2.6 If $A = \{3, 5, 7\}$ and $B = \{0, 1\}$, list the elements in $A \otimes B$ and in $B \otimes A$.

Question 2.2.7 Suppose A is the set of points satisfying the inequality $x^2 + y^2 \le 1$. Let B be the set $B = \{z : 0 \le z \le 1\}$. Geometrically, what is the Cartesian cross product $A \otimes B$?

Question 2.2.8 A ballplayer has four official at bats during a game. For each at bat the team statistician records whether the player made a hit (h) or an out (o). List the corresponding sample space. What outcomes are in the event "player makes at least three hits in a row"? How many outcomes would be in S if the hit category was subdivided into singles, doubles, triples, and home runs?

Question 2.2.9 In the home, the amount of radiation emitted by a color television set does not pose a health problem of any consequence, but the same may not be true in department stores, where as many as 15 or 20 sets are turned on at the same time in a relatively confined area. Suppose measurements are taken on the radiation levels found on the sales floors of four department stores. Characterize the sample space. (*Note:* Even though the meter reading will necessarily be quantized, the number of possible radiation levels is so large that for all practical purposes the range for each measurement can be thought of as the entire set of nonnegative real numbers.) The National Council on Radiation Protection has set the safety limit in these situations at 0.5 milliroentgens per hour (40). Characterize the event A, "stores 1, 2, and 4 are in compliance with the NCRP limit but store 3 is not."

===

Whether the sample space appropriate for a given problem refers to a single experiment or is a Cartesian cross product resulting from the simultaneous or sequential performance of several subexperiments, we will often find it necessary to work with events that are formed in various ways from other events. Thus, the craps shooter to whom we alluded earlier will win on the initial roll if the shooter throws either a 7 or an 11. In the language of Boolean algebra, that event—"shooter rolls a 7 or an 11"—is the *union* of two simple events—"shooter rolls a 7" and "shooter rolls an 11." If E denotes the former event and A and B the latter two, we write $E = A \cup B$. The next several definitions and examples review some of the algebra of sets that we will find useful.

Definition 2.2.1. Let A and B be any two events defined over the same sample space, S. Then:

1. The *intersection* of A and B, written $A \cap B$, is the event whose outcomes belong to both A and B.
2. The *union* of A and B, written $A \cup B$, is the event whose outcomes belong to either A or B or both.

Example 2.2.8

A single card is drawn from a poker deck. Let A be the event "ace is selected":

$$A = \{\text{ace of hearts, ace of diamonds, ace of clubs, ace of spades}\}$$

Let B be the event "heart is drawn":

$$B = \{\text{two of hearts, three of hearts, } \ldots \text{, king of hearts, ace of hearts}\}$$

Then

$$A \cap B = \{\text{ace of hearts}\}$$

and

$$A \cup B = \{\text{two of hearts, three of hearts, } \ldots \text{, ace of hearts,}$$
$$\text{ace of diamonds, ace of clubs, ace of spades}\}$$

(Let C be the event "club is drawn." Which cards are in $B \cup C$? in $B \cap C$?)

Example 2.2.9

Let A be the set of x-values for which $x^2 + 2x = 8$ and B be the set for which $x^2 + x = 6$. Find $A \cap B$ and $A \cup B$.

Since the first equation factors into $(x + 4)(x - 2) = 0$, its solution set is $A = \{-4, 2\}$. Similarly, the second equation can be written $(x + 3)(x - 2) = 0$, so that $B = \{-3, 2\}$. Therefore,

$$A \cap B = \{2\}$$

and

$$A \cup B = \{-4, -3, 2\}$$

(What would $A \cap B$ and $A \cup B$ be if the two equations were actually *inequalities: $x^2 + 2x \leq 8$ and $x^2 + x \leq 6$?*)

Example 2.2.10

An electronic system has four components divided into two pairs. The two components of a pair are wired in parallel while the pairs themselves are in series (see Figure 2.2.6). Let A_{ij} denote the event "ith component in jth pair fails," $i = 1, 2; j = 1, 2$. Let A be the event "system fails." How might we express A in terms of the A_{ij}'s?

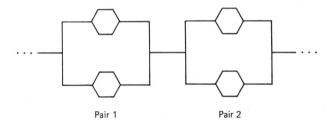

Pair 1   Pair 2   **Figure 2.2.6**

The system will fail if either pair 1 or pair 2 (or both) fails. Thus if P_1 and P_2 denote the events "pair 1 fails" and "pair 2 fails," respectively, then $A = P_1 \cup P_2$. But because it is wired in parallel, a pair fails only if both its components fail. That is, a pair failure is equivalent to an *intersection* of component failures. Putting all this information together gives us our expression for A:

$$A = (A_{11} \cap A_{21}) \cup (A_{12} \cap A_{22})$$

Example 2.2.11

Suppose the Cleveland Indians and the Boston Red Sox are playing a four-game weekend series in Municipal Stadium. Let A be the event "Cleveland wins final game" and B the event "Cleveland wins series." List the outcomes making up the intersection $A \cap B$.

Observe that the sample space here is a Cartesian cross product of the sample spaces for each of the four games. That is,

$$S = \{s_{i_1}, s_{i_2}, s_{i_3}, s_{i_4}\}_{i_1, i_2, i_3, i_4}$$

where, for any j, the set of s_{i_j}'s consists of two outcomes—c_j: "Cleveland wins jth game" and b_j: "Boston wins jth game." Of the $2^4 = 16$ members in S, a total of four comprise the desired intersection:

$$A \cap B = \{(c_1, c_2, c_3, c_4), (c_1, c_2, b_3, c_4), (c_1, b_2, c_3, c_4), (b_1, c_2, c_3, c_4)\}$$

(What outcomes would be in $A \cup B$?)

Definition 2.2.2. Events A and B defined over the same sample space are said to be *mutually exclusive* if they have no outcomes in common—that is, if $A \cap B = \emptyset$, where $\emptyset$ is the null set.

Example 2.2.12

Consider a single throw of two dice. Define A to be the event that the *sum* of the faces showing is odd. Let B be the event that the two faces themselves are odd. Then the intersection is clearly empty, the sum of two odd numbers necessarily being even. In symbols, $A \cap B = \emptyset$. (Recall the event $B \cap C$ asked for in Example 2.2.8.)

Definition 2.2.3. Let A be any event defined on a sample space, S. The *complement* of A, written A^C, is the event consisting of all the outcomes in S other than those contained in A.

Example 2.2.13

Let A be the set of (x, y)-values for which $x^2 + y^2 < 1$. Draw a sketch showing the region in the xy-plane corresponding to A^C.

From algebra, we recognize that $x^2 + y^2 < 1$ describes the interior of a circle of radius 1 centered at the origin. Figure 2.2.7 indicates the obvious complement.

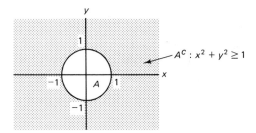

Figure 2.2.7

Question 2.2.10 Consider the electrical circuit shown at the top of page 15. Let A_i denote the event that switch i fails to close, $i = 1, 2, 3, 4$. Let A be the event "circuit is not completed." Express A in terms of the A_i's.

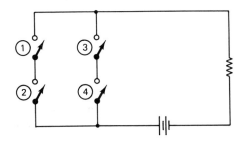

Question 2.2.11 Find $A \cup B$ and $A \cap B$ if

$$A = \{(x, y) : 0 < x < 3, 0 < y < 3\}$$

and

$$B = \{(x, y) : 2 < x < 4, 2 < y < 4\}$$

Question 2.2.12 Let A, B, and C denote any three events defined on S. Write down an expression that represents each of the following statements.
a. Exactly one of the events A, B, and C occurs.
b. Exactly one of the events A, B, and C does not occur.
c. At least two of the events occur.

Question 2.2.13 Define $A = \{x : 0 \leq x \leq 1\}$, $B = \{x : 0 \leq x \leq 3\}$, and $C = \{x : -1 \leq x \leq 2\}$. Draw a diagram showing each of the following sets of points.
a. $A^C \cap B \cap C$
b. $A^C \cup (B \cap C)$
c. $A \cap B \cap C^C$
d. $((A \cup B) \cap C^C)^C$

Question 2.2.14 In poker, a five-card hand is called a *straight* if the denominations of the cards are consecutive—for example, a 4 of hearts, 5 of spades, 6 of spades, 7 of hearts, and 8 of clubs. A hand is called a *flush* if all 5 cards are in the same suit—such as a 3, 6, 10, jack, and king of diamonds. If A denotes the set of straights and B the set of flushes, how many outcomes are in the intersection $A \cap B$? Assume that an ace can count high or low—that is, an ace, 2, 3, 4, and 5 constitutes a straight and so does a 10, jack, queen, king, and ace. (*Hint:* Picture the deck of 52 cards as a 4×13 matrix, where the rows correspond to the 4 suits and the columns represent the 13 denominations.)

Question 2.2.15 Let A and B be any two events defined on a sample space S. Then S is the union of A, $A^C \cap B$, and what other mutually exclusive event?

Question 2.2.16 Construct a counterexample to show that

$$A \cup (B \cap C) \neq (A \cup B) \cap C$$

Question 2.2.17 How could events A and B be characterized if (a) $A \cap B = B$, (b) $A \cup B = A$?

Question 2.2.18 Consolidated Industries has come under considerable pressure to eliminate its discriminatory hiring practices. Company officials have agreed that during the next five years, 60% of their new employees will be females and 30% will be black. One out of four new employees, though, will be white males. What percentage of black females are they committed to hiring?

Question 2.2.19 Let $A_i = \{x : 0 \leq x < 1/i\}$, $i = 1, 2, \ldots, k$. Describe each set.

a. $\bigcup\limits_{i=1}^{k} A_i$

b. $\bigcap\limits_{i=1}^{k} A_i$

Question 2.2.20 Define the three events

A: all females
B: all math professors
C: all professors with tenure

Describe in words the indicated sets.

a. $A \cap B^C \cap C$
b. $A \cap B \cap C^C$
c. $(A^C \cap B \cap C^C) \cup (A \cap B \cap C^C) \cup (A^C \cap B \cap C)$
d. $(A \cap B^C \cap C) \cup (A \cap B \cap C)$
e. $(A \cup B) \cap C^C$

Venn diagrams provide an easy and often very useful method for picturing the sorts of relationships that can exist among events. Figure 2.2.8 shows Venn diagrams illustrating an intersection, a union, a complement, and two events that are mutually exclusive. In each case, the interior of a set is intended to represent the outcomes it includes.

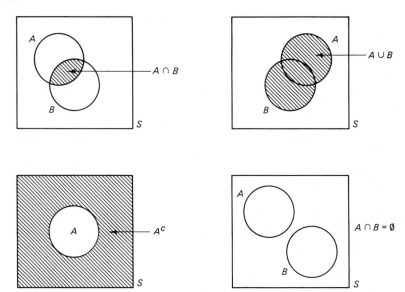

Figure 2.2.8 Venn diagrams.

Comment

Venn diagrams are especially helpful for visualizing more-complicated event relationships. Let A and B be any two events defined over the same sample space S. Convince yourself that:

1. *Exactly one* (of the two) occurs is $(A \cap B^C) \cup (B \cap A^C)$.

2. *At most one* (of the two) occurs is $(A \cap B)^C$.

(How would the event "at least one occurs" be written?)

The final example in this section shows two ways of "verifying" identities involving events. The first is nonrigorous and uses Venn diagrams; the second is a formal approach, in which every outcome in the left-hand side of the presumed identity is shown to belong to the right-hand side, and vice versa. The particular identity at which we look here is a very useful distributive property for intersections.

Example 2.2.14

Let A, B, and C be any three events defined over the same sample space S. Show that

$$A \cap (B \cup C) = (A \cap B) \cup (A \cap C) \qquad (2.2.1)$$

Figure 2.2.9 indicates the set formed by intersecting A with the union of B and C. Similarly, Figure 2.2.10 shows the union of the intersection of A and B with the intersection of A and C. It would appear at this point that the identity is true: The rightmost diagrams in Figures 2.2.9 and 2.2.10 *do* show the same shaded region. Still, what we have done is not a definitive proof. How we drew the events in the first place may not do justice to the problem *in general,* in which case the diagrams would be a misleading oversimplification. Can we be certain, for example, that Equation 2.2.1 remains true if, say, C is mutually exclusive of A or B? Or if B is a proper subset of A?

As an alternative to drawing an entire set of Venn diagrams and verifying

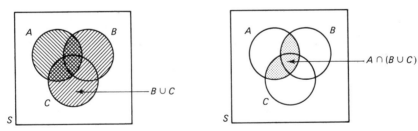

Figure 2.2.9

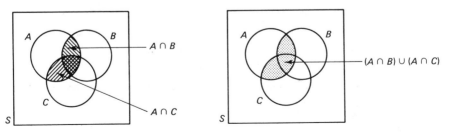

Figure 2.2.10

the identity for each one, we can instead turn to the algebraic approach of showing that (1) $A \cap (B \cup C)$ is contained in $(A \cap B) \cup (A \cap C)$ and (2) $(A \cap B) \cup (A \cap C)$ is contained in $A \cap (B \cup C)$. Let $s \in A \cap (B \cup C)$. Then $s \in A$ and $s \in B \cup C$. But if $s \in B \cup C$, then either $s \in B$ or $s \in C$. If $s \in B$, then $s \in A \cap B$ and $s \in (A \cap B) \cup (A \cap C)$. Likewise, if $s \in C$, it follows that $s \in (A \cap B) \cup (A \cap C)$. Therefore, every sample outcome in $A \cap (B \cup C)$ is also contained in $(A \cap B) \cup (A \cap C)$. Going the other way, assume that $s \in (A \cap B) \cup (A \cap C)$. Therefore, either $s \in A \cap B$ or $s \in A \cap C$ (or both). Suppose $s \in A \cap B$. Then $s \in A$ and $s \in B$, in which case $s \in A \cap (B \cup C)$. The same conclusion holds if $s \in A \cap C$. Thus, every sample outcome in $(A \cap B) \cup (A \cap C)$ is in $A \cap (B \cup C)$. It follows that $A \cap (B \cup C)$ and $(A \cap B) \cup (A \cap C)$ are identical.

Question 2.2.21 A fashionable country club has 100 members, 30 of whom are lawyers. Rumor has it that 25 of the club members are liars and that 55 are neither lawyers nor liars. What proportion of the lawyers are liars?

Question 2.2.22 Verify the associative laws for unions and intersections.
 a. $A \cup (B \cup C) = (A \cup B) \cup C$
 b. $A \cap (B \cap C) = (A \cap B) \cap C$

Question 2.2.23 Example 2.2.14 proved one of the distributive laws for unions and intersections. Show that the other one if also true—that is, verify that

$$A \cup (B \cap C) = (A \cup B) \cap (A \cup C)$$

Question 2.2.24 The following information was collected on six students responding to a campus newspaper announcement asking for volunteers to help with a psychology experiment.

ID	Parents divorced?	Left-handed?	Number of years required to finish high school
U	No	Yes	3
V	Yes	No	4
W	No	No	3
X	Yes	No	7
Y	No	Yes	2
Z	No	No	4

Define the following events:

A: "student's parents are divorced"
B: "student is left-handed"
C: "student completed high school in 3 years or less"

Draw a Venn diagram showing the three events A, B, and C and the location of each of the six outcomes. Which individuals are in the following sets?
 a. $A \cup B$
 b. $A \cap B \cap C^c$

c. $A^C \cap B^C \cap C^C$

d. $A^C \cup C^C$

Question 2.2.25 Let A and B be any two events defined on S. Show that A is the null set if and only if

$$B = (A \cap B^C) \cup (A^C \cap B)$$

Question 2.2.26 If $A = C \otimes D$ and $B = E \otimes F$, does $A \cap B = (C \cap E) \otimes (D \cap F)$?

Question 2.2.27 At a convention attended by 100 country music buffs, a trade magazine sought to get the fans' opinions of Waylon Jennings, Merle Haggard, and Willie Nelson. Twenty said they liked Jennings but not Haggard or Nelson, 22 were strongly partial to Haggard, and 28 listened only to Nelson. Fourteen liked Jennings *and* Haggard; 18, Haggard and Nelson; 12, Jennings and Nelson; 8, all the performers. It was also reported that 4 did not like any of the three. Comment on the accuracy of the results.

Question 2.2.28 Let A and B be any two events. Use Venn diagrams to show that (a) the complement of their intersection is the union of their complements,

$$(A \cap B)^C = A^C \cup B^C$$

and (b) the complement of their union is the intersection of their complements,

$$(A \cup B)^C = A^C \cap B^C$$

Question 2.2.29 Prove the two statements in Question 2.2.28 formally.

═══════════════════════════════

2.3 THE PROBABILITY FUNCTION

Having introduced the twin concepts of experiment and sample space in the previous section, we are now ready to pursue in a formal way the all-important problem of assigning a probability to an experiment's outcome—and, more generally, to an event. The backdrop for our discussion will be the unions, intersections, and complements of set theory.

Over the years, the mathematical definition of probability has undergone a considerable metamorphosis, but the most recent formulations have not invalidated their predecessors. As a consequence, there are, today, four very different ways of defining what is to be meant by the probability of an event. Historically, the first definition evolved quite naturally out of the gambling context in which so many of the early problems studied by Pascal, Fermat, and others were posed. Imagine an experiment, or game, having n possible outcomes—*and suppose that those outcomes are all equally likely.* Then if some event A were satisfied by m of those n, it reasonably follows that the probability of A (written $P(A)$) should be set equal to m/n. This is called the *classical,* or *a priori,* definition of probability. Recall Example 2.2.1. If two fair dice are tossed, there are $n = 36$ possible outcomes (all equally likely); of those 36, a total of $m = 6$ satisfy the event A "a 7 appears," so we write $P(A) = \frac{6}{36}$, or $\frac{1}{6}$.

There are many situations where the classical definition of probability is quite sufficient. We will see some in this chapter and many others in Chapter 3. Still, it has obvious limitations. What if the outcomes of the experiment are not equally likely?

Or the number of outcomes is not finite? Questions such as these provided the impetus for the formation of a more general and more experimentally oriented definition of probability. The first formal statement of what became known as the *empirical,* or *a posteriori,* definition of probability is often credited to the twentieth-century German mathematician Richard von Mises, but the basic notion was implicit in the work of many other probabilists at least a century earlier.

Consider a sample space S and any event A defined on S. If our experiment were performed *one* time, either A or A^C would be the outcome. If it were performed n times, the resulting set of sample outcomes would be members of A on m occasions, m being some integer between 0 and n, inclusive. Hypothetically, we could continue this process an infinite number of times. As n gets larger, the ratio m/n will fluctuate less and less (we will make that statement more precise in Chapter 7 when we study limit theorems). The number to which m/n converges is called the *empirical probability* of A: that is $P(A) = \lim_{n \to \infty} m/n$ (see Figure 2.3.1).

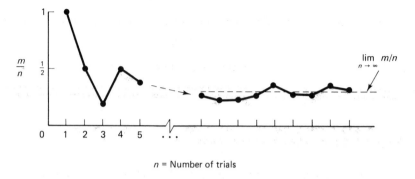

n = Number of trials

Figure 2.3.1

While the approach of von Mises does shore up some of the inadequacies seen in the classical definition of probability, it is not without problems of its own. The limit on which it focuses is not the usual one in standard analysis. For example, if a fair coin were to be tossed repeatedly with A being the event "head appears," it seems reasonable to expect that $P(A) = \lim_{n \to \infty} m/n$ would equal $\frac{1}{2}$. But it is still possible for the coin to come up tails every time, in which case $m/n = 0$ for all n. Also, there is a bit of an inconsistency in extolling $\lim_{n \to \infty} m/n$ as a way of defining a probability *experimentally* when the very act of repeating an experiment under identical conditions an infinite number of times is physically impossible. And left unanswered is the question of how large n must be to give a good approximation for $\lim_{n \to \infty} m/n$.

The next attempt at defining probability was entirely a product of the twentieth century. Modern mathematicians have shown a keen interest in developing subjects axiomatically—for example, consider Hilbert's work with Euclidean geometry. It was to be expected, then, that probability would come under a similar scrutiny and be defined not as a ratio or as the limit of a ratio, but simply as a function that behaved in accordance with a prescribed set of axioms. The major breakthrough on this front came in 1933 when the great Russian mathematician Andrei Kolmogorov published

Grundbegriffe der Wahrscheinlichkeitsrechnung (*Foundations of the Theory of Probability*). Kolmogorov's work was a masterpiece of mathematical elegance—it reduced the behavior of the probability function to a set of just three or four simple postulates, three if the sample space is limited to a finite number of outcomes and four if S is infinite.

If S has a finite number of members, Kolmogorov showed that a necessary and sufficient set of axioms for P are these three.

Axiom 1.   Let A be any event defined over S. Then $P(A) \geq 0$.

Axiom 2. $P(S) = 1$.

Axiom 3.   Let A and B be any two mutually exclusive events defined over S. Then

$$P(A \cup B) = P(A) + P(B)$$

When S has an infinite number of members, a fourth axiom is needed. It appears in *Grundbegriffe* as follows.

Axiom 4.   If $A_1 \supset A_2 \supset A_3 \supset \cdots$ and $\bigcap_{n=1}^{\infty} A_n = \emptyset$, then $\lim_{n \to \infty} P(A_n) = 0$.

Equivalently, Axiom 4 can be rephrased as an infinite analog of Axiom 3:

Axiom 4′.   Let $A_1, A_2, \ldots,$ be events defined over S. If $A_i \cap A_j = \emptyset$ for each $i \neq j$, then

$$P\left(\bigcup_{i=1}^{\infty} A_i \right) = \sum_{i=1}^{\infty} P(A_i)$$

From these few axioms, we can derive all the other properties of the P function.

One final way of quantifying the concept of probability needs to be mentioned before we investigate some of the basic results involving P—results that hold true regardless of which definition is invoked. Even more recent than the construction of an axiom system for probability have been the efforts to define P in *subjective* terms, as a person's measure of belief that some given event will occur. For example, suppose we ask: What is the probability that a nuclear war will break out in the Mid East sometime in the next 5 years? It is impossible to cast such a question meaningfully in a strictly empirical framework. Any number with which we might come up would necessarily be our own personal (subjective) assessment of the situation, based on the various countries' past history, extrapolations of their current policies, and so on. For the kinds of problems with which we will deal, though, the subjective definition of probability is inappropriate. Where this approach does become important is in an application of statistics known as Bayesian inference (see (7) or (79) for more details).

Now, let us return to Kolmogorov's axioms and use them to derive some basic properties of the probability function. Of the seven at which we will look, Theorems 2.3.1, 2.3.5, 2.3.6, and 2.3.7 will prove to be especially important later.

Theorem 2.3.1. $P(A^C) = 1 - P(A)$.

Proof. By Axiom 2 and Definition 2.2.3,

$$P(S) = 1 = P(A \cup A^C)$$

But A and A^C are mutually exclusive, so

$$P(A \cup A^C) = P(A) + P(A^C)$$

and the result follows.

Theorem 2.3.2. $P(\emptyset) = 0$.

Proof. Since $\emptyset = S^C$, $P(\emptyset) = P(S^C) = 1 - P(S) = 0$.

Theorem 2.3.3. If $A \subset B$, then $P(A) \leq P(B)$.

Proof. Note that the event B may be written in the form

$$B = A \cup (B \cap A^C)$$

where A and $(B \cap A^C)$ are mutually exclusive. Therefore,

$$P(B) = P(A) + P(B \cap A^C)$$

which implies that $P(B) \geq P(A)$, since $P(B \cap A^C) \geq 0$.

Theorem 2.3.4. For any event A, $P(A) \leq 1$.

Proof. The proof follows immediately from Theorem 2.3.3 because $A \subset S$ and $P(S) = 1$.

Theorem 2.3.5. Let $A_1, A_2, \ldots, A_n$ be events defined over S. If $A_i \cap A_j = \emptyset$ for $i \neq j$, then

$$P\left(\bigcup_{i=1}^{n} A_i \right) = \sum_{i=1}^{n} P(A_i)$$

Proof. The proof is a straightforward induction argument, with Axiom 3 being the starting point.

Theorem 2.3.6. $P(A \cup B) = P(A) + P(B) - P(A \cap B)$.

Proof. The Venn diagram for $A \cup B$ certainly suggests that the statement of the theorem is true (recall Figure 2.2.8). More formally, from Axiom 3 we have that

$$P(A) = P(A \cap B^C) + P(A \cap B)$$

and

$$P(B) = P(B \cap A^C) + P(A \cap B)$$

Adding these two equations gives

$$P(A) + P(B) = [P(A \cap B^C) + P(B \cap A^C) + P(A \cap B)] + P(A \cap B)$$

By Theorem 2.3.5, the sum in the brackets is $P(A \cup B)$. If we subtract $P(A \cap B)$ from both sides of the equation, the result follows.

The next result is a generalization of Theorem 2.3.6 that considers the probability of a union of n events. We have elected to retain the two-event case, $P(A \cup B)$, as a separate theorem simply for pedagogical reasons.

Theorem 2.3.7. Let $A_1, A_2, \ldots, A_n$ be any n events defined on S. Then

$$P\left(\bigcup_{i=1}^{n} A_i\right) = \sum_{i=1}^{n} P(A_i) - \sum_{i<j} P(A_i \cap A_j)$$

$$+ \sum_{i<j<k} P(A_i \cap A_j \cap A_k) - \sum_{i<j<k<l} P(A_i \cap A_j \cap A_k \cap A_l)$$

$$+ \cdots + (-1)^{n+1} \cdot P(A_1 \cap A_2 \cap \cdots \cap A_n)$$

Proof. We will defer a formal proof until Chapter 3. It should be noted, though, that for $n = 2$, both Theorems 2.3.6 and 2.3.7 give the same result. And for $n = 3$, a Venn diagram corroborates the statement (see Question 2.3.2).

Comment

The probability functions with which we will deal are broadly categorized into two groups: (1) those associated with a sample space that is either finite or countably infinite and (2) those associated with a sample space whose outcomes are uncountably infinite. The first are referred to as *discrete;* the second are said to be *continuous.* In the next two sections we will examine in some detail the properties of these two types of probability functions and how they are applied. As a preview of that material, we conclude this section with two examples that make use of several of the properties just derived, as well as some of the set theory results from Section 2.2.

Example 2.3.1

Two cards are drawn successively from a deck of 52. What is the probability that the second card is higher in rank than the first?

Suppose we draw the first card and remove it from the deck without looking at it. Define S to be the sample space associated with the *second* card. Every element in S belongs to exactly one of three mutually exclusive subsets:

A: the set of cards having a higher rank than the first card

B: the set of cards having a lower rank than the first card
C: the set of cards having the same rank as the first card

By symmetry, it must be true that $P(A) = P(B) = p$. Also, of the 51 cards that can be selected on the second draw, three have the same rank as the first card, so $P(C) = \frac{3}{51}$. Since A, B, and C are mutually exclusive and $S = A \cup B \cup C$, it follows from Axioms 2 and 3 that

$$P(S) = P(A) + P(B) + P(C)$$

and

$$1 = p + p + \tfrac{3}{51}$$

From the latter we find that $p = \frac{8}{17}$.

Example 2.3.2

Imagine a premed student being summarily rejected by all 126 U.S. medical schools. Desperate, he sends his vitae to the two least-selective foreign schools he can think of, the two branch campuses (X and Y) of Swampwater Tech. Based on the success his friends have had there, he estimates that his probability of being accepted at X is 0.7, and at Y, 0.4. He also suspects there is a 75% chance that at least one of his applications will be rejected. What is the probability that at least one of the schools will accept him?

Let A be the event "branch X accepts him" and B be the event "branch Y accepts him." We are given that $P(A) = 0.7$, $P(B) = 0.4$, and $P(A^C \cup B^C) = 0.75$; what we want to find is $P(A \cup B)$.

From Theorem 2.3.6,

$$P(A \cup B) = P(A) + P(B) - P(A \cap B)$$

We can get a value for the probability of the intersection by appealing to Question 2.2.28. Since $A^C \cup B^C = (A \cap B)^C$,

$$P(A \cap B) = 1 - P\{A \cap B)^C\} = 1 - 0.75 = 0.25$$

Substituting, then, into the expression for the union, we find that our premed's chances are not all that bad—he has an 85% chance of getting at least one acceptance:

$$P(A \cup B) = 0.7 + 0.4 - 0.25 = 0.85$$

A curious verbal interpretation of this result can be gotten if we generalize the problem slightly and treat $P(A \cup B)$ as a function of $P(A^C \cup B^C)$. Since

$$P(A \cap B) = 1 - P\{(A \cap B)^C\} = 1 - P(A^C \cup B^C)$$

we can write the union as

$$P(A \cup B) = P(A) + P(B) - 1 + P(A^C \cup B^C)$$

But this latter expression implies that for $P(A)$ and $P(B)$ fixed, as the probability of at least one rejection increases, so does the probability of at least one accept-ance! Or, to take a numerical example, if $P(A) = 0.7$ and $P(B) = 0.4$, it is more reassuring to be told that we have a 90% chance of at least one rejection than to be told we have a 70% chance of at least one rejection (see Question 2.3.1).

Question 2.3.1 For the conditions stated in Example 2.3.2, compute $P(A \cup B)$ for $P(A^C \cup B^C) = 0.7$ and for $P(A^C \cup B^C) = 0.9$. (As a convenient model, take S to be the points along $[0, 1]$ and let the different events be subintervals of $[0, 1]$. For any event A, define $P(A) = $ (length of A)/(length of S) = length of A.)

Question 2.3.2 Draw a Venn diagram to "verify" Theorem 2.3.7 for the case of three events A, B, and C.

Question 2.3.3 If $P(A) = \frac{1}{3}$, $P(B) = \frac{1}{2}$, and $P(A \cup B) = \frac{3}{4}$, find each probability.
   a. $P(A \cap B)$
   b. $P(A^C \cup B^C)$
   c. $P(A^C \cap B)$

Question 2.3.4 An experiment has two possible outcomes: The first occurs with probability p and the second occurs with probability p^2. Find p.

Question 2.3.5 We say that the *odds* are a to b for E if $P(E) = a/(a + b)$. Suppose the odds for E are 3 to 1 and the odds for $E \cup F$ are 4 to 1. Find upper and lower bounds for $P(F)$.

Question 2.3.6 Use a Venn diagram to show that

$$P\{A \cap B^C) \cup (A^C \cap B)\} = P(A) + P(B) - 2P(A \cap B)$$

Give a short verbal description of the event "$(A \cap B^C) \cup (A^C \cap B)$."

Question 2.3.7 The runoff election for a seat on the Public Service Commission is between three candidates (A, B, and C). According to the pollsters, the probabilities of A, B, and C winning are 0.4, 0.5, and 0.1, respectively. Two days before the election, candidate C drops out. What would be a revised estimate of A's chances of winning? What assumption are you making?

Question 2.3.8 Suppose S has a finite number of outcomes, all equally likely. Let $n(A)$ denote the number of outcomes in the event A, where $A \subset S$. Define

$$P(A) = \frac{n(A)}{n(S)}$$

Show that $P(A)$ satisfies Axioms 1 through 3.

Question 2.3.9 For any events A and B, show that

$$P(A \cap B) \geq 1 - P(A^C) - P(B^C)$$

Question 2.3.10 Express the following probabilities in terms of $P(A)$, $P(B)$, and $P(A \cap B)$.
   a. $P(A^C \cup B^C)$
   b. $P(A^C \cap (A \cup B))$

Question 2.3.11 For any set of events $A_1, A_2, \ldots, A_k$, prove that

$$P\left(\bigcup_{i=1}^{k} A_i \right) \leq \sum_{i=1}^{k} P(A_i)$$

Question 2.3.12 Let $A_1, A_2, \ldots$ be a countable sequence of events. Show that

$$P\left(\bigcap_{i=1}^{\infty} A_i\right) \geq 1 - \sum_{i=1}^{\infty} P(A_i^C)$$

Question 2.3.13 Let $A_1, A_2, \ldots, A_n$ be a set of n events $(n > 1)$. Use induction to prove *Bonferroni's inequality:*

$$\sum_{i=1}^{n} P(A_i) - \sum_{i<j} P(A_i \cap A_j) \leq P\left(\bigcup_{i=1}^{n} A_i\right) \leq \sum_{i=1}^{n} P(A_i)$$

Question 2.3.14 Let $S = \{x : 0 \leq x \leq 1\}$. Define

$$A_1 = \{x : \tfrac{1}{3} < x < \tfrac{2}{3}\}$$

$$A_2 = \{x : \tfrac{1}{9} < x < \tfrac{2}{9} \cup \tfrac{7}{9} < x < \tfrac{8}{9}\}$$

$$A_3 = \{x : \tfrac{1}{27} < x < \tfrac{2}{27} \cup \tfrac{7}{27} < x < \tfrac{8}{27} \cup \tfrac{19}{27} < x < \tfrac{20}{27} \cup \tfrac{25}{27} < x < \tfrac{26}{27}\}$$

In general, let A_n = union of the middle third (open) intervals in $(A_1 \cup A_2 \cup \cdots \cup A_{n-1})^C$. Let $A = \bigcup_{i=1}^{\infty} A_i$. If $P(A_i)$ = length of A_i, show that $P(A) = 1$. (*Note:* In analysis, A^C is known as *Cantor's ternary set.*)

2.4 DISCRETE PROBABILITY FUNCTIONS

Suppose the sample space for a given experiment has either a finite or a countably infinite number of outcomes. Then any function P such that

1. $0 \leq P(s)$ for each $s \in S$
2. $\sum_{\text{all } s \, \in \, S} P(s) = 1$

is said to be a *discrete probability function*. Applied to any event A, it follows that

$$P(A) = \sum_{\text{all } s \, \in \, A} P(s)$$

The next several examples look at some representative discrete probability functions and the kinds of problems to which they can give rise. Much more will be said about sample spaces of this type in Chapter 3.

Example 2.4.1

To encipher, or *code,* a message, each letter of the original text is replaced by some other letter, according to a prescribed plan. Thus, if the substitution scheme called for the new letters to be shifted one place to the left of the old, a word like *DOG* would become *CNF.* To break such a simple cipher (see Question 2.4.1), all we need to know is the letter frequency—that is, the probability function—of the language being coded. English, for example, generates a sample space of 26 outcomes, the probabilities at issue being $P(A)$, $P(B), \ldots, P(Z)$. Estimates for these probabilities are credited to Dewey (20), who made a tally of some 438,023 letters of English text and found the relative frequencies shown in Table 2.4.1.

TABLE 2.4.1

Letter	Relative frequency	Letter	Relative frequency
E	0.1268	F	0.0256
T	0.0978	M	0.0244
A	0.0788	W	0.0214
O	0.0776	Y	0.0202
I	0.0707	G	0.0187
N	0.0706	P	0.0186
S	0.0634	B	0.0156
R	0.0594	V	0.0102
H	0.0573	K	0.0060
L	0.0394	X	0.0016
D	0.0389	J	0.0010
U	0.0280	Q	0.0009
C	0.0268	Z	0.0006

Given the information in Table 2.4.1, we can answer questions about relative likelihoods of letters, as well as about events composed of subsets of letters. An example of the first would be the observation that E's in English text are approximately 80 times as common as X's: $P(E)/P(X) = 0.1268/0.0016 = 79.3$. For an example of the second, suppose our experiment consists of choosing a single letter and we define the event "a vowel is chosen." If Y is considered a vowel,

$$P(\text{vowel is chosen}) = P(A) + P(E) + P(I) + P(O) + P(U) + P(Y)$$

$$= 0.0788 + 0.1268 + 0.0707 + 0.0776 + 0.0280 + 0.0202$$

$$= 0.4021$$

A similar approach could be used to determine the probability that a consonant is chosen—that is, adding the relative frequencies of the 20 consonants—but Theorem 2.3.1 suggests an easier solution. Since the events "vowel is chosen" and "consonant is chosen" are complementary,

$$P(\text{consonant is chosen}) = 1 - P(\text{vowel is chosen})$$

$$= 1 - 0.4021$$

$$= 0.5979$$

(What is the probability that a letter chosen at random is either a vowel or in the first half of the alphabet?)

Example 2.4.2

A fair coin is to be tossed until a head comes up for the first time. What are the chances of that happening on an odd-numbered toss?

Note, first of all, that the sample space here is substantially different than the one we encountered in Example 2.4.1. There, S was finite; here, the numbers of outcomes in both S and in the event for which we are looking are countably infinite (we could go on tossing forever!). But other than having to deal with summations where the indices range from one to infinity, going from a finite to a countably infinite sample space poses no particular mathematical problems. (As we will see in the next section, though, significant changes in our definition of the probability function *do* occur when the sample space becomes uncountably infinite.)

Suppose we let $P(k)$ denote the probability that the first head appears on the kth toss. Since the coin was presumed to be fair, $P(1) = \frac{1}{2}$. Furthermore, we would expect half of the coins that showed a tail on the first toss to come up heads on the second, so, intuitively, $P(2) = \frac{1}{4}$. In general, $P(k) = (\frac{1}{2})^k$, $k = 1$, 2, ... (see Question 2.4.4).

Let E be the event "first head appears on an odd-numbered toss." Then

$$P(E) = P(1) + P(3) + P(5) + \cdots$$

$$= \sum_{i=0}^{\infty} \left(\frac{1}{2}\right)^{2i+1}$$

$$= \frac{1}{2} \sum_{i=0}^{\infty} \left(\frac{1}{4}\right)^{i}$$

Recall the formula for the sum of a geometric series: If $0 < x < 1$,

$$\sum_{k=0}^{\infty} x^k = \frac{1}{1-x}$$

Applying that result to $P(E)$ gives

$$P(E) = \frac{1}{2} \cdot \left(\frac{1}{1 - \frac{1}{4}}\right)$$

$$= \frac{2}{3}$$

A similar computation would show that the probability of the first head appearing on an even-numbered toss is $\frac{1}{3}$.

An important special case of a discrete probability function (when S is finite) is the *equally likely model:* If S contains n outcomes, such a function assigns $P(s) = 1/n$, for all s. If A is an event composed of m outcomes, then $P(A) = m/n$. The next example shows the application of this particular model to a Cartesian product sample space.

Example 2.4.3

Two gamblers, A and B, each choose an integer from 1 to m (inclusive) at random. What is the probability that the two numbers they pick do not differ by more than n?

It will be easier if we approach this problem via its complement. Let x and y denote the numbers selected by A and B, respectively. The complement has two cases, depending on whether $x < y$ or $x > y$. Let us first suppose that $x < y$. Then, for a given x, the values of y such that $y - x > n$ are $y = x + n + 1$, $y = x + n + 2$, . . . , and $y = m$—altogether, a range of $m - n - x$ choices. Summing over x, we find that the total number of (x, y)-pairs such that $y - x > n$ reduces to the sum of the first $m - n - 1$ integers:

$$\sum_{x=1}^{m-n-1} (m - n - x) = \sum_{i=1}^{m-n-1} i = \frac{(m - n - 1)(m - n)}{2}$$

By symmetry, the same number of (x, y)-pairs satisfies the second case: $x > y$ and $x - y > n$. Thus the total number of (x, y)-selections such that $|x - y| > n$ is $(m - n - 1)(m - n)$.

The sample space S that corresponds to these pairs is a Cartesian product, $S = S_1 \otimes S_2$, where the m outcomes in S_1 are A's possible choices and the m outcomes in S_2 are B's. Thus S contains m^2 outcomes, all equally likely by assumption. It follows that

$$P(|x - y| \le n) = 1 - \frac{(m - n - 1)(m - n)}{m^2}$$

(Intuitively, how would the solution here be modified if the problem called for choosing two real numbers from the unit interval?)

Example 2.4.4

Among the most important of all the discrete probability models is the *Poisson distribution*. The sample space for the Poisson is the set of all nonnegative integers, with the probabilities assigned to the sample outcomes being given by

$$P(s) = \frac{e^{-\lambda} \lambda^s}{s!}, \qquad s = 0, 1, 2, \ldots \qquad (2.4.1)$$

where λ is some positive constant. That $P(s)$ is a legitimate probability function follows readily, since

$$\sum_{\text{all } s} P(s) = \sum_{s=0}^{\infty} \frac{e^{-\lambda} \lambda^s}{s!} = e^{-\lambda} \sum_{s=0}^{\infty} \frac{\lambda^s}{s!}$$
$$= e^{-\lambda} e^{\lambda} = e^0 = 1$$

What makes the Poisson so important is that it describes remarkably well the way in which many real-world phenomena behave. For example, consider the number of daily fatalities among senior citizens in a large city. Over a period of 3 years (1096 days) in London, records showed that a total of 903 deaths occurred among males 85 years of age and older (88). Columns 1 and 2 of Table

2.4.2 give the breakdown of those 903 deaths according to the number occurring on a given day.

TABLE 2.4.2

(1) Number of deaths, s	(2) Number of days	(3) Empirical probability	(4) Poisson probability
0	484	0.442	0.440
1	391	0.357	0.361
2	164	0.150	0.148
3	45	0.041	0.040
4	11	0.010	0.008
5	1	0.001	0.003
6+	0	0	0.000

Note that by dividing a given entry in the second column by the overall number of days, 1096, we get the after-the-fact probability that on a given day there will be that many deaths. Thus, $\frac{164}{1096} = 0.150$, implying that the probability was 0.150 that on a given day in London exactly two elderly men would die.

The Poisson distribution provides a good model here in the sense that it can closely approximate the entire "observed" probability distribution associated with the integers 0 through 5. To see that, consider the particular case of Equation 2.4.1 where λ is replaced by 0.82 (the reason for the choice of that particular value will be discussed later):

$$P(s) = \frac{e^{-0.82}(0.82)^s}{s!}, \qquad s = 0, 1, 2, 3, 4, 5, \ldots$$

By substituting the various values of s into $P(s)$, we can generate a set of theoretical probabilities (see column 4 of Table 2.4.2). If the entries in column 4 match up well with those in column 3, we are justified in saying that $P(s)$ is a good model for the phenomenon of daily fatalities among elderly males in London. In this case they do.

In Chapter 6 we will look at a number of other applications of the Poisson distribution and examine its mathematical properties more thoroughly. Among all the probability functions that we will encounter, whether discrete or continuous, the Poisson is among the handful that would be considered the most important.

Comment

In this instance the agreement between the empirical probabilities (column 3) and the Poisson probabilities (column 4) is so close that it seems eminently reasonable to argue that the latter provides a good model for the former. Oftentimes, though, there may be some serious question as to whether a given $P(s)$ adequately describes the probabilistic behavior of the sample outcomes. Or, put another way, how much disagreement are we willing to tolerate between observed and expected probabilities before we *reject* the proposed model? This is a legitimate question, but one too "statistical" for us to examine in any detail here.

We conclude this section with an interesting puzzle. It concerns a very simple finite sample space over which is defined a very simple discrete probability function. What is intriguing about the problem is the fact that the *obvious* answer to the question it poses turns out to be the *wrong* answer!

Example 2.4.5

Three prisoners, A, B, and C, all with identical records, are up for parole and the Corrections Board decides to free two of them without initially revealing which two. Because the Board's release choices are (A, B), (A, C), and (B, C), it seems clear that $P(\text{A is released}) = \frac{2}{3}$. As it happens, though, A has befriended one of the jailers who knows the Board's decision. Being understandably curious about his fate, A considers asking the jailer to tell him which prisoner, other than himself, will be released (he deems it too bold to inquire about his own fate directly). After thinking about it, though, he decides not to pursue the matter because it seems that by asking the jailer for information he will have diminished his own chances for freedom. For example, if the jailer says, "B will be the other prisoner released," then the Board will be setting free either A and B or B and C, suggesting that A's chances have now fallen to $\frac{1}{2}$. Is A's reasoning correct?

The answer is no, however plausible A's arguments may sound to the contrary. The fallacy lies in his perception of the relevant sample space. It *is* true that initially—before the jailer is brought into the problem—the sample space consists of three (equally likely) outcomes, each a possible combination of prisoners to be released:

$$S = \{(A, B), (A, C), (B, C)\}$$

When the jailer enters the picture, though, the sample space changes to include not only the identities of the prisoners released but also the corresponding pronouncement of which inmate other than A is to be set free. Thus the former S is replaced by a new S' consisting of *four* elements:

$$S' = \{(A, B, \text{jailer says B}), (A, C, \text{jailer says C})$$

$$(B, C, \text{jailer says B}), (B, C, \text{jailer says C})\}$$

We must be careful here to observe that the outcomes in S' are *not* equally likely. Specifically, the two outcomes (B, C, jailer says B) and (B, C, jailer says C) now share the probability value of $\frac{1}{3}$ previously associated with (B, C) (why?). Thus the first two outcomes in S' each have probability $\frac{1}{3}$, while the last two each have probabilty $\frac{1}{6}$.

Now, consider A's status if the jailer says B. The two possible outcomes in S' that could have occurred are (A, B, jailer says B) and (B, C, jailer says B). But the former is twice as likely as the latter, implying that the probability of A being released is still $\frac{2}{3}$. Similarly, given that the jailer says C, the probability of (A, C, jailer says C) being the actual outcome is also $\frac{2}{3}$. Thus, no matter what the jailer says, A's chances for release are exactly what they were at the outset—$\frac{2}{3}$.

Comment

What the prisoner's errant reasoning underscores is the importance of having a well-defined sample space. Fallacies in probability arguments are far from uncommon, a claim to which future chapters will bear ample testimony. Taking the time to examine a problem's sample space carefully is often the best way to unravel an apparent ambiguity.

Question 2.4.1 Use Table 2.4.1 to translate the following quotation:

> HVOCZHVODXDVIN VMZ GDFZ AMZIXCHZI;
> RCVOZQZM TJP NVT OJ OCZH,
> OCZT OMVINGVOZ DIOJ OCZDM JRI GVIBPVBZ
> VIY AJMOCRDOC DO DN NJHZOCDIB ZIODMZGT YDAAZMZIO.
> BJZOCZ

Question 2.4.2 Ace-six flats are a type of crooked dice where the cube is shortened in the one-six direction. Suppose an ace-six die has $P(1) = P(6) = \frac{1}{4}$ and $P(2) = P(3) = P(4) = P(5) = \frac{1}{8}$. If two such dice are rolled, what is the probability the sum of the faces will be a 7? Compare your answer with the probability of rolling a 7 with two fair dice.

Question 2.4.3 Consider a countably infinite sample space consisting of all the positive integers except 1. Define

$$P(s) = c\left(\tfrac{2}{3}\right)^s, \qquad s = 2, 3, \ldots$$

What is the probability the outcome of the experiment will be less than or equal to 4? What is the probability the outcome will be an even number?

Question 2.4.4 Verify that the function derived in Example 2.4.2 sums to 1. That is, show that

$$\sum_{k=1}^{\infty} \left(\frac{1}{2}\right)^k = 1$$

Question 2.4.5 Suppose a rather limited lottery is set up that contains only 12 tickets, numbered 1 through 12. One ticket is to be drawn. Let A be the event "number on ticket is divisible by 2" and let B be the event "number on ticket is divisible by 3." Find $P(A \cup B)$.

Question 2.4.6 Recall the game of *odd person out* described in Question 2.2.5. If m players each toss a fair coin once, what is the probability no one will be an odd person out?

Question 2.4.7 A die is loaded in such a way that the probability of any particular face's showing is directly proportional to the number on that face. What is the probability a number greater than 4 appears?

Question 2.4.8 Three fair dice are rolled, one red, one green, and one blue. What is the probability the three faces are all different and the blue equals the sum of the red and the green?

Question 2.4.9 Records were kept during World War II of the number and the location of bombs falling in south London. The particular part of the city studied was divided into 576 areas, each covering $\frac{1}{4}$ km². The numbers of areas experiencing x hits, $x = 0, 1, 2, 3, 4, 5$, are listed in the following table (11).

Number of hits, x	Frequency
0	229
1	211
2	93
3	35
4	7
5	1
	576

Compute the *expected* frequencies to see how well the Poisson model applies. Use

$$P(x) = \frac{e^{-0.93}(0.93)^x}{x!}, \qquad x = 0, 1, 2, \ldots$$

Question 2.4.10 Show that

$$f(x) = \frac{1}{1 + \lambda}\left(\frac{\lambda}{1 + \lambda}\right)^x, \qquad x = 0, 1, 2, \ldots, \lambda > 0$$

qualifies as a probability function. (To statisticians, this is known as *Pascal's distribution*—it can be shown to be a special case of the *negative binomial distribution,* an important family of probability functions we will encounter in Chapter 6. Physicists know it as *Furry's distribution.*)

Question 2.4.11 Show that

$$f(x) = \left(\frac{1}{2\alpha}\right)e^{-|x-\mu|/\alpha}, \qquad \alpha > 0, -\infty < x < \infty$$

(also known as *Laplace's distribution*) qualifies as a probability function.

Question 2.4.12 Jean D'Alembert, the renowned eighteenth-century French mathematician, was once asked the following question: What is the probability of getting at least one head in two tosses of a fair coin? D'Alembert answered $\frac{2}{3}$, his argument being that there are three possible outcomes, H, TH, and TT, two of which—H and TH—satisfy the event "at least one head." Discuss his answer and his method of solution.

2.5 CONTINUOUS PROBABILITY FUNCTIONS

We will call S a *continuous* sample space if it contains an interval of real numbers, either bounded or unbounded. Associated with such an S will be a *continuous probability function, f*: If A is an event defined over S, f is a real-valued function having the property that $P(A) = \int_A f(x)\, dx$. Analogous to what was imposed on P in the previous section, f must satisfy the following two conditions:

1. $0 \le f(x)$ for each $x \in S$
2. $\int_S f(x)\, dx = 1$

Comment

The distinction between discrete and continuous probability functions needs to be kept clearly in mind. In the discrete case, $P(s)$ equals the probability that the outcome of the experiment will be s. In the continuous case, $f(x)$ does *not* represent the probability of the outcome being x. Rather, the function f describes the outcomes of an experiment in the sense that the probability of any event A will be the integral of $f(x)$ over all those points that make up A—that is,

$$P(A) = \int_A f(x)\, dx$$

Should A be just a single point, such as x', then $P(A) = 0$, since

$$\int_{x'}^{x'} f(x)\, dx = 0$$

although $f(x')$ may very well be positive.

Example 2.5.1

Suppose an enemy aircraft flies directly over the Alaskan pipeline and fires a single air-to-surface missile. If the missile hits anywhere within 20 ft of the pipeline, major structural damage will be incurred and oil flow will be disrupted. Suppose that with a certain sighting device, the probability function describing the missile's point of impact is given by

$$f(x) = \begin{cases} \dfrac{60 + x}{3600} & \text{for } -60 < x < 0 \\[2ex] \dfrac{60 - x}{3600} & \text{for } 0 \le x < 60 \\[2ex] 0 & \text{elsewhere} \end{cases}$$

where x is the perpendicular distance from the pipeline to the point of impact.

Let A be the event "flow is disrupted." Then the probability of A is the area under $f(x)$ above the interval $(-20, +20)$:

$$P(A) = \int_{-20}^{20} f(x)\, dx = \int_{-20}^{0} \frac{60 + x}{3600}\, dx + \int_{0}^{20} \frac{60 - x}{3600}\, dx$$

$$= \left[\frac{60x}{3600} + \frac{x^2}{7200} \right]_{-20}^{0} + \left[\frac{60x}{3600} - \frac{x^2}{7200} \right]_{0}^{20}$$

$$= \frac{1200}{3600} - \frac{400}{7200} + \frac{1200}{3600} - \frac{400}{7200}$$

$$= 0.55$$

Figure 2.5.1 is a graph of $f(x)$ and shows the area representing $P(A)$. (What is the probability the missile will land at least 40 ft from the pipeline?)

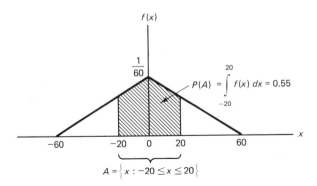

$$P(A) = \int_{-20}^{20} f(x)\, dx = 0.55$$

$$A = \{x : -20 \leq x \leq 20\}$$

Figure 2.5.1

Example 2.5.2

In the kinetic theory of gases, the distance X that a molecule travels before colliding with another molecule is described probabilistically by an exponential function,

$$f(x) = \left(\frac{1}{\lambda}\right) e^{-x/\lambda}, \qquad x > 0$$

where λ is come constant greater than 0 (see Figure 2.5.2). Physicists denote the average distance between collisions as the *mean free path* and define it by the expression,

$$\mu = \text{mean free path} = \int_0^\infty x \cdot f(x)\, dx$$

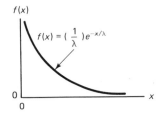

$$f(x) = \left(\frac{1}{\lambda}\right) e^{-x/\lambda}$$

Figure 2.5.2

What is the probability that the distance a molecule travels between consecutive collisions is less than half its mean free path?

First, note that the numerical value for the mean free path is just λ:

$$\mu = \int_0^\infty x \cdot \left(\frac{1}{\lambda}\right) e^{-x/\lambda}\, dx$$

$$= \lambda \int_0^\infty y e^{-y}\, dy \qquad \left(\text{where } y = \frac{x}{\lambda}\right)$$

$$= \lambda$$

the final step resulting from an integration by parts. Therefore, the probability of the event A, "intercollision distance is less than half the mean free path," is given by

$$P(A) = \int_0^{\lambda/2} \left(\frac{1}{\lambda}\right) e^{-x/\lambda} \, dx \qquad (2.5.1)$$

Making the substitution $y = x/\lambda$, we find that a molecule has a 39% chance of colliding with another molecule before it can traverse half its mean free path:

$$P(A) = \int_0^{1/2} e^{-y} \, dy = \left. -e^{-y} \right|_0^{1/2} = 1 - e^{-1/2} = 0.39$$

Comment

The *median*, x_m, associated with a continuous probability function is that value of x having the property that

$$\int_{-\infty}^{x_m} f(x) \, dx = \int_{x_m}^{\infty} f(x) \, dx = \frac{1}{2}$$

Is the mean free path equal to x_m? If not, draw a diagram of $f(x)$ showing the relative positions of x_m and μ.

Example 2.5.3

It can be shown that

$$f(x) = \frac{1}{(\alpha - 1)! \, \beta^\alpha} x^{\alpha - 1} e^{-x/\beta}, \qquad x > 0, \; \beta > 0, \; \alpha \text{ a positive integer}$$

is a probability function (we will formally do so in Chapter 6). Assuming it is, evaluate

$$\int_0^\infty x^5 e^{-3x} \, dx$$

This problem is similar to some we encountered in Section 2.4 in the sense that it makes use of the fact that the probability associated with the entire sample space is 1. Here, we note that if $f(x)$ is a legitimate probability function,

$$\int_0^\infty f(x) \, dx = \int_0^\infty \frac{1}{(\alpha - 1)! \, \beta^\alpha} x^{\alpha - 1} e^{-x/\beta} \, dx = 1$$

which implies that

$$\int_0^\infty x^{\alpha - 1} e^{-x/\beta} \, dx = (\alpha - 1)! \, \beta^\alpha$$

Let $\alpha = 6$ and $\beta = \frac{1}{3}$. Then

$$\int_0^\infty x^{\alpha - 1} e^{-x/\beta} \, dx = \int_0^\infty x^5 e^{-3x} \, dx = (6 - 1)! \, (\tfrac{1}{3})^6 = 0.165$$

(Try evaluating this integral without making use of its relationship to a probability function.)

Example 2.5.4

The lifetime of a piece of equipment is a variable whose sample space is clearly continuous. If x denotes the "age" at which the equipment fails, we write $S = \{x : 0 \leq x < \infty\}$. Many empirical studies have shown that lifetimes of electrical equipment are often described by the same exponential probability function that was introduced in Example 2.5.2:

$$f(x) = \begin{cases} \left(\dfrac{1}{\lambda}\right)e^{-x/\lambda} & \text{for } x > 0 \\ 0 & \text{elsewhere} \end{cases}$$

For reasons that will be explored in a later chapter, the factor λ is assigned a numerical value equal to the equipment's observed life expectancy.

Davis (17) "fit" the exponential model to the lifetimes of V805 transmitter tubes, components that were once standard in aircraft radar systems. The results for some 903 tubes are summarized in Table 2.5.1.

TABLE 2.5.1

Lifetime (hours)	Number of tubes	Relative frequency	Probability from model
0–40	166	0.184	0.201
40–80	151	0.167	0.160
80–120	132	0.146	0.128
120–160	98	0.108	0.103
160–200	73	0.081	0.082
200–240	45	0.050	0.066
240–280	53	0.059	0.052
280–320	40	0.044	0.042
320–360	23	0.025	0.033
360–400	26	0.029	0.027
400–440	24	0.026	0.021
440–480	9	0.010	0.017
480–520	9	0.010	0.014
520–560	8	0.009	0.011
560–600	9	0.010	0.009
600–700	17	0.019	0.015
700–800	8	0.009	0.008
800–up	12	0.013	0.011

For these particular tubes, the life expectancy is 179 h, so the probability function being proposed is

$$f(x) = \begin{cases} \left(\frac{1}{179}\right)e^{-x/179} & \text{for } x > 0 \\ 0 & \text{elsewhere} \end{cases}$$

Suppose the event A were defined to be "a tube fails in 40 h or less." The corresponding probability would be the integral of $f(x)$ over the interval $(0, 40]$:

$$P(A) = \int_0^{40} (\tfrac{1}{179})e^{-x/179}\, dx$$

$$= \int_0^{40/179} e^{-u}\, du = 1 - e^{-40/179}$$

$$= 0.201$$

That is, the probability function predicts that 20% of the tubes will fail in 40 h or less. What actually happened was that 166 out of 903, or 18.4%, failed that soon. The last two columns of Table 2.5.1 show that the agreement between the empirical and postulated probability functions is quite good—although not as good as what we found in Example 2.4.4. (According to the model, how many of the 903 tubes should have survived to within 50 h (plus or minus) of the 179-h average?)

Example 2.5.5

By far the most important of all probability functions—important because of both its widespread applicability as well as its theoretical properties—is the *normal distribution*. This is the familiar bell-shaped curve. The sample space for the normal distribution is the entire real line; its probability function is given by

$$f(x) = \frac{1}{\sqrt{2\pi}\sigma} e^{-1/2[(x-\mu)/\sigma]^2}, \quad -\infty < x < \infty, \ -\infty < \mu < \infty, \ \sigma > 0$$

Depending on the values assigned to μ and σ, $f(x)$ takes on a variety of shapes and locations along the x-axis (see Figure 2.5.3).

As an example of a phenomenon that can be modeled by the normal curve, consider the data in Table 2.5.2, which give a recent year's traffic death rates (per 100 million motor vehicle miles, mvm) for each of our 50 states (58).

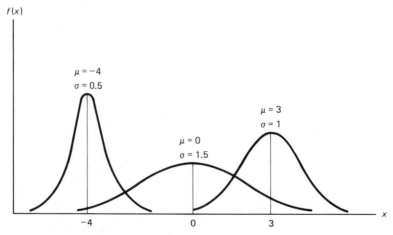

Figure 2.5.3 Normal curves.

TABLE 2.5.2 TRAFFIC DEATHS PER 100 MILLION MVM

Ala	6.4	La	7.1	Ohio	4.5
Alaska	8.8	Maine	4.6	Okla	5.0
Ariz	6.2	Mass	3.5	Ore	5.3
Ark	5.6	Md	3.9	Pa	4.1
Cal	4.4	Mich	4.2	RI	3.0
Color	5.3	Minn	4.6	SC	6.5
Conn	2.8	Miss	5.6	SD	5.4
Del	5.2	Mo	5.6	Tenn	7.1
Fla	5.5	Mont	7.0	Tex	5.2
Ga	6.1	NC	6.2	Utah	5.5
Hawaii	4.7	ND	4.8	Va	4.5
Idaho	7.1	Nebr	4.4	Vt	4.7
Ill	4.3	Nev	8.0	WVa	6.2
Ind	5.1	NH	4.6	Wash	4.3
Iowa	5.9	NJ	3.2	Wisc	4.7
Kans	5.0	NM	8.0	Wy	6.5
Ky	5.6	NY	4.7		

If we group these data into classes, 2.0–2.9, 3.0–3.9, and so on, we get the histogram shown in Figure 2.5.4. Superimposed over the histogram is the particular normal curve having $\mu = 5.3$ and $\sigma = 1.3$ (again, we will have to defer to a later chapter our reasons for choosing these particular values for μ and σ). Quite clearly, the normal curve works well here—that is, it describes very adequately the state-to-state variability in traffic fatality rates. We will have much more to say about this singularly important distribution in Chapters 6 and 7.

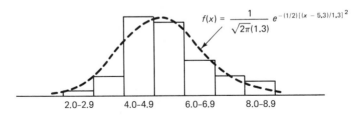

$$f(x) = \frac{1}{\sqrt{2\pi}(1.3)} e^{-(1/2)[(x-5.3)/1.3]^2}$$

2.0-2.9   4.0-4.9   6.0-6.9   8.0-8.9

Figure 2.5.4 Traffic deaths per 100 million mvm.

Question 2.5.1 In actuarial science, one of the simplest models used for describing mortality is

$$f(t) = k \cdot t^2 (100 - t)^2, \quad 0 \le t \le 100$$

where t denotes the age at which a person dies.
a. Find k.
b. Let A be the event "person lives past 60." Find $P(A)$.
c. What value does the model give for a person's median life expectancy?

Question 2.5.2 The length of a cotter pin that is part of a wheel assembly is supposed to be 6 cm. However, the machine that stamps out the parts makes them $6 + x$ cm long, where x varies from pin to pin according to the probability function

$$f(x) = k(x + x^2), \qquad 0 \le x \le 2$$

where k is a constant. If a pin is longer than 7 cm, it is unusable. What proportion of cotter pins produced by this machine will be unusable?

Question 2.5.3 In England from 1875 to 1951, the interval t (in days) between consecutive mining accidents resulting in at least ten fatalities was well described (57) by the exponential probability function $f(t) = (\frac{1}{241})e^{-t/241}$, $t > 0$. Estimate the probability that the gap between consecutive accidents would be somewhere between 50 and 100 days, inclusive.

Question 2.5.4 A batch of small-caliber ammunition is accepted as satisfactory if one shell selected at random is fired and lands within 2 ft of the center of a target. Assume that for a given batch, the probability function describing r, the distance from the target center to a shell's point of impact, is

$$f(r) = \frac{2re^{-r^2}}{1 - e^{-9}}, \qquad 0 < r < 3$$

Find the probability the batch will be accepted.

Question 2.5.5 Suppose the probability function describing the life of an ordinary 60-W bulb is

$$f(x) = (\frac{1}{1000})e^{-x/1000}, \qquad x > 0$$

a. What is the probability that such a bulb will burn more than 1000 h?
b. Find the bulb's *median* life (recall the comment following Example 2.5.2).

Question 2.5.6 Assume the reaction time of motorists over the age of 70 to a certain visual stimulus is described by a continuous probability function of the form

$$f(x) = xe^{-x}, \qquad x > 0$$

where x is measured in seconds. Let A be the event "motorist requires longer than 1.5 s to react." Find $P(A)$.

Question 2.5.7 Consider a probability function of the form

$$f(x) = ax^2e^{-bx}, \qquad k > 0, 0 < x$$

Integrate by parts twice to show that $a = k^3/2$.

Question 2.5.8 Given that

$$f(x) = \frac{1}{\sqrt{2\pi\sigma}}e^{-1/2[(x-\mu)/\sigma]^2}, \qquad -\infty < x < \infty$$

is a probability function for any number μ and any positive number σ, evaluate

$$\int_0^\infty e^{-4x^2}\, dx$$

Question 2.5.9 Let x be a variable described by the particular exponential probability function for which λ is equal to 1—that is,

$$f(x) = e^{-x}, \qquad x > 0$$

What is the probability that the first digit to the right of the decimal point in x is a 1? (*Hint:* If A is the event "first digit to the right of the decimal point is a 1," then

$$A = \bigcup_{n=0}^\infty [n + 0.1, n + 0.2)$$

The simplest f that can be defined over a continuous (and bounded) sample space is the *uniform*, or *rectangular,* probability function. Such a function is constant wherever it has nonzero probability:

$$f(x) = \begin{cases} \dfrac{1}{b-a} & \text{for } x \text{ in the interval } (a, b) \\ 0 & \text{elsewhere} \end{cases}$$

(see Figure 2.5.5). Thus, the probability of x lying in any subinterval in (a, b) of length t is the same, regardless of where in (a, b) the interval is located.

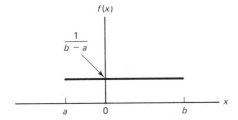

Figure 2.5.5

The uniform function is being singled out here for several reasons. First, its simplicity makes it the easiest of all probability functions with which to work. There are many situations in which a theorem can be stated and proved in general terms for arbitrary $f(x)$, but where the only tractable examples of that theorem (that is, applications where results can be obtained without having to resort to numerical integration) occur when $f(x)$ is set equal to $1/(b-a)$ for $x \in (a, b)$. We will see the uniform function reappearing in this role any number of times in the chapters ahead. A second reason has to do with the familiar phrase *at random.* If a discrete sample space has n outcomes and we say that one of them is picked at random, we mean that the probability function associated with S is the equally likely model $P(s) = 1/n$ for all s. If the sample space is an interval, though, picking a point x at random implies that $f(x) = 1/(b-a)$ for $a < x < b$.

Example 2.5.6

Suppose the electronic sensors set up to time bobsled races are accurate to the nearest tenth of a second. If the Olympic record for a particular course is 17.18 s and this year's Czechoslovakian team finished with a recorded time of 17.20 s, what is the probability that the world mark was actually eclipsed? Assume the timing errors are uniformly distributed.

What "accurate to the nearest 0.1 s" and "uniformly distributed" imply in this context is that the *true* time, x, achieved by the Czechoslovakian team could have been anything from 17.15 s to 17.25 s (17.20 ± 0.05) and that $f(x) = 10$ for any x in that interval (see Figure 2.5.6). Furthermore, if A is the event "world mark was broken," then

$$P(A) = \int_{x<17.18} f(x) \, dx = \int_{17.15}^{17.18} 10 \, dx = 0.3$$

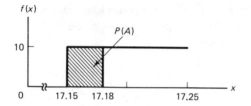

f(x)

10

P(A)

0   17.15 17.18   17.25   x   **Figure 2.5.6**

(Does the assumption that $f(x)$ be uniform seem reasonable? If not, what might we expect $f(x)$ to look like?)

Parzen (66) gives the following example of a uniform probability function that points out a rather curious property of the real numbers.

Example 2.5.7

Suppose x is a number randomly selected from the interval $(0, 1)$. That is, $f(x) = 1$ for $0 < x < 1$ and $f(x)$ is zero elsewhere. What is the probability that the second digit in $\sqrt{x}$ is equal to k, $k = 0, 1, 2, \ldots , 9$? We will show that even though x is uniform, the second digit in $\sqrt{x}$ is not.

Let A_k denote the set of numbers in $(0, 1)$ whose square roots have a second digit equal to k. We are looking for $P(A_k)$. Since

$$\sqrt{x} = 0 \cdot a_1 a_2 a_3 \cdots$$

it follows that

$$10\sqrt{x} = a_1 \cdot a_2 a_3 \cdots$$

and, for some m, $m = 0, 1, \ldots , 9$,

$$m + \frac{k}{10} \leq 10\sqrt{x} \leq m + \frac{k + 1}{10}$$

Writing the inequality in terms of x rather than $10\sqrt{x}$, we find that

$$\frac{1}{100}\left(m + \frac{k}{10}\right)^2 \leq x \leq \frac{1}{100}\left(m + \frac{k + 1}{10}\right)^2$$

More significantly, the *length* of the interval surrounding x is $(1/10,000) \times (20m + 2k + 1)$:

$$\frac{1}{100}\left(m + \frac{k + 1}{10}\right)^2 - \frac{1}{100}\left(m + \frac{k}{10}\right)^2 = \frac{1}{10,000}(20m + 2k + 1)$$

It was presumed at the outset that x was uniform over $(0, 1)$. Therefore, $P(A_k)$ is simply the sum—over m—of lengths of the form $(1/10,000) \times (20m + 2k + 1)$. That is,

$$P(A_k) = \sum_{m=0}^{9} \frac{1}{10,000}(20m + 2k + 1)$$

$$= 0.091 + 0.002k$$

(2.5.2)

which is *not* a uniform probability—the likelihood of the second digit in the square root being k increases as k increases. Table 2.5.3 lists the A_k probabilities resulting from Equation 2.5.2.

TABLE 2.5.3

Probability of second digit in $\sqrt{x}$ being k

k	Probability
0	0.091
1	0.093
2	0.095
3	0.097
4	0.099
5	0.101
6	0.103
7	0.105
8	0.107
9	0.109

(How would you expect the probability distribution for the second digit in $\sqrt{x}$ to compare with that for the *third*?)

A third reason for emphasizing the uniform probability function is that it can easily be extended to describe situations where two or more quantities are simultaneously varying at random. We will have much more to say about problems of this sort in Chapter 4, but we can preview some of that material here at an intuitive level.

Suppose the outcome of an experiment is an ordered *pair* of real numbers (x, y), where the coordinates are restricted to a rectangular area given by $S = \{(x, y) : 0 \le x \le a, 0 \le y \le b\}$ (see Figure 2.5.7). Let A_1 and A_2 be any two subsets

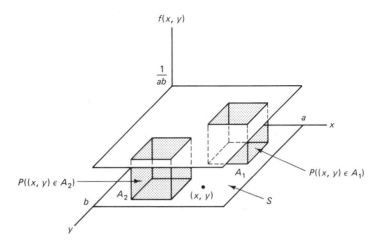

Figure 2.5.7

of S such that the *area* of A_1 is equal to the *area* of A_2. If it is also true that

$$P((x, y) \in A_1) = P((x, y) \in A_2) \qquad (2.5.3)$$

then x and y are said to be *uniformly distributed over the region S.*

In one-variable problems where the sample space is continuous, we have seen that the probability associated with an interval A is a *single* integral: $P(x \in A) = \int_A f(x)\, dx$. Generalizing that idea to two variables, it seems reasonable to associate the probability that (x, y) lies in some area A with the *double* integral of some $f(x, y)$ over the corresponding set of x's and y's: $P((x, y) \in A) = \iint_A f(x, y)\, dx\, dy$. Geometrically, this means that probabilities are now being represented as *volumes* under the surface $f(x, y)$, as opposed to *areas* under the curve $f(x)$. Functions like $f(x, y)$ are called *joint probability functions.* They need to be everywhere nonnegative and, of course, $\iint_S f(x, y)\, dx\, dy$ must equal 1.

For the joint *uniform* distribution, Equation 2.5.3 implies that $f(x, y)$ is a surface everywhere parallel to the xy-plane. Also, the condition that $\iint_S f(x, y)\, dx\, dy = 1$ implies that the height of the $f(x, y)$ plane is $1/ab$. Formally, then, we can define the joint uniform probability function over S by writing

$$f(x, y) = \begin{cases} \dfrac{1}{ab} & 0 \le x \le a;\ 0 \le y \le b \\ 0 & \text{elsewhere} \end{cases}$$

Note that the constancy of $f(x, y)$ over S allows us to express any double integral of $f(x, y)$ as a ratio of areas:

$$P((x, y) \in A) = \iint_A f(x, y)\, dx\, dy = \frac{\text{area of } A}{\text{area of } S}$$

The next several examples are all situations that reduce to the problem of selecting points at random in some prescribed region of the xy-plane. In each instance, the appropriate probability function is the joint uniform and the answer is gotten by taking a ratio of areas.

Example 2.5.8

A carnival operator wants to set up a ringtoss game. Players will throw a ring of diameter d onto a grid of squares, the side of each square being of length s (see Figure 2.5.8). If the ring lands entirely inside a square, the player wins a prize. To ensure a profit, the operator must keep the player's chances of winning down to something less than one in five. How small can the operator make the ratio d/s?

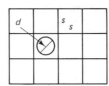

Figure 2.5.8

First, it will be assumed that the player is required to stand far enough away so that no skill is involved and the ring is falling at random on the grid. From Figure 2.5.9, we see that in order for the ring not to touch any side of the square, the ring's center must be somewhere in the interior of a smaller square, each side of which is a distance $d/2$ from one of the grid lines. Since the area of a grid square is s^2 and the area of an interior square is $(s - d)^2$, the probability of a winning toss can be written as a simple ratio:

$$P(\text{ring touches no lines}) = \frac{(s - d)^2}{s^2}$$

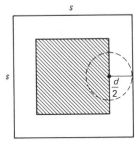

Figure 2.5.9

But the operator requires that

$$\frac{(s - d)^2}{s^2} \leq 0.20$$

Solving for d/s gives

$$\frac{d}{s} \geq 1 - \sqrt{0.20} = 0.55$$

That is, if the diameter of the ring is at least 55% as long as the side of one of the squares, the player will have no more than a 20% chance of winning.

Example 2.5.9

Two friends agree to meet, on the University Commons "sometime around 12:30." But neither of them is particularly punctual—or patient. What will actually happen is that each will arrive at random sometime in the interval from 12:00 to 1:00. If one arrives and the other is not there, the first person will wait 15 min or until 1:00, whichever comes first, and then leave. What is the probability the two will get together?

To simplify notation, we can represent the time period from 12:00 to 1:00 as the interval from 0 to 60 min. Then if x and y denote the two arrival times, the sample space is the 60×60 square shown in Figure 2.5.10. Furthermore, the event M, "the two friends meet," will occur if and only if $|x - y| \leq 15$ or, equivalently, if and only if $-15 \leq x - y \leq 15$. These inequalities appear as the shaded region in Figure 2.5.10.

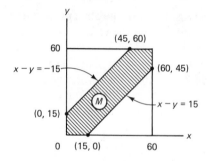

$$(45, 60)$$

$x - y = -15$

$(60, 45)$

$(0, 15)$ M $x - y = 15$

0 $(15, 0)$ 60

Figure 2.5.10

Notice that the areas of the two triangles above and below M are each equal to $\frac{1}{2}(45)(45)$. It follows that the two friends have a 44% chance of meeting:

$$P(M) = \frac{\text{area of } M}{\text{area of } S}$$

$$= \frac{(60)^2 - 2[\frac{1}{2}(45)(45)]}{(60)^2}$$

$$= 0.44$$

(How long would each have to wait for the other in order to raise their chances of meeting to 75%?)

Example 2.5.10

Consider a line AB divided into two parts by a point C, where the length of segment AC is greater than or equal to the length of segment CB (see Figure 2.5.11). Suppose a point X is chosen at random along the segment AC and a point Y is chosen at random along the segment CB. Let x and y denote the distances of X and Y from A and B, respectively. What is the probability the three segments AX, XY, and YB will fit together to form a triangle?

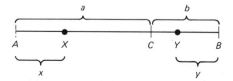

Figure 2.5.11

The key here is to recognize that three conditions must be met if the segments are to form a triangle—each segment must be shorter than the sum of the other two:

1. $x < (a + b - x - y) + y$
2. $a + b - x - y < x + y$
3. $y < x + (a + b - x - y)$

Intuitively, it seems clear that the probability of the segments forming a triangle will be greatest when C is the midpoint of AB: As b gets smaller, y tends to get smaller, and the likelihood of conditions (1) and (2) both being true diminishes. To make that argument precise, we need to determine what proportion of the sample space $S = \{(x, y) : 0 \le x \le a, \ 0 \le y \le b\}$ is included in the (x, y)-values satisfying conditions (1), (2), and (3).

Note that

1. $x < (a + b - x - y) + y \ \Rightarrow \ x < \dfrac{a + b}{2}$

2. $a + b - x - y < x + y \ \Rightarrow \ x + y > \dfrac{a + b}{2}$

3. $y < x + (a + b - x - y) \ \Rightarrow \ y < \dfrac{a + b}{2}$

The (x, y)-values satisfying all three of these inequalities make up the interior of the triangle shown in Figure 2.5.12. Call that interior E. It follows that the probability of the segments forming a triangle will equal the area of $E \cap S$ divided by the area of S:

$$P(\text{segments form triangle}) = \frac{\dfrac{1}{2} b \cdot \left(\dfrac{a + b}{2} - \dfrac{a - b}{2} \right)}{ab}$$

$$= \frac{b}{2a}$$

As expected, the probability is greatest when C is midway between A and B, and it decreases fairly rapidly as C approaches B.

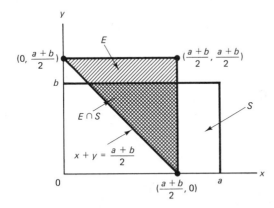

Figure 2.5.12

The next example is among the most-frequently cited of all geometric probability problems. What is particularly interesting about it is that it offers us a rather unusual way of estimating π. Known as *Buffon's needle problem*, it dates back to the eighteenth century.

Comment

Georges-Louis Leclerc, Comte de Buffon (1707–1788), and Montesquieu, Voltaire, and Rousseau were the four major literary figures of the French Enlightenment. Buffon's monumental work was the 44-volume *Histoire Naturelle, Generale et Particuliere,* which established him as one of the foremost theoretical biologists and geologists of the century. A man of eclectic interests, he also studied mathematics extensively, although his contributions in that area were not quite so far-reaching. Probably his best-known work in mathematics today is the needle problem described here, although in his own century he was more famous for having authored a French translation of Newton's *Method of Fluxions.*

Example 2.5.11

Imagine dropping a needle *s* inches long onto a grid of horizontal lines, each *s* inches apart. What is the probability the needle will cross one of the lines?
Figure 2.5.13 shows a typical result. Let

$$y = \text{distance from needle's center to nearest grid line}$$

$$\theta = \text{angle formed by the needle and the vertical}$$

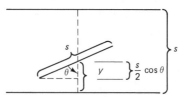

Figure 2.5.13

The sample space for the (θ, y)-values is then a rectangle of length π and height $s/2$:

$$S = \left\{ (\theta, y) : \frac{-\pi}{2} \leq \theta \leq \frac{\pi}{2};\ 0 \leq y \leq \frac{s}{2} \right\}$$

Furthermore, because the needle is being dropped at random, we can take the joint probability function for θ and y to be the joint uniform:

$$f(\theta, y) = \begin{cases} \dfrac{2}{\pi s} & -\dfrac{\pi}{2} \leq \theta \leq \dfrac{\pi}{2};\ 0 \leq y \leq \dfrac{s}{2} \\ 0 & \text{elsewhere} \end{cases}$$

Notice from Figure 2.5.13 that the needle touches a line if and only if $y \leq (s/2) \cos \theta$. The probability of that inequality being true is the ratio of the shaded area shown in Figure 2.5.14 to the total area of S, $\pi s/2$. Numerically,

$$P\left(y \leq \left(\frac{s}{2}\right) \cos \theta\right) = \int_{-\pi/2}^{\pi/2} \int_{0}^{(s/2)\cos\theta} \left(\frac{2}{\pi s}\right) dy\, d\theta = \int_{-\pi/2}^{\pi/2} \left(\frac{\cos\theta}{\pi}\right) d\theta$$

$$= \frac{2}{\pi} \tag{2.5.4}$$

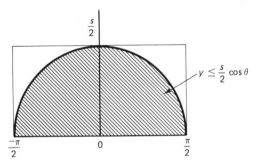

$y \leq \frac{s}{2} \cos \theta$

Figure 2.5.14

That is, each time we drop the needle under these conditions, the probability is $2/\pi$, or about 64%, that it will touch a line.

Equation 2.5.4 is a curious result—using it, we can come up with an unusual method for estimating π. Suppose a needle is dropped on the grid not once, but n times. And suppose that out of those n "trials," it crosses a line on m different occasions. Then the *empirical* proportion of crossings, m/n, is an obvious estimate for the *true* proportion, $2/\pi$. But $m/n \doteq 2/\pi$ implies that

$$\pi \doteq \frac{2n}{m}$$

Thus, if we tossed a needle 100 times and it crossed a line 60 times, the resulting estimate for π would be 3.03:

$$\pi \doteq \frac{2(100)}{60} = 3.03$$

(As might be expected, this is hardly one of the more efficient ways to estimate π. It can be shown (5) that something on the order of 1,000,000 trials would be necessary if we wanted to be reasonably certain that $2n/m$ had a value within 0.001 of the true value of π!)

Question 2.5.10 The time it takes a commuter to travel from home to the nearest train station is uniformly distributed between 15 and 20 minutes. The train leaves at exactly 7:30 A.M. Find the probability the commuter catches the train if she leaves home at 7:12 A.M.

Question 2.5.11 If 20 points are picked at random from the unit square, how many would we expect to find lying outside the inscribed circle of radius $\frac{1}{2}$?

Question 2.5.12 A point is chosen at random from the interior of a circle whose equation is $x^2 + y^2 \leq 4$. What is $f(x, y)$, the joint probability function for x and y? Use $f(x, y)$ to find $P(A)$, where $A = \{(x, y) : 1 \leq x^2 + y^2 \leq 3\}$.

Question 2.5.13 Suppose x and y are randomly chosen from the unit square. Let A be the set of (x, y)-values for which $x + y \leq 1$. Let B be the set of (x, y)-values for which $xy \leq \frac{2}{9}$. Find $P(A \cap B)$.

Question 2.5.14 A real number x is chosen at random from the interval $(0, 3)$. A second number, y, is chosen at random from the interval $(0, 4)$. Find the 80th percentile of the sum of the two numbers. (*Note:* In general, the pth percentile of $x + y$ would be that number—say, z_p—such that $P(x + y \le z_p) = p/100$.)

Question 2.5.15 Suppose the coefficient b in the quadratic equation

$$3x^2 + bx + 3 = 0$$

is chosen at random from the interval $(-10, 10)$. What is the probability the equation has complex roots?

Question 2.5.16 If a point is selected at random from the interior of the following circle of radius $4r$, what is the probability it lies in one of the shaded areas?

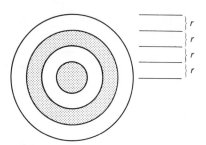

Question 2.5.17 A random number is chosen from the interval $(0, 10)$. What is the probability that a circle whose radius is that number has an area less than $\pi/4$?

Question 2.5.18 A point is chosen at random from the interior of a right triangle with base b and height h. What is the probability the point is in the lower half of the triangle?

Question 2.5.19 Two points X and Y ($X < Y$) are chosen at random along the interval AB. What is the probability the three line segments AX, XY, and YB can form a triangle?

Question 2.5.20 A right triangle is formed, where the length of the base is a number chosen at random from the interval $(0, b)$ and the height is a number chosen at random from the interval $(0, h)$. Find the 75th percentile of the area of the triangle (see Question 2.5.14).

Question 2.5.21 The volume V of a frustum of a right circular cone is given by $V = (\frac{1}{3})\pi h (R_1^2 + R_2^2 + R_1 R_2)$. Suppose a point is selected at random from the interior of a frustum whose height is 10 in. and whose upper and lower radii are 2 in. and 5 in., respectively. What is the probability the point selected is in the upper half of the frustum?

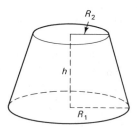

2.6 CONDITIONAL PROBABILITY

In this section we will see that the probability of an event A may have to be recomputed if we know for certain that some other event, B, has already occurred. That probabilities *should* change in the light of additional information is certainly not unreasonable and is an easily demonstrated concept. Consider a fair die being tossed with A defined as the event "the number 6 appears." Clearly, $P(A) = \frac{1}{6}$. But suppose the die has already been tossed by someone who refuses to tell us whether or not A occurred but does enlighten us to the point of confirming that B occurred, where B is the event "an even number appears." What are the chances of A now? The answer, of course, is obvious: There are three equally likely even numbers making up the event B, one of which satisfies the event A, so the updated probability is $\frac{1}{3}$.

Notice that the effect of additional information, such as the knowledge that B has occurred, is to revise—indeed, to *shrink*—the original sample space S to a new set of outcomes, S'. Here, the original S contains six outcomes; the conditional sample space contains three (see Figure 2.6.1).

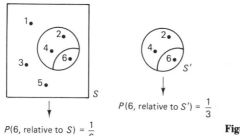

$P(6, \text{relative to } S) = \dfrac{1}{6}$

$P(6, \text{relative to } S') = \dfrac{1}{3}$

Figure 2.6.1

The symbol $P(A \mid B)$, read "the probability of A given B," is used to denote a conditional probability. Specifically, $P(A \mid B)$ refers to the probability that A *will occur* given that B has *already occurred*. It will prove convenient to have a formula for $P(A \mid B)$ that can be evaluated in terms of the original S rather than the revised S'. To motivate that formula, we need to consider a situation just slightly more general than the dice problem just mentioned. Suppose S is a finite sample space with n outcomes and P is the equally likely probability function. Suppose A and B are two events containing a and b outcomes, respectively, and let c denote the number of outcomes in the intersection of A and B. Then, from what we have just seen, the conditional probability of A given B is the ratio of c to b. But c/b can be written as the ratio of still two other ratios—namely,

$$\frac{c}{b} = \frac{c/n}{b/n}$$

which is a form that suggests the next definition.

Definition 2.6.1. Let A and B be any two events defined on S such that $P(B) > 0$. The conditional probability of A, assuming B has already occurred, is written $P(A|B)$ and is given by

$$P(A|B) = \frac{P(A \cap B)}{P(B)}$$

Example 2.6.1

Consider families with two children and assume the four possible outcomes—(younger is a boy, older is a boy), (younger is a boy, older is a girl), (younger is a girl, older is a boy), and (younger is a girl, older is a girl)—are all equally likely. What is the probability that both children are boys *given that at least one is a boy?* As a first thought, it may seem like the answer should be $\frac{1}{2}$, but it is not. The event "at least one is a boy" reduces the original sample space from four outcomes to three: $S' = \{$(younger is a boy, older is a girl), (younger is a girl, older is a boy), (younger is a boy, older is a boy)$\}$. Only one of these, though, is a two-boy family, so the correct probability is $\frac{1}{3}$. Reassuringly, Definition 2.6.1 gives the same answer. If A is the event "both children are boys" and B is "at least one child is a boy," then

$$P(A|B) = \frac{P(A \cap B)}{P(B)} = \frac{P(A)}{P(B)}$$

$$= \frac{\frac{1}{4}}{\frac{3}{4}} = \frac{1}{3}$$

(Note that $P(A \cap B) = P(A)$ here because $A \subset B$.)

Example 2.6.2

Recall the probability function describing mortality that was given in Question 2.5.1: If t is the age at which a person dies,

$$f(t) = 3 \times 10^{-9} \cdot t^2 (100 - t)^2, \qquad 0 \le t \le 100$$

What is the probability a person will die between the ages of 80 and 85 given that that person has lived to be at least 70?

Let A be the event "$80 \le t \le 85$" and let B be the event "$t > 70$." Then

$$P(\text{person dies between the ages of 80}$$
$$\text{and 85 given that that person}$$
$$\text{has lived to be at least 70}) = P(A|B)$$

$$= \frac{P(A \cap B)}{P(B)}$$

$$= \frac{\int_{80}^{85} 3 \times 10^{-9} \cdot t^2 (100 - t)^2 \, dt}{\int_{70}^{100} 3 \times 10^{-9} \cdot t^2 (100 - t)^2 \, dt}$$

$$= \frac{0.0313}{0.1631}$$

$$= 0.19$$

(Compare the conditional probability $P(80 \le t \le 85 \,|\, t \ge 70)$ with the *unconditional* probability $P(80 \le t \le 85)$. Which do you expect to be larger?)

Question 2.6.1 Suppose we ignored the ages of the children in two-child families and merely distinguished three family *types*: (boy, boy), (girl, boy), and (girl, girl). Would the conditional probability of both children being boys given that least one is a boy be different than the $\frac{1}{3}$ that was calculated in Example 2.6.1? Explain.

Question 2.6.2 One card is drawn from a standard deck.
    a. What is the probability the card is a king given the card is a club?
    b. What is the probability the card is a club given the card is a king?
    c. What is the probability the card is a 10 if we know it is not a face card?

Question 2.6.3 A sample space S consists of the integers 1 to n, inclusive. Each has an associated probability proportional to its magnitude. One integer is chosen at random. What is the probability that 1 was chosen given that the number selected was in the first m integers?

Question 2.6.4 If $P(A) = a$ and $P(B) = b$, show that

$$P(A \,|\, B) \ge \frac{a + b - 1}{b}$$

Question 2.6.5 If $P(A \,|\, B) < P(A)$, show that $P(B \,|\, A) < P(B)$.

Question 2.6.6 Let S be a sample space and P be a probability function defined on the events of S. Suppose H is a subset of S with $P(H) > 0$. Define a probability function P_H on S by $P_H(A) = P(A \,|\, H)$, for any $A \subset S$. Show that P_H satisfies Axioms 1, 2, and 3 of Section 2.3.

Question 2.6.7 Show that $P(B \,|\, A) > P(B)$ implies that $P(B^C \,|\, A) < P(B^C)$.

Question 2.6.8 Suppose two fair dice are tossed. What is the probability the sum equals 10 given it exceeds 8?

Definition 2.6.1 can be rewritten to give a useful formula for the probability of the intersection of two events: If $P(A \,|\, B) = P(A \cap B)/P(B)$, then

$$P(A \cap B) = P(A \,|\, B)P(B) \tag{2.6.1}$$

A similar formulation holds for intersections of higher order. For example, the probability of the intersection of events A, B, and C can be written as the product of three probabilities, two conditional and one unconditional:

$$P(A \cap B \cap C) = P(A \cap (B \cap C)) = P(A \mid B \cap C)P(B \cap C)$$
$$= P(A \mid B \cap C)P(B \mid C)P(C)$$

(2.6.2)

The next several examples illustrate Equations 2.6.1 and 2.6.2.

Example 2.6.3

Suppose an urn contains $r = 1$ red chips and $w = 1$ white chips. One is drawn at random. If the chip selected is red, that chip along with $k = 2$ additional red chips are put back into the urn. If it is white, the chip is simply returned to the urn, then a second chip is drawn. What is the probability that both selections are red?

If we let R_1 be the event "red chip is selected on first draw" and R_2 be "red chip is selected on second draw," it should be clear that

$$P(R_1) = \tfrac{1}{2} \quad \text{and} \quad P(R_2 \mid R_1) = \tfrac{3}{4}$$

Substituting these probabilities into Equation 2.6.1 gives

$$P(R_1 \cap R_2) = P(\text{both chips are red}) = P(R_1)P(R_2 \mid R_1)$$
$$= (\tfrac{1}{2}) \cdot (\tfrac{3}{4}) = \tfrac{3}{8}$$

Comment

What we have described in Example 2.6.3 can serve as a probability model for the spread of contagious diseases. Think of the red chips as, say, measles cases. The constants r and w characterize the initial status of the population at risk, while k reflects the strength of the contagion.

Example 2.6.4

The possibility of importing liquefied natural gas (LNG) from Algeria has been suggested as one way of easing the energy crunch in the United States. Complicating matters, though, is that LNG is highly volatile and poses an enormous safety hazard. Any major spill occurring near a U.S. port could result in a fire of catastrophic proportions. The question, therefore, of the *likelihood* of a spill becomes critical input for the policymakers who will have to decide whether or not to go ahead with the proposal.

Two numbers need to be taken into account: (1) the probability that a tanker will have an accident near a port and (2) the probability that a major spill will develop given that an accident has happened. Although no significant spills of LNG have yet occurred anywhere in the world, these probabilities can be approximated from records kept on similar tankers transporting less dangerous cargo. On the basis of such data, it has been estimated (22) that the probability that an LNG tanker will have an accident on any one trip is 8/50,000. Similarly, given that an accident *has* occurred, it is suspected that only 3 times out of

15,000 will the damage be severe enough to produce a major spill. Thus, the single-trip probability for a major LNG disaster is 3.2×10^{-8}:

$$P(\text{accident occurs and spill develops}) = P(\text{spill} \mid \text{accident})P(\text{accident})$$

$$= \left(\frac{3}{15,000}\right)\left(\frac{8}{50,000}\right)$$

$$= 0.000000032$$

Example 2.6.5

A man has n keys on a chain, one of which opens the door to his apartment. Having celebrated a little too much one evening, he returns home, only to find himself unable to distinguish one key from another. Resourceful, he works out a fiendishly clever plan: He will choose a key at random and try it. If it fails to open the door, he will discard it and choose one of the remaining $n - 1$ keys at random, and so on. Clearly, the probability that he gains entrance with the first key he selects is $1/n$. What is the probability the door opens with the second key he tries?

It would be tempting here to answer $1/(n - 1)$, but in this case our intuition would be in error. Actually, $1/(n - 1)$ is a right answer, but to the wrong question. To see why, let K_i, $i = 1, 2, \ldots, n$, denote the event "ith key tried opens door." Then $P(K_1)$ is certainly $1/n$, but the event "second key tried opens door" can occur only if the first key *does not* open the door. That is,

$$P(K_2) = P(K_2 \cap K_1^C) \qquad (2.6.3)$$

Applying Equation 2.6.1 to the right-hand side of Equation 2.6.3, we see that the probability that the second key tried opens the door is the same as the probability for the first key, $1/n$:

$$P(K_2 \cap K_1^C) = P(K_2 \mid K_1^C)P(K_1^C)$$

$$= \left(\frac{1}{n - 1}\right)\left(\frac{n - 1}{n}\right)$$

$$= \frac{1}{n}$$

Thus, the ratio $1/(n - 1)$ does answer a *conditional* probability question (what is $P(K_2 \mid K_1^C)$?), but the query as originally posed is asking about something *unconditional*—K_2. (Extend this argument to show that for *any* i, $P(K_i) = 1/n$.)

Example 2.6.6

The highways connecting two resort areas at A and B are shown in Figure 2.6.2: There is a direct route through the mountains and a more-circuitous route going through a third resort area at C in the foothills. Travel between A and B during the winter months is not always possible, the roads sometimes being closed due

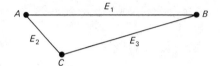

Figure 2.6.2

to snow and ice. Suppose we let E_1, E_2, and E_3 denote the events that highways AB, AC, and CB are passable, respectively, and we know from past years that on a typical winter day,

$$P(E_1) = \tfrac{2}{5} \qquad\qquad P(E_3 | E_2) = \tfrac{4}{5}$$
$$P(E_2) = \tfrac{3}{4} \qquad P(E_1 | E_2 \cap E_3) = \tfrac{1}{2}$$
$$P(E_3) = \tfrac{2}{3}$$

What is the probability that a traveler will be able to get from A to B?

If E denotes the event that we *can* get from A to B, then

$$E = E_1 \cup (E_2 \cap E_3)$$

It follows that

$$P(E) = P(E_1) + P(E_2 \cap E_3) - P[E_1 \cap (E_2 \cap E_3)]$$

Applying Equation 2.6.1 three times gives

$$P(E) = P(E_1) + P(E_3|E_2)P(E_2) - P[E_1|(E_2 \cap E_3)]P(E_2 \cap E_3)$$
$$= P(E_1) + P(E_3|E_2)P(E_2) - P[E_1|(E_2 \cap E_3)]P(E_3|E_2)P(E_2)$$
$$= \tfrac{2}{5} + (\tfrac{4}{5})(\tfrac{3}{4}) - (\tfrac{1}{2})(\tfrac{4}{5})(\tfrac{3}{4})$$
$$= 0.7$$

(Which route should a traveler starting from A try first to maximize the chances of getting to B?)

Example 2.6.7

Part of the support structure for a small pier consists of two oak logs sunk into the riverbank on either side of a wooden platform. It is possible that neither, either, or both of the logs will settle. On the basis of past experience, a realistic estimate for the chances of each log settling is 0.05. Also, if one settles, the probability of the other doing the same is 0.7. What is the probability the pier will be level?

If A is the event "first log settles," B is the event "second log settles," and E is the event "pier is not level," then

$$E = (A \cap B^C) \cup (A^C \cap B)$$

But $A \cap B^C$ and $A^C \cap B$ are mutually exclusive, so

$$P(E) = P(A \cap B^C) + P(A^C \cap B)$$

Replacing the intersection probabilities with conditional probabilities gives

$$P(E) = P(B^C|A)P(A) + P(A^C|B)P(B)$$

Since $P(A|B) = P(B|A) = 0.7$, it follows that $P(A^C|B) = P(B^C|A) = 1 - 0.7 = 0.3$, so

$$P(E) = (0.3)(0.05) + (0.3)(0.05)$$

$$= 0.03$$

Thus, there is a 97% chance the pier will be level. (Determine $P(E)$ a second way, by first finding $P(E^C)$.)

Example 2.6.8

A sandwich vendor parks her truck near the University Library each night to sell snacks, soft drinks, and coffee. She estimates that 20% of the students returning to their dormitories buy snacks, 30% soft drinks, and 10% cups of coffee. She has also observed that *if* a student buys a snack, there is a 50% chance he or she will also buy a soft drink and a 60% chance he or she will buy some coffee. No one, though, ever buys a soft drink *and* coffee. Approximately how many of the 200 students walking by the truck on a typical night will buy something?

Let A, B, and C represent the events "student buys snack," "student buys soft drink," and "student buys coffee," respectively. We first need to find $P(A \cup (B \cup C))$. That probability, multiplied by 200, would be the obvious estimate of the number of customers the vendor can expect to serve on any given night. By Theorem 2.3.6,

$$P(\text{student buys something}) = P(A \cup (B \cup C))$$

$$= P(A) + P(B \cup C) - P(A \cap (B \cup C))$$

Since B and C are mutually exclusive, $P(B \cap C) = 0$, so

$$P(A \cup (B \cup C)) = P(A) + P(B) + P(C) - P(A \cap (B \cup C))$$

From Equation 2.2.1,

$$P(A \cap (B \cup C)) = P((A \cap B) \cup (A \cap C))$$

$$= P(A \cap B) + P(A \cap C)$$

(Why is $P((A \cap B) \cap (A \cap C)) = 0$?) If these latter two intersection probabilities are written in conditional form,

$$P(A \cup (B \cup C)) = P(A) + P(B) + P(C) - [P(B|A)P(A) + P(C|A)P(A)]$$

It follows, then, from the estimates the vendor has for the right-hand-side probabilities that she can expect 76 of the students to buy something:

$$200 \cdot P(A \cup (B \cup C))$$

$$= 200 \cdot [0.20 + 0.30 + 0.10 - ((0.50)(0.20) + (0.60)(0.20))]$$

$$= 200 \cdot (0.38)$$

$$= 76$$

(How would the vendor's business be affected if a student's decision to buy either a soft drink or a cup of coffee was not influenced by whether or not he or she had already brought a snack?)

Question 2.6.9 Home-security experts estimate that an untrained housedog has a 70% probability of detecting an intruder and, given detection, a 50% chance of scaring the intruder away (1). What is the probability Fido successfully thwarts a burglar?

Question 2.6.10 A certain kind of avocado is grown in two valleys in southern California. Both areas are sometimes infested with a red aphid that can damage the fruit. Let A be the event "valley X is infested" and B be "valley Y is infested." Suppose $P(A) = \frac{2}{5}$, $P(B) = \frac{3}{4}$, and $P(A \cup B) = \frac{4}{5}$. If state inspectors find aphids on a shipment of avocados coming from valley Y, what is the probability that farmers in valley X are experiencing a similar problem?

Question 2.6.11 Recall Example 2.6.5. How might you deduce that $P(K_i) = 1/n$, $i = 1, 2, \ldots, n$, on purely intuitive grounds? (*Hint:* Imagine a key chain with the keys numbered 1 through n.)

Question 2.6.12 Suppose a probability function is given by

$$f(x) = \left(\frac{1}{\lambda}\right)e^{-x/\lambda}, \qquad x > 0$$

Show that the conditional probability of x exceeding some value $s + t$ given that x has already exceeded t is equal to the probability that x exceeds s. What does this imply about "wearout" if $f(x)$ is being used to model equipment failure, as in Example 2.5.4?

Question 2.6.13 An urn contains two white chips and one black chip. One is drawn at random and replaced with an additional chip of the same color. The same procedure is repeated two more times. Find the probabilities associated with the eight points in the sample space.

Question 2.6.14 Suppose A and B are mutually exclusive and $P(A \cup B) \neq 0$. Find an expression for $P(A \mid A \cup B)$ in terms of $P(A)$ and $P(B)$.

Question 2.6.15 If $B \subset A$, $C \subset A$, and $P(C) > 0$, show that

$$\frac{P(B)}{P(C)} = \frac{P(B \mid A)}{P(C \mid A)}$$

Question 2.6.16 Three actors audition for parts in a forthcoming TV series. On the basis of their past experience and the caliber of competition they face, A has a 40% chance of being hired, B a 50% chance, and C a 30% chance. If exactly two of the three are cast, what is the probability that A was rejected?

Question 2.6.17 Two fair dice are tossed. What is the probability at least one shows a 5 given that their sum is even?

Question 2.6.18 Foster, who is something of a troublemaker, is taking a probability course and after reading the urn problem of Example 2.6.3 disagrees with the proposed solution. He believes the answer should be $\frac{2}{6}$, not $\frac{3}{8}$. Suppose, says Foster, we label the initial chips in the urn r_1 and w_1 and the two red chips that may be added r_2 and r_3. Then the sample space contains six outcomes,

$$S = \{(r_1, r_1), (r_1, r_2), (r_1, r_3), (r_1, w_1), (w_1, w_1), (w_1, r_1)\}$$

all of which are equally likely since the sampling is done at random. So, by inspection, $P(R_1 \cap R_2) = P\{(r_1, r_1), (r_1, r_2), (r_1, r_3)\} = \frac{3}{6}$. Where has Foster gone wrong, or is he right?

Equation 2.6.1 can be extended in still another way. Suppose S is a sample space partitioned by a set of mutually exclusive and exhaustive events $A_1, A_2, \ldots, A_n$—that is, every sample outcome in S belongs to one and only one A_i (see Figure 2.6.3). Let B be some other event defined over S. What we will derive is a formula for the unconditional probability for B in terms of the n conditional probabilities $P(B|A_i)$. This will prove to be a fundamentally important result.

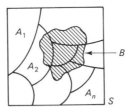

Figure 2.6.3

Theorem 2.6.1. Let $\{A_i\}_{i=1}^{n}$ be a set of events defined over S such that $S = \bigcup_{i=1}^{n} A_i$, $A_i \cap A_j = \emptyset$ for $i \neq j$, and $P(A_i) > 0$ for $i = 1, 2, \ldots, n$. For any event B,

$$P(B) = \sum_{i=1}^{n} P(B|A_i)P(A_i)$$

Proof. Because of the first two conditions imposed on the A_i's,

$P(B) = P(B \cap S)$

$\qquad = P\left(B \cap \left(\bigcup_{i=1}^{n} A_i\right)\right) = P\left(\bigcup_{i=1}^{n} (B \cap A_i)\right)$ (Recall Example 2.2.14.)

$\qquad = \sum_{i=1}^{n} P(B \cap A_i)$ (Why?)

Apply Equation 2.6.1 to each of the intersection probabilities and the result follows:

$$P(B) = \sum_{i=1}^{n} P(B|A_i)P(A_i)$$

Example 2.6.9

The starship *Enterprise* is planning a surprise attack against the Klingons at 0100 hours. The possibility of encountering a radiation storm, though, is causing Captain Kirk and Mr. Spock to reassess their strategy. According to Spock's calculations, the probability of the *Enterprise* encountering a severe radiation storm at 0100 hours is 0.6, a moderate storm, 0.3, and no storm, 0.1. Captain Kirk feels the probability of the attack's being a success is 0.8 if there is a severe storm, 0.7 if there is a moderate storm, and 0.2 if there is no storm. Spock claims the attack would be a tactical misadventure if its probability of success were not at least 0.7306. Should they attack?

No! Let A_1, A_2, and A_3 denote the events "severe radiation storm," "moderate radiation storm," and "no radiation storm," respectively; let B be the event "attack is a success." Substituting directly into Theorem 2.6.1, we see that the proposed mission's chances for success are not sufficiently promising:

$$P(B) = P(B|A_1)P(A_1) + P(B|A_2)P(A_2) + P(B|A_3)P(A_3)$$
$$= (0.8)(0.6) + (0.7)(0.3) + (0.2)(0.1)$$
$$= 0.71$$

Kirk and Spock should let the Klingons alone and go find some Romulans to pick on!

Example 2.6.10

A toy manufacturer buys ball bearings from three different suppliers—50% of his total order comes from supplier 1, 30% from supplier 2, and the rest from supplier 3. Past experience has shown that the quality control standards of the three suppliers are not all the same. Of the ball bearings produced by supplier 1, 2% are defective, while suppliers 2 and 3 produce defective bearings 3% and 4% of the time, respectively. What proportion of the ball bearings in the toy manufacturer's inventory are defective?

Let A_i be the event "bearing came from supplier i," $i = 1, 2, 3$. Let B be the event "bearing in toy manufacturer's inventory is defective." Then

$$P(A_1) = 0.5, \qquad P(A_2) = 0.3, \qquad P(A_3) = 0.2$$

and

$$P(B|A_1) = 0.02, \qquad P(B|A_2) = 0.03, \qquad P(B|A_3) = 0.04$$

Combining these probabilities according to Theorem 2.6.1 gives

$$P(B) = (0.02)(0.5) + (0.03)(0.3) + (0.04)(0.2)$$
$$= 0.027$$

meaning the manufacturer can expect 2.7% of his stock of ball bearings to be defective.

Example 2.6.11

Consider the following con game. Three cards are put into a hat. Card 1 is white on both sides, card 2 is red on both sides, and card 3 is red on one side and white on the other. A card is drawn from the hat and placed on a table, without the underside being seen. Suppose the side showing is red. The con man argues that what has been drawn is clearly not the white-white card, so the underside is equally likely to be red or white. He offers to take "red" at even money. It all sounds fair enough, but if you watched a series of games like this being played, the con man will win $\frac{2}{3}$ of the time rather than half the time. Why?

A correct analysis of this game recognizes that a conditional probability is involved. Let A_i be the event of drawing card i, $i = 1, 2, 3$. Then $P(A_1) = P(A_2) = P(A_3) = \frac{1}{3}$. Let D be the event "red side is down" and U the event "red side is up." The bet as proposed is on the occurrence of D given the occurrence of U. With the problem reduced to these terms, the solution is a simple application of Definition 2.6.1 and Theorem 2.6.1:

$$P(D|U) = \frac{P(D \cap U)}{P(U)} = \frac{P(D \cap U)}{P(U|A_1)P(A_1) + P(U|A_2)P(A_2) + P(U|A_3)P(A_3)}$$

$$= \frac{\frac{1}{3}}{0(\frac{1}{3}) + 1(\frac{1}{3}) + \frac{1}{2}(\frac{1}{3})}$$

$$= \frac{2}{3}$$

(Note the similarity between this problem and Example 2.6.1. What makes the answer go contrary to our common sense here is that we are not in effect drawing a card, which is what the statement of the question deliberately leads us to believe; rather, we are drawing the *side* of a card. Write out the sample space and convince yourself without appealing formally to Theorem 2.6.1 that $\frac{2}{3}$ is the right answer.)

Question 2.6.19 At the University Computing Center, 35% of all programs submitted are written in COBOL; the remaining 65% are in FORTRAN. Suppose that 10% of the COBOL programs and 15% of the FORTRAN programs compile on their first run. What is the probability the next program submitted will compile on its first run?

Question 2.6.20 Recall the survival lottery described in Example 2.2.4. What is the probability of release associated with the prisoner's optimal strategy?

Question 2.6.21 In an upstate congressional race, the incumbent Republican (R) is running against three Democrats (D_1, D_2, D_3) seeking the nomination. Political pundits estimate that D_1, D_2, and D_3 have probabilities 0.35, 0.40, and 0.25, respectively, of being nominated. Furthermore, they feel that R's chance of winning the general election over D_1 is 0.40; over D_2, 0.35; and over D_3, 0.60. What is the probability that R will win the election?

Question 2.6.22 Urn I contains two red chips and four white chips; urn II contains three red and one white. A chip is drawn at random from urn I and transferred to urn II. Then a chip is drawn from urn II. What is the probability the chip drawn from urn II is red?

Question 2.6.23 The table summarizes the prediction record of a TV weather forecaster over the past several years. Entries in the table are intersection probabilities.

		Forecast		
		Sunny	Cloudy	Rain
Actual Weather	Sunny	0.3	0.05	0.05
	Cloudy	0.04	0.2	0.02
	Rain	0.1	0.04	0.2

a. What proportion of the days were actually sunny?

b. How often was the weather forecaster wrong?

c. What was the probability of rain on the days the forecast was for sunny weather?

Question 2.6.24 Urn I contains three red chips and one white chip. Urn II contains two red chips and two white chips. One chip is drawn from each urn and transferred to the other urn. Then a chip is drawn from the first urn. What is the probability the chip ultimately drawn from urn I is red?

Question 2.6.25 Show that the following argument is incorrect. A study has shown that 7 out of 10 people will call heads in calling a coin toss. But heads occurs only 5 times out of 10, on the average. Thus, it is to your advantage to let the other person call the toss.

Question 2.6.26 A telephone solicitor is responsible for canvassing three suburbs. In the past, 60% of the completed calls to Belle Meade have resulted in contributions, compared to 55% for Oak Hill and 35% for Antioch. Her list of telephone numbers includes 1000 households from Belle Meade, 1000 from Oak Hill, and 2000 from Antioch. Suppose she picks a number at random from the list and places the call. What is the probability she gets a donation?

Question 2.6.27 A card is drawn from the top of a standard deck and dealt face down. Then the next card is turned over. What is the probability that that second card is an ace? Do the problem two ways: first, using Theorem 2.6.1 and second, using intuition.

Question 2.6.28 Two points are chosen at random along the unit interval. What is the probability the distance between them is less than q?

Question 2.6.29 Define a probability function P' by

$$P'(A) = \frac{P(A \cap B)}{P(B)}$$

Show that P' satisfies Axioms 1–3 of Section 2.3.

Question 2.6.30 An urn contains four white chips and three red chips. Two are drawn out at random and not replaced. Then a third is selected. What is the probability the third chip will be white?

Question 2.6.31 Hebephrenia, catatonia, and paranoia are three mental disorders related to schizophrenia. The data in the table were collected (76) to see what influence, if any, heredity has in determining the type of schizophrenic a person is likely to become. The subjects were 160 children and young adults with mental disorders. Each had a relative with a diagnosed mental condition.

		Diagnosis of index case		
		Hebephrenia	Catatonia	Paranoia
Diagnosis of Relative	Hebephrenia	49	6	5
	Catatonia	10	31	10
	Paranoia	21	15	13

a. Which is more likely, two similar disorders in the same family or two dissimilar disorders?

b. Given that the index case is something other than paranoia, what is the probability the relative will be something other than catatonic?

c. Use Theorem 2.6.1 to find P(relative is hebephrenic).

Question 2.6.32 Show that

$$P(A|B^C) = \frac{P(A) - P(A \cap B)}{1 - P(B)}$$

Question 2.6.33 Show that

$$P(A \cup B|C) = P(A|C) + P(B|C) - P(A \cap B|C)$$

Question 2.6.34 An urn contains five white chips, four black chips, and three red chips. Four chips are drawn sequentially and without replacement. What is the probability of obtaining the sequence (white, red, white, black)?

Question 2.6.35 Under what conditions does $P(A) = P(A|B) + P(A|B^C)$?

Question 2.6.36 Let $A_1, A_2, \ldots, A_n$ be any set of events defined on a sample space S such that $P(A_1) > 0, P(A_1 \cap A_2) > 0, \ldots, P(A_1 \cap A_2 \cap \cdots \cap A_n) > 0$. Use induction to prove the generalization of Theorem 2.6.2:

$$P(A_1 \cap A_2 \cap \cdots \cap A_n) = P(A_1)P(A_2|A_1)P(A_3|A_1 \cap A_2) \cdots$$

$$P(A_n|A_1 \cap A_2 \cap \cdots \cap A_{n-1})$$

Question 2.6.37 Consider a set of n urns, each containing one red chip and two white chips. A chip is drawn at random from the first urn: If it is white, it is transferred to the second urn; otherwise, it is put back in urn 1 and nothing is transferred. Then a chip is drawn from urn 2 and transferred to urn 3 *if it is white,* and so on. What is the probability a white chip is drawn from the nth urn? (*Hint:* Get an expression for $p_n = P$(white chip is drawn from urn n) in terms of p_{n-1}.)

2.7 BAYES' THEOREM

The theorem in this section has an interesting history. Its first explicit statement, coming in 1812, was due to Laplace, but its name derives from the Reverend Thomas Bayes, whose 1763 paper (published posthumously) had already outlined the result. On one level, the theorem is a relatively minor extension of the definition of conditional probability. When viewed from a loftier perspective, though, it takes on some rather profound philosophical implications. These implications, in fact, have precipitated a schism among practicing statisticians: Bayesians analyze data one way, non-Bayesians, another.

Our concern with the result will have nothing to do with its statistical interpretation. We will use it simply as the Reverend Bayes originally intended, as a formula for evaluating a certain kind of "inverse" probability. If we know $P(B|A_i)$ for all i, the theorem enables us to compute conditional probabilities "in the other direction"—that is, we can use the $P(B|A_i)$'s to find $P(A_j|B)$.

> **Theorem 2.7.1 (Bayes').**   Let $\{A_i\}_{i=1}^{n}$ be a set of n events, each with positive probability, that partition S in such a way that $\cup_{i=1}^{n} A_i = S$ and $A_i \cap A_j = \emptyset$ for $i \neq j$. For any event B (also defined on S), where $P(B) > 0$,
>
> $$P(A_j|B) = \frac{P(B|A_j)P(A_j)}{\sum\limits_{i=1}^{n} P(B|A_i)P(A_i)}$$
>
> for any $1 \leq j \leq n$.

Proof. From Definition 2.6.1,

$$P(A_j|B) = \frac{P(A_j \cap B)}{P(B)} = \frac{P(B|A_j)P(A_j)}{P(B)}$$

But Theorem 2.6.1 allows the denominator to be written as

$$\sum_{i=1}^{n} P(B|A_i)P(A_i)$$

and the result follows.

Example 2.7.1

Urn I contains two white chips (w_1, w_2) and one red chip (r_1); urn II has one white chip (w_3) and two red chips (r_2, r_3). One chip is drawn at random from urn I and transferred to urn II. Then one chip is drawn from urn II (see Figure 2.7.1). Suppose a red chip is selected from urn II. What is the probability the chip *transferred* was white?

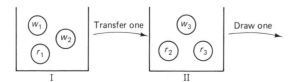

                    I   II   **Figure 2.7.1**

Let A_1 and A_2 denote the events "red chip is transferred from urn I" and "white chip is transferred from urn I." Let B be the event "red chip is drawn from urn II." What we are asking for is $P(A_2|B)$. We will find it instructive here to examine the problem at its most elementary level by enumerating the sample space. If, for example, (r_i, w_j) denotes the particular outcome of transferring the ith red chip from urn I and then drawing the jth white chip from urn II, the sample space reduces to a set of 12 such points, all equally likely. Figure 2.7.2 shows S together with the events A_1, A_2, and B. By inspection, $P(A_2|B) = \frac{4}{7}$.

To do the problem more formally, we note that $P(A_1) = \frac{1}{3}$, $P(A_2) = \frac{2}{3}$, $P(B|A_1) = \frac{3}{4}$, and $P(B|A_2) = \frac{2}{4}$. Therefore, by Theorem 2.7.1,

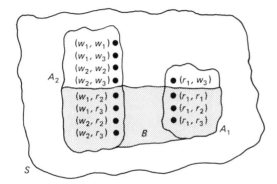

S

Figure 2.7.2

$$P(A_2|B) = \frac{P(B|A_2)P(A_2)}{P(B|A_1)P(A_1) + P(B|A_2)P(A_2)}$$

$$= \frac{(\frac{2}{4})(\frac{2}{3})}{(\frac{3}{4})(\frac{1}{3}) + (\frac{2}{4})(\frac{2}{3})}$$

$$= \frac{4}{7}$$

the same answer obtained by direct enumeration.

Example 2.7.2

A weather satellite is sending a binary code of 0s and 1s describing a developing tropical storm. Channel noise, though, can be expected to introduce a certain amount of transmission error. Suppose 70% of the message being relayed is 0s and there is an 80% chance of a given 0 or 1 being received properly. If a 1 is received, what is the probability that a 0 was sent?

Let B_i denote the event "i was sent," $i = 0, 1$, and let A_i denote the event "i is received," $i = 0, 1$. We want to find $P(B_0|A_1)$. According to what we are given,

$$P(B_0) = 0.7 \qquad (\Rightarrow P(B_1) = 0.3)$$
$$P(A_0|B_0) = 0.8 \qquad (\Rightarrow P(A_1|B_0) = 0.2)$$
$$P(A_1|B_1) = 0.8 \qquad (\Rightarrow P(A_0|B_1) = 0.2)$$

From Theorem 2.7.1, then, the probability of a 0 having been sent given that a 1 was received is 0.37:

$$P(B_0|A_1) = \frac{P(A_1|B_0)P(B_0)}{P(A_1|B_0)P(B_0) + P(A_1|B_1)P(B_1)}$$

$$= \frac{(0.2)(0.7)}{(0.2)(0.7) + (0.8)(0.3)}$$

$$= 0.37$$

(What are the chances that a 1 was sent given that a 0 is received?)

Example 2.7.3

State College is playing Backwater A & M for the conference football championship. If Backwater's first-string quarterback is healthy, A & M has a 75% chance of winning. If they have to start their backup quarterback, their chances of winning drop to 40%. The team physician says there is a 70% chance the first-string quarterback will play.

  a. What is the probability that Backwater will win the game?
  b. Suppose you miss the game but read in the headlines of Sunday's paper that Backwater won. What is the probability the second-string quarterback started?

The answer to part (a) comes from Theorem 2.6.1. Let B be the event "Backwater wins" and let A_1 and A_2 be the events, "first-string quarterback starts" and "second-string quarterback starts," respectively. Then

$$P(B) = P(B|A_1)P(A_1) + P(B|A_2)P(A_2)$$
$$= (0.75)(0.70) + (0.40)(0.30)$$
$$= 0.645$$

For part (b) we need to appeal to Bayes' theorem—the question is asking for $P(A_2|B)$. But

$$P(A_2|B) = \frac{P(B|A_2)P(A_2)}{P(B|A_1)P(A_1) + P(B|A_2)P(A_2)}$$

$$= \frac{(0.40)(0.30)}{(0.75)(0.70) + (0.40)(0.30)}$$

which reduces to 0.186. That is, given that Backwater won, there is approximately a 19% chance it was the backup quarterback who played. (What is the probability Backwater wins *and* the second-string quarterback plays?)

Bayes' theorem has been applied with considerable success to the problem of diagnosing medical conditions—specifically, to estimating the probability that a patient has a certain disease given that a particular diagnostic procedure says he or she does. The next example is one such application and it makes a significant point: When the disease for which we are looking is very rare, the number of incorrect diagnoses can be alarmingly high.

Example 2.7.4

Consider the problem of screening for cervical cancer (74). Let C be the event "a woman has the disease" and B be the event "positive biopsy"—that is, B occurs when the diagnostic procedure indicates that a woman *does* have cervical cancer. We will assume that $P(C) = 0.0001$, $P(B|C) = 0.90$ (the test correctly identifies 90% of all the women who do have the disease), and

$P(B|C^C) = 0.001$ (the test gives one false positive, on the average, out of every 1000 patients). Find $P(C|B)$, the probability a woman actually does have cervical cancer given the biopsy says she does.

While the method of solution here is straightforward, the actual numerical answer is not at all what we would expect. From Theorem 2.7.1,

$$P(C|B) = \frac{P(B|C)P(C)}{P(B|C)P(C) + P(B|C^C)P(C^C)}$$

$$= \frac{(0.9)(0.0001)}{(0.9)(0.0001) + (0.001)(0.9999)}$$

$$= 0.08$$

That is, only 8% of those women identified as having the disease actually do! Table 2.7.1 shows the strong dependence of $P(C|B)$ on $P(C)$ and $P(B|C^C)$.

TABLE 2.7.1

$P(C)$	$P(B|C^C)$	$P(C|B)$
0.0001	0.001	0.08
	0.0001	0.47
0.001	0.001	0.47
	0.0001	0.90
0.01	0.001	0.90
	0.0001	0.99

In light of these figures, the practicality of large-scale screening programs directed at diseases with low prevalence is open to question, particularly when the diagnostic procedure itself may be a health hazard, as would be the case in using annual chest X rays to look for tuberculosis.

Question 2.7.1 Suppose 0.5% of all the students seeking treatment at a school infirmary are eventually diagnosed as having mononucleosis. Of those who do have mono, 90% complain of a sore throat. But 30% of those who do not have mono also claim to have sore throats. If a student comes to the infirmary and says he or she has a sore throat, what is the probability the student has mono?

Question 2.7.2 During a power blackout, 100 persons are arrested on suspicion of looting. Each is given a polygraph test. From past experience, it is known that the polygraph is 90% reliable when administered to a guilty suspect and 98% reliable when given to someone who is innocent. Suppose that of the 100 persons taken into custody, only 12 were actually involved in any wrongdoing. What is the probability that a given suspect is innocent given that the polygraph says he or she is guilty?

Question 2.7.3 A dashboard warning light is supposed to flash red if a car's oil pressure is too low. On a certain model, the probability of the light flashing when it should is 0.95; 2% of the time, though, it flashes for no apparent reason. If there is a 10% chance the oil pressure really is low, what is the probability a driver needs to be concerned if the warning light goes on?

Question 2.7.4 A biased coin, twice as likely to come up heads as tails, is tossed once. If it shows heads, a chip is drawn from urn I, which contains three white chips and four red chips; if it shows tails, a chip is drawn from urn II, which contains six white chips and three red chips. Given that a white chip was drawn, what is the probability the coin came up tails?

Question 2.7.5 Your next-door neighbor has a rather old and tempermental burglar alarm. If someone breaks into his house, the probability of the alarm sounding is 0.95. In the last 2 years, though, it has gone off on five different nights, each time for no apparent reason. Police records show that the chances of a home being burglarized in your community on any given night are 2 in 10,000. If your neighbor's alarm goes off tomorrow night, what is the probability his house is being burglarized?

Question 2.7.6 Two sections of a senior probability course are being taught. From what she has heard about the two instructors listed, Francesca estimates that her chances of passing the course are 0.85 if she gets professor X and 0.60 if she gets professor Y. The section into which she is put is determined by the registrar. Suppose that her chances of being assigned to professor X are 4 out of 10. Fifteen weeks later we learn that Francesca did, indeed, pass the course. What is the probability she was enrolled in professor X's section?

Question 2.7.7 A liquor store owner is willing to cash personal checks for amounts up to $50, but she has become wary of customers who wear sunglasses. Fifty percent of checks written by persons wearing sunglasses bounce. In contrast, 98% of the checks written by persons not wearing sunglasses clear the bank. She estimates that 10% of her customers wear sunglasses. If the bank returns a check and marks it "insufficient funds," what is the probability it was written by someone wearing sunglasses?

Question 2.7.8 A loaded die is tossed for which

$$P(i \text{ appears}) = ki, \qquad i = 1, 2, \ldots, 6$$

If an i is rolled, a fair coin is tossed i times. Given that at least one head has appeared, what is the probability a 2 was rolled?

Question 2.7.9 The governor of a western state has decided to come out strongly for prison reform and has proposed a new pardon program. Its guidelines are simple: If a prisoner is related to a member of the governor's staff, he has a 90% chance of being pardoned; if he is not a relative, the chances for a pardon are 0.01. Suppose that 40% of all inmates are related to someone on the governor's staff. If a prisoner is pardoned, what is the probability he was related to someone in the administration?

Question 2.7.10 Bart is planning to murder his rich uncle Basil in hopes of claiming his inheritance a bit early. Hoping to take advantage of his uncle's predilection for immoderate desserts, Bart has put rat poison in the cherries flambé and cyanide in the chocolate mousse. The probability of the rat poison being fatal is 0.60; the cyanide, 0.90. Based on other dinners he has had with his uncle, Bart estimates that Basil has a 50% chance of asking for the cherries flambé, a 40% chance of ordering the chocolate mousse, and a 10% chance of either requesting something else or skipping dessert altogether. Given that Basil did, indeed, suffer a premature demise, what is the probability it was the chocolate mousse that did him in?

Question 2.7.11 Josh takes a 20-question multiple choice exam where each question has 5 answers. Some of the answers he knows, while others he gets right just by making lucky guesses. Suppose the conditional probability of his knowing the answer to a randomly selected question given that he got it right is 0.92. For how many of the 20 questions was he prepared?

Question 2.7.12 Recently the U.S. Senate Committee on Labor and Public Welfare investigated the feasibility of setting up a national screening program to detect child abuse. A team of consultants estimated the following probabilities: (1) 1 child in 90 is abused, (2) a physician can detect an abused child 90% of the time, and (3) a screening program would incorrectly label 3% of all nonabused children as abused. What is the probability a child is actually abused given the screening program diagnoses him or her as such? How does this probability change if the incidence of abuse is 1 in 1000? One in 50?

2.8 INDEPENDENCE

The previous two sections have dealt with the problem of reevaluating the probability of a given event in light of the additional information that some other event has already occurred. It sometimes is the case, though, that the probability of the given event remains unchanged regardless of the outcome of the second event—that is, $P(A|B) = P(A) = P(A|B^C)$. Events sharing this property are said to be *independent*. Definition 2.8.1 gives a necessary and sufficient condition for two events to be independent.

> **Definition 2.8.1.** Two events A and B are said to be *independent* if $P(A \cap B) = P(A) \cdot P(B)$.

Comment

The fact that the probability of the intersection of two independent events is equal to the product of their individual probabilities follows immediately from our first definition of independence, that $P(A|B) = P(A)$. Recall that the definition of conditional probability holds true for *any* two events A and B (provided $P(B) > 0$):

$$P(A|B) = \frac{P(A \cap B)}{P(B)}$$

But $P(A|B)$ can equal $P(A)$ only if $P(A \cap B)$ factors into $P(A)$ times $P(B)$.

The first set of examples in this section address the problem of establishing whether or not two (or more than two) events are independent. The final set, Examples 2.8.6–2.8.10, show how probabilities involving intersections are greatly simplified if we know that the events in question are independent.

Example 2.8.1

Let K be the event "drawing a king" from a standard poker deck and D be the event "drawing a diamond." Then, by Definition 2.8.1, K and D are independent because the probability of their intersection—drawing a king of diamonds—is equal to $P(K) \cdot P(D)$:

$$P(K \cap D) = \tfrac{1}{52} = \tfrac{1}{13} \cdot \tfrac{1}{4} = P(K)P(D)$$

Example 2.8.2

Suppose A and B are independent events. Does it follow that A^C and B^C are also independent? That is, does $P(A \cap B) = P(A)P(B)$ guarantee that $P(A^C \cap B^C) = P(A^C)P(B^C)$?

The answer is yes, the proof being accomplished by equating two different expressions for $P(A^C \cup B^C)$. First, by Theorem 2.3.6,

$$P(A^C \cup B^C) = P(A^C) + P(B^C) - P(A^C \cap B^C) \qquad (2.8.1)$$

But the union of two complements is also the complement of their intersection (recall Question 2.2.28). Therefore,

$$P(A^C \cup B^C) = 1 - P(A \cap B) \qquad (2.8.2)$$

Combining Equations 2.8.1 and 2.8.2, we get

$$1 - P(A \cap B) = 1 - P(A) + 1 - P(B) - P(A^C \cap B^C)$$

Since A and B are independent, $P(A \cap B) = P(A) \cdot P(B)$, so

$$P(A^C \cap B^C) = 1 - P(A) + 1 - P(B) - (1 - P(A)P(B))$$

$$= (1 - P(A))(1 - P(B))$$

$$= P(A^C)P(B^C)$$

the latter factorization implying that A^C and B^C are, themselves, independent. (If A and B are independent, are A and B^C independent?)

It is not immediately obvious how to extend Definition 2.8.1 to *three* events. To call A, B, and C independent, should we require that the probability of the three-way intersection factors into the product of the three original probabilities,

$$P(A \cap B \cap C) = P(A)P(B)P(C) \qquad (2.8.3)$$

or should we impose the definition we already have on the three *pairs* of events:

$$P(A \cap B) = P(A)P(B)$$

$$P(B \cap C) = P(B)P(C) \qquad (2.8.4)$$

$$P(A \cap C) = P(A)P(C)$$

As the next two examples show, neither condition by itself is sufficient. If three events satisfy Equations 2.8.3 *and* 2.8.4, we will call them independent (or mutually independent), but Equation 2.8.3 does not imply Equation 2.8.4, nor does Equation 2.8.4 imply Equation 2.8.3.

Example 2.8.3

Suppose two fair dice (one red and one green) are thrown, with events A, B, and C defined as follows:

A: "a 1 or a 2 shows on the red die"

B: "a 3, 4, or 5 shows on the green die"

C: "the dice total is 4, 11, or 12"

By direct enumeration, it is a simple matter to show that $P(A) = \frac{1}{3}$, $P(B)$ $= \frac{1}{2}$, $P(C) = \frac{1}{6}$, $P(A \cap B) = \frac{1}{6}$, $P(A \cap C) = \frac{1}{18}$, $P(B \cap C) = \frac{1}{18}$, and $P(A \cap B \cap C) = \frac{1}{36}$. As a result, Equation 2.8.3 is satisfied:

$$P(A \cap B \cap C) = \frac{1}{36} = P(A)P(B)P(C) = (\tfrac{1}{3})(\tfrac{1}{2})(\tfrac{1}{6})$$

However, Equation 2.8.4 is not:

$$P(B \cap C) = \frac{1}{18} \neq P(B)P(C) = (\tfrac{1}{2})(\tfrac{1}{6}) = \frac{1}{12}$$

Example 2.8.4

A roulette wheel has 36 numbers colored red or black according to the pattern indicated in Figure 2.8.1. Let R be the event "red number appears," E be the event "even number appears," and T be the event "total is ≤ 18." Then $P(R) = P(E) = P(T) = \frac{1}{2}$, $P(R \cap E) = \frac{1}{4}$, $P(E \cap T) = \frac{1}{4}$, and $P(R \cap T) = \frac{1}{4}$. Clearly, R, E, and T are all pairwise independent (that is, Equation 2.8.4 holds), and yet the probability of the three-way intersection does not factor in accordance with Equation 2.8.3:

$$P(R \cap E \cap T) = \frac{4}{36} = \frac{1}{9} \neq P(R)P(E)P(T) = (\tfrac{1}{2})^3 = \frac{1}{8}$$

1	2	3	4	5	6	7	8	9	10	11	12	13	14	15	16	17	18
R	R	R	R	R	B	B	B	B	R	R	R	R	B	B	B	B	B
36	35	34	33	32	31	30	29	28	27	26	25	24	23	22	21	20	19

Figure 2.8.1 Roulette wheel pattern.

The upshot of Examples 2.8.3 and 2.8.4 is that for n events to be independent, the probabilities of *all* possible intersections must factor into the product of the probabilities of the component events. Definition 2.8.2 gives the formal statement.

Definition 2.8.2. Events $A_1, A_2, \ldots, A_n$ are said to be *independent* if for every set of indices $i_1, i_2, \ldots, i_k$ between 1 and n (inclusive),

$$P(A_{i_1} \cap A_{i_2} \cap \cdots \cap A_{i_k}) = P(A_{i_1})P(A_{i_2}) \cdots P(A_{i_k})$$

Example 2.8.5

Suppose a fair coin is flipped three times. Let H_1 be the event "a head on the first flip," T_2 be the event "a tail on the second flip," and H_3 be the event "a head on the third flip." Then

$$P(H_1) = P(T_2) = P(H_3) = \tfrac{1}{2}$$

Also,

$$P(H_1 \cap T_2) = P(\text{HTH, HTT}) = \tfrac{2}{8} = \tfrac{1}{4} = \tfrac{1}{2} \cdot \tfrac{1}{2} = P(H_1)P(T_2)$$

Similarly,

$$P(H_1 \cap H_3) = \tfrac{1}{4} = \tfrac{1}{2} \cdot \tfrac{1}{2} = P(H_1)P(H_3)$$

and

$$P(T_2 \cap H_3) = \tfrac{1}{4} = \tfrac{1}{2} \cdot \tfrac{1}{2} = P(T_2)P(H_3)$$

Finally,

$$P(H_1 \cap T_2 \cap H_3) = P(\text{HTH}) = \tfrac{1}{8} = \tfrac{1}{2} \cdot \tfrac{1}{2} \cdot \tfrac{1}{2} = P(H_1)P(T_2)P(H_3)$$

By Definition 2.8.2, then, events H_1, T_2, and H_3 are independent.

Question 2.8.1 A large company is responding to an affirmative-action commitment by setting up hiring quotas by race and sex for office personnel. So far they have agreed to employ the 120 people indicated in the table. How many black women must they include if they want to be able to claim that the race and sex of the people on their staff are independent? Is that claim equivalent to saying that they hire people independent of their race and sex? Explain.

	White	Black
Male	50	30
Female	40	

Question 2.8.2 Spike is not a terribly bright student. His chances of passing chemistry are 0.35; mathematics, 0.40; and both, 0.12. Are the events "Spike passes chemistry" and "Spike passes mathematics" independent? What is the probability he fails both subjects?

Question 2.8.3 Suppose two events A and B, each having nonzero probability, are mutually exclusive. Are they also independent?

Question 2.8.4 How many probability equations need to be verified to establish the mutual independence of four events?

Question 2.8.5 The probability of a passenger having a bomb on board a plane is very small. The probability of *two* passengers having bombs on board the same plane is even smaller. Does it follow that your chances of being on board a plane someone else intends to blow up are lessened if you bring a bomb on board yourself? Explain.

Question 2.8.6 Let events A, B, and C be mutually independent. Show that $A \cup B$ and C are also independent.

Question 2.8.7 In a roll of a pair of dice (one red and one green), let A be the event "red die shows a 3, 4, or 5"; let B be the event "green die shows a 1 or a 2"; and let C be the event "dice total is 7." Show that A, B, and C are independent.

Question 2.8.8 In a roll of a pair of dice (one red and one green), let A be the event "odd number on the red die," let B be the event "odd number on the green die," and let C be the event "odd sum." Show that any pair of these events are independent, but that A, B, and C are not mutually independent.

Question 2.8.9 Suppose S contains just three points and we define three events A_1, A_2, and A_3 on S such that $0 < P(A_i) < 1$, $i = 1, 2, 3$. Prove that A_1, A_2, and A_3 cannot be pairwise independent.

Question 2.8.10 Show that

$$P(A \mid B \cap C) = P(A)$$

for any set of mutually independent events A, B, and C.

Question 2.8.11 If $A_1, A_2, \ldots, A_n$ are independent events, show that

$$P(A_1 \cup A_2 \cup \cdots \cup A_n) = 1 - [1 - P(A_1)][1 - P(A_2)] \cdots [1 - P(A_n)]$$

Question 2.8.12 Four persons are applying for a single civil service position: one is white, one is male, one is a veteran, and one is a white, male veteran. Let A be the event "person hired is white"; let B be the event "person hired is male"; let C be the event "person hired is a veteran." If the four candidates actually have the same number of "rating" points (so the final selection is done at random), show that A, B, and C are pairwise independent but not mutually independent.

Question 2.8.13 Suppose A and B are independent events and $P(C) > 0$. Prove that $P(A \cap B \mid C)$ is not necessarily equal to $P(A \mid C) \cdot P(B \mid C)$.

Question 2.8.14 For independent events $A_1, A_2, \ldots, A_n$, show that

$$P(A_1 \cup A_2 \cup \cdots \cup A_n) \geq 1 - e^{-[P(A_1) + \cdots + P(A_n)]}$$

While the previous examples in this section have focused on the problem of investigating whether or not a set of events are independent (by examining the conditions spelled out in Definition 2.8.2), there are many situations where the independence of $A_1, A_2, \ldots, A_n$ follows immediately from physical considerations. In these cases we can turn the definition around and use it to provide us with an easy method for evaluating probabilities of intersections. The next set of examples in this section—and all those in the next—will illustrate this very important problem-solving technique.

Example 2.8.6

Suppose one of the genes associated with the control of carbohydrate metabolism exhibits two alleles—a dominant allele, W, and a recessive allele, w. If the probabilities of the WW, Ww, and ww genotypes in the present generation are p, q, and r, respectively, for both males and females, what are the chances that an individual in the *next* generation will be a ww?

Let A_w denote the event that an offspring receives a w allele from its father; let B_w denote the event that it receives the recessive allele from its mother. What we are being asked to find is $P(A_w \cap B_w)$.

According to the information given,

$$p = P(\text{parent has genotype } WW) = P(WW)$$

$$q = P(\text{parent has genotype } Ww) = P(Ww)$$

$$r = P(\text{parent has genotype } ww) = P(ww)$$

If an offspring is equally likely to receive either of its parent's alleles, the probabilities of A_w and B_w can easily be computed using Theorem 2.6.1:

$$P(A_w) = P(A_w|WW)P(WW) + P(A_w|Ww)P(Ww) + P(A_w)|ww)P(ww)$$

$$= 0 \cdot p + \tfrac{1}{2} \cdot q + 1 \cdot r$$

$$= r + \frac{q}{2} = P(B_w)$$

We are not provided here with any explicit statement regarding the relationship between A_w and B_w, but it certainly seems reasonable to presume the two are independent. (Why?) Under that hypothesis, the intersection probability for which we are looking factors into the two components we have just computed:

$$P(A_w \cap B_w) = P(\text{offspring has genotype } ww)$$

$$= P(A_w) \cdot P(B_w)$$

$$= \left(r + \frac{q}{2} \right)^2$$

(Generalize the expression for $P(A_w)$ to include the case where there is a nonzero probability t that the dominant allele mutates to the recessive allele.)

Comment

The model for allele segregation that we have used here, together with the independence assumption, is called *random Mendelian mating*.

Example 2.8.7

An insurance company has three clients—one in Alaska, one in Missouri, and one in Vermont—whose estimated chances of living to the year 2000 are 0.7, 0.9, and 0.3, respectively. What is the probability that by the end of 1999 the company will have had to pay death benefits to exactly one of the three?

Let $A, M,$ and V denote the events of the Alaska client, the Missouri client, and the Vermont client, respectively, surviving through 1999. Then the event E: "exactly one dies" can be written as the union of three intersections:

$$E = (A \cap M \cap V^C) \cup (A \cap M^C \cap V) \cup (A^C \cap M \cap V)$$

But each of the intersections is mutually exclusive of the other two, so

$$P(E) = P(A \cap M \cap V^C) + P(A \cap M^C \cap V) + P(A^C \cap M \cap V)$$

Furthermore, there is no reason to believe that for all practical purposes the fates of the three are not independent. That being the case, each of the intersection probabilities reduces to a product, and we can write

$$P(E) = P(A)P(M)P(V^C) + P(A)P(M^C)P(V) + P(A^C)P(M)P(V)$$

$$= (0.7)(0.9)(0.7) + (0.7)(0.1)(0.3) + (0.3)(0.9)(0.3)$$

$$= 0.543$$

(What is the probability that by the end of 1999 the company will have had to pay death benefits to *at least* one of the three?)

Comment

"Declaring" events independent for reasons other than those prescribed in Definition 2.8.2 is a necessarily subjective endeavor. Here we might feel fairly certain that a random person dying in Alaska will not affect the survival chances of a random person residing in Missouri (or Vermont). But there may be special circumstances that invalidate that sort of argument. For example, what if the three individuals in question were mercenaries fighting in an African border war and were all crew members assigned to the same helicopter? In general, all we can do is look at each situation on an individual basis and try to make a reasonable judgment as to whether the occurrence of one event is likely to influence the outcome of another.

Example 2.8.8

Diane and Lew are health physicists employed by the Department of Public Health. One of their duties is being on call during nonworking hours to handle any nuclear-related incidents that might endanger the public safety. Each carries a pager that can be activated by personnel at Civil Defense Headquarters. Lew is a conscientious worker and is within earshot of his pager 80% of the time. Not nearly as reliable, Diane is capable of responding to a pager only 40% of the time. (a) If Lew and Diane report into Civil Defense independently, what is the probability that at least one of them could be contacted in the event of a nuclear emergency? (b) Suppose Mike, who has a 60% chance of hearing his pager go off, is added to the emergency team. How much would his presence increase the team's response probability?

Consider Diane and Lew first. Let D and L denote the events "Diane responds to alert" and "Lew responds to alert," respectively. What we are looking for is $P(D \cup L)$. From Theorem 2.3.6,

$$P(D \cup L) = P(D) + P(L) - P(D \cap L)$$

By assumption, $P(D) = 0.40$ and $P(L) = 0.80$, and since D and L are independent, $P(D \cap L) = P(D)P(L) = (0.4)(0.8) = 0.32$. Therefore,

$$P(D \cup L) = 0.4 + 0.8 - 0.32 = 0.88$$

If Mike is added to the team, the team's response probability derives from the three-way union $D \cup L \cup M$, where M is the event "Mike responds to alert." Applying Theorem 2.3.7 and presuming the independence of D, L, and M, we can write

$$P(D \cup L \cup M) = P(D) + P(L) + P(M) - P(D \cap L) - P(D \cap M)$$
$$- P(L \cap M) + P(D \cap L \cap M)$$

$$= P(D) + P(L) + P(M) - P(D)P(L) - P(D)P(M)$$
$$- P(L)P(M) + P(D)P(L)P(M)$$
$$= 0.4 + 0.8 + 0.6 - (0.4)(0.8) - (0.4)(0.6)$$
$$- (0.8)(0.6) + (0.4)(0.8)(0.6)$$
$$= 0.952$$

The addition of Mike, then, would raise the team's response probability from 0.88 to 0.952. Whether or not an increase of that magnitude (0.072) is sufficiently large to justify the added cost of a third person would depend on the consequences of a team's *not* responding. Of course, one obvious compromise would be to remove Diane from the team and replace her with Mike. Working together, the two men would have a 92% chance of responding to an emergency. (Derive the three-person response probability an easier way by using the fact that the two events $(D \cup L \cup M)^C$ and $D^C \cap L^C \cap M^C$ are equivalent.)

The past several examples have focused on events for which an a priori assumption of independence was eminently reasonable. We must be careful, though, not to invoke this property too quickly; as the comment following Example 2.8.7 suggested, there are situations where events may *seem* to be independent but are not. The next example is one such instance.

Example 2.8.9

A crooked gambler has nine dice in her coat pocket: Three are fair and six are biased. The biased ones are loaded in such a way that the probability of rolling a 6 is $\frac{1}{2}$. She takes out one die at random and rolls it twice. Let A be the event "6 appears on the first roll" and B be the event "6 appears on the second roll." Are A and B independent?

Our intuition here would most probably answer yes—but, appearances notwithstanding, this is not a typical dice problem. Repeated throws of a die *do* qualify as independent events *if* the probabilities associated with the different faces are known. In this situation, though, those probabilities are *not* known and depend in a random way on which die the gambler draws from her pocket.

To see formally what effect having two different dice has on the relationship between A and B, we must appeal to Theorem 2.6.1. Let F and L denote the events "fair die selected" and "loaded die selected," respectively. Then

$$P(A \cap B) = P(6 \text{ on first roll} \cap 6 \text{ on second roll})$$
$$= P(A \cap B | F)P(F) + P(A \cap B | L)P(L)$$

Conditional on either F or L, A and B are independent, so

$$P(A \cap B) = \left(\tfrac{1}{6}\right)\left(\tfrac{1}{6}\right)\left(\tfrac{3}{9}\right) + \left(\tfrac{1}{2}\right)\left(\tfrac{1}{2}\right)\left(\tfrac{6}{9}\right)$$
$$= \tfrac{19}{108}$$

Similarly,

$$P(A) = P(A|F)P(F) + P(A|L)P(L)$$
$$= (\tfrac{1}{6})(\tfrac{3}{9}) + (\tfrac{1}{2})(\tfrac{6}{9})$$
$$= \tfrac{7}{18} = P(B)$$

But note that

$$P(A \cap B) = \tfrac{19}{108} = \tfrac{57}{324} \neq P(A) \cdot P(B) = (\tfrac{7}{18})(\tfrac{7}{18}) = \tfrac{49}{324}$$

proving that A and B are *not* independent. (For what values of $P(F)$ and $P(L)$ will the difference between $P(A \cap B)$ and $P(A) \cdot P(B)$ be greatest?)

Among the recent legal trends in the United States has been the increasing use of probability in the courtroom as a way of formalizing and quantifying arguments centering on the likelihood (or unlikelihood) of a given sequence of events. Discrimination suits, for example, often take such an approach. But, given the lack of mathematical training that is characteristic of most lawyers, judges, and juries, whether probability clarifies problems or obscures them is still an open question.

The final example in this section looks at a court trial where the prosecution used the probability of an intersection as a measure of culpability. The numbers sounded convincing and the jury returned a guilty verdict. Later, though, on an appeal, the defense countered with a *second* probability argument and the verdict was overturned. (We will discuss the case for the defense in Chapter 4.)

Example 2.8.10

In 1964, the purse of a woman shopping in Los Angeles was snatched by a young, blond female wearing a pony tail. The thief fled on foot but was seen shortly thereafter getting into a yellow automobile driven by a black male who had a mustache and a beard. A police investigation subsequently turned up a suspect, one Janet Collins, who was blond, wore a pony tail, and associated with a black male who drove a yellow car and had a mustache. An arrest was made.

Having no tangible evidence and no reliable witnesses, the prosecutor sought to build his case on the unlikelihood of Ms. Collins and her companion sharing these characteristics and not being the guilty parties. First, the bits of evidence that were available were assigned probabilities. It was estimated, for example, that the probability of a female wearing a pony tail in Los Angeles was $\tfrac{1}{10}$. Table 2.8.1 lists the probabilities quoted for the six "facts" agreed on by the victim and the eyewitnesses (23).

TABLE 2.8.1

Characteristic	Probability
Yellow automobile	$\tfrac{1}{10}$
Man with a mustache	$\tfrac{1}{4}$
Woman with a pony tail	$\tfrac{1}{10}$
Woman with blond hair	$\tfrac{1}{3}$
Black man with beard	$\tfrac{1}{10}$
Interracial couple in car	$\tfrac{1}{1000}$

The prosecutor multiplied these six numbers together and claimed that the product, $(\frac{1}{10}) \cdot (\frac{1}{4}) \cdots (\frac{1}{1000})$, or 1 in 12 million, was the probability of the intersection—that is, the probability that a random couple would fit this description. A probability of 1 in 12 million is so small, he argued, that the only reasonable decision is to find the defendants guilty. The jury agreed and handed down a verdict of second-degree robbery. Later, though, the Supreme Court of California *disagreed*. Ruling on an appeal, the higher court reversed the decision, claiming the probability argument was incorrect and misleading (see Question 2.8.18).

Question 2.8.15 School-board officials are debating whether to require all high school seniors to take a proficiency exam before graduating. A student passing all three parts (mathematics, language skills, and general knowledge) would be awarded a diploma; otherwise, he or she would receive only a certificate of attendance. A practice test given to this year's 9500 seniors resulted in the numbers of failures given in the table.

Subject area	Number of students failing
Mathematics	3325
Language skills	1900
General knowledge	1425

If "student fails mathematics," "student fails language skills," and "student fails general knowledge" are independent events, what proportion of next year's seniors can be expected to fail to qualify for a diploma? Does independence seem a reasonable assumption in this situation?

Question 2.8.16 On his way to work, a commuter encounters four traffic signals. The distance between each of the four is great enough that the probability of getting a green light at any intersection is independent of what happened at any prior intersection. If each light is green (or yellow) for 40 s of every minute, what is the probability the driver has to stop at least three times?

Question 2.8.17 A 50-lb load (*L*) is hung on the support structure shown in the figure. Compression studies indicate that a load of that magnitude will cause arms *a*, *b*, and *c* to crack with probabilities 0.05, 0.04, and 0.03, respectively. What is the probability the support will hold? (Assume that (1) the arms fail independently and (2) if at least one arm cracks, the entire structure will collapse.)

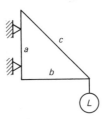

Question 2.8.18 If you were counsel for the defense, how would you counter the prosecution's probability argument given in Example 2.8.10?

Question 2.8.19 Two fair dice are rolled. What is the probability the number showing on one will be twice the number appearing on the other?

Question 2.8.20 Consider the five-switch circuit shown in the figure. If all switches operate independently and $P(\text{switch closes}) = p$, what is the probability the circuit is completed?

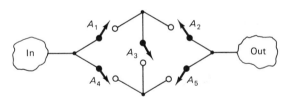

Question 2.8.21 Urn I has three red chips, two black chips, and five white chips; urn II has two red, four black, and three white chips. One chip is drawn at random from each urn. What is the probability both chips are the same color?

Question 2.8.22 Suppose that for both men and women the distribution of blood types in the general population can be characterized by the figures shown in the table. What is the probability that a man and woman getting married will have different blood types? What assumption are you making? Is it reasonable? Explain.

Blood type	Probability
A	0.40
B	0.10
AB	0.05
O	0.45

Question 2.8.23 Three biased dice are tossed, where $P(i \text{ appears}) = ki, i = 1, 2, \ldots, 6$. What is the probability the same face appears all three times?

Question 2.8.24 Trip, Brandon, and Dante play a round-robin tennis tournament where each of them plays the other two exactly once. If $P(\text{Trip beats Brandon}) = 0.6$, $P(\text{Trip beats Dante}) = 0.8$, and $P(\text{Brandon beats Dante}) = 0.9$, who is more likely to win the tournament, Trip or Brandon?

Question 2.8.25 Each of m urns contains three red chips and four white chips. A total of r samples with replacement is taken from each urn. What is the probability at least one red chip is drawn from at least one urn?

Question 2.8.26 A coin for which $P(\text{heads}) = p$ is to be tossed twice. For what value of p will the probability of the event "same side comes up twice" be minimized?

Question 2.8.27 Two myopic deer hunters fire rifles simultaneously and independently at a nearby rooster. The probability of hunter A's shot killing the rooster is 0.2; of hunter B's, 0.3. Suppose the rooster is hit and killed by only one bullet. What is the probability that hunter B fired the fatal shot?

Question 2.8.28 An equilateral triangle is inscribed in a circle of radius r. Three points are selected at random from the interior of the circle. What is the probability exactly two of the three are outside the triangle?

Question 2.8.29 Three points, X_1, X_2, and X_3, are chosen at random in the interval $(0, a)$. Three more points, Y_1, Y_2, and Y_3, are chosen at random in the interval $(0, b)$. Let A be the event "X_2 is between X_1 and X_3." Let B be the event "$Y_1 < Y_2 < Y_3$." Find $P(A \cap B)$.

Question 2.8.30 Amanda tells the truth 80% of the time. Bitsy tells the truth 75% of the time. What is the probability Amanda is telling the truth given that Bitsy says she (Amanda) is lying? Assume the events "Amanda is telling the truth" and "Bitsy is telling the truth" are independent.

Question 2.8.31 In a certain corporation, protocol for making major decisions follows the flow chart shown in the figure. Any proposal is first screened by A. If A approves it, the document is forwarded to B, C, and D. If either B or D concurs, it goes to E. If either E or C says yes, it moves on to F for a final reading. Only if F is also in agreement does the measure pass. Suppose A, C, and F each has a 50% chance of saying yes, whereas B, D, and E will each concur with probability 0.70. If everyone comes to a decision independently, what is the probability a proposal will pass?

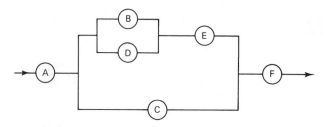

2.9 REPEATED INDEPENDENT TRIALS

We now take up a broad class of problems that represent a special case of the independence notions introduced in the previous section. Algebraically, these problems often make use of the familiar summation result that for $0 < r < 1$, $\sum_{k=0}^{\infty} r^k = 1/(1 - r)$. Conceptually, they hinge on the definition of the Cartesian cross product given in Section 2.2. We begin by reviewing the latter, but in a slightly more general framework than used in our earlier discussion.

We have already pointed out that it is not unusual for an experiment to be made up of a series of n subexperiments, each performed under essentially the same conditions. What needs to be considered in such cases is the relationship between the sample spaces of the individual subexperiments and the sample space of the overall experiment. They are not the same.

Recall the game of craps, whose rules were given in Question 2.2.4. The one or more rolls of two dice that comprise each game qualify as subexperiments. For the shooter to win with, say, a point of 9 on the fourth roll, he or she must roll a 9 on the first throw, something other than a 9 or a 7 on the second and third throws, and a 9 on the fourth. If E_1, E_2, E_3, and E_4 denote those particular outcomes,

$$P(\text{shooter wins on fourth roll with a point of 9}) = P(E_1 \cap E_2 \cap E_3 \cap E_4)$$

On intuitive grounds, we would like to argue that because the rolls of the dice are independent (for obvious physical reasons), the probability of the intersection factors according to Equation 2.9.1:

$$P(E_1 \cap E_2 \cap E_3 \cap E_4) = P(E_1)P(E_2)P(E_3)P(E_4) \qquad (2.9.1)$$

In point of fact, Equation 2.9.1 *is* correct, but it does need some justification. Definition 2.8.2, which gives an expression for the probability of the intersection of independent events, does not apply here because the sample space for the intersection is different than the sample spaces for the individual E_i's. To shore up the mathematics, we must appeal to the notion of a *Cartesian product*.

Phrasing the problem more generally, suppose an experiment is repeated n times under identical conditions. Let $E_1, E_2, \ldots, E_n$ denote events defined on the 1st, 2nd, $\ldots$, nth trials, respectively. Let $S_k = \{s_i^{(k)}\}_i$ denote the sample space for the kth trial, $k = 1, 2, \ldots, n$. (For simplicity, we are assuming the S_k's are discrete.) The Cartesian product of $S_1, S_2, \ldots, S_n$ is defined to be the set of n-tuples,

$$S = \{(s_{i_1}^{(1)}, s_{i_2}^{(2)}, \ldots, s_{i_n}^{(n)})\}_{i_1, i_2, \ldots, i_n}$$

(In mathematics texts, this is often written $S = S_1 \otimes S_2 \otimes \cdots \otimes S_n$.)

Set up in this way, S serves as a common denominator for events related to individual trials, as well as to groups of trials. For example, suppose E_j is just a single outcome in S_j—$E_j = \{s_{j*}^{(j)}\}$. Then, relative to S, E_j can be written

$$E_j = \{(s_{i_1}^{(1)}, \ldots, s_{i_{j-1}}^{(j-1)}, s_{j*}^{(j)}, s_{i_{j+1}}^{(j+1)}, \ldots, s_{i_n}^{(n)})\}_{i_1, \ldots, i_{j-1}, i_{j+1}, \ldots, i_n}$$

At the other extreme, we can use S to express the intersection of n (single-outcome) E_j's:

$$E_1 \cap E_2 \cap \cdots \cap E_n = \{(s_{1*}^{(1)}, s_{2*}^{(2)}, \ldots, s_{n*}^{(n)})\}$$

This latter formulation, together with the next definition, leads us to the desired computing formula for $P(E_1 \cap E_2 \cap \cdots \cap E_n)$.

Definition 2.9.1. Let $E_1, E_2, \ldots, E_n$ denote the outcomes of a series of n trials. If, for all j, the probability of any given outcome on the jth trial does not depend on the outcomes of the preceding $j - 1$ trials, the trials are said to be *independent*.

Comment

The analog of Definition 2.8.1 holds for repeated independent trials:

$$P(E_1 \cap E_2 \cap \cdots \cap E_n) = P(E_1)P(E_2) \cdots P(E_n)$$

That we can factor intersections in this fashion greatly simplifies the problems we are about to consider.

The first example in this section offers some interesting advice on battle tactics. It concerns a three-cornered pistol duel, in which the participants fire sequentially at whomever they please. Because of the constraints put on the problem—specifically, on the duelists' shooting abilities—it can be shown that the welfare of the combatant who is allowed to shoot first is best served if he deliberately elects to fire his initial shot into the ground! Hearing it for the first time, we would most certainly dismiss such a strategy as hopelessly suicidal—in reality, however, its optimality is quite easy to demonstrate.

In the second example we take another look at the rules governing the game of craps and show that the probability of the shooter winning can be reduced to an exercise in independent trials and mutually exclusive events. Next is an early effort by Laplace in using the concept of independence to find the probability of an intersection. Under certain assumptions, Laplace's result has an unusual application: It gives us a formula for estimating the probability that the sun will rise!

Example 2.9.1

Andy, Bob, and Charley have gotten into a disagreement over a female acquaintance, Donna, and decide to settle their dispute with a three-cornered pistol duel. Of the three, Andy is the worst shot, hitting his target only 30% of the time. Charley, a little better, is on-target 50% of the time, while Bob never misses (see Figure 2.9.1). The rules they agree to are simple: They are to fire at the targets of their choice in succession, and cyclically, in the order Andy, Bob, Charley, Andy, Bob, Charley, and so on until only one of them is left standing. (On each "turn," they get only one shot. If a combatant is hit, he no longer participates, either as a shooter or as a target.)

Andy

P(hits target) = 0.3

Bob

P(hits target) = 1.0

Charley

P(hits target) = 0.5 **Figure 2.9.1**

As Andy loads his revolver he mulls over his options (his objective, of course, is clear—to maximize his probability of survival). According to the rules, he has his choice of shooting at either Bob or Charley, but he quickly rules out the latter as being counterproductive to his future well-being. If he shot at Charley and had the misfortune of hitting him, it would then be Bob's turn, and Bob would have no recourse but to shoot at Andy. From Andy's point of view, this would be a decidedly grim turn of events, since Bob never misses. It seems clear, then, that Andy's only viable option is to shoot at Bob. This leaves two scenarios: (1) He might shoot at Bob and hit him, or (2) he might shoot at Bob and miss.

Consider the first possibility. If Andy hits Bob, Charley will proceed to shoot at Andy, Andy will shoot back at Charley, and so on, until one of them hits the other. Let CH_i and CM_i denote the events "Charley hits Andy with ith shot" and "Charley misses Andy with ith shot," respectively. Define AH_i and AM_i analogously. Then Andy's chances of survival (given that he has killed Bob) reduce to a countably infinite union of intersections:

$$P(\text{Andy survives}) = P((CM_1 \cap AH_1) \cup (CM_1 \cap AM_1 \cap CM_2 \cap AH_2)$$
$$\cup \ (CM_1 \cap AM_1 \cap CM_2 \cap AM_2 \cap CM_3 \cap AH_3) \cup \cdots)$$

Note that each intersection is mutually exclusive of all the others and its component events are independent. Therefore,

$$P(\text{Andy survives}) = P(CM_1)P(AH_1) + P(CM_1)P(AM_1)P(CM_2)P(AH_2)$$
$$+ \ P(CM_1)P(AM_1)P(CM_2)P(AM_2)P(CM_3)P(AH_3) + \cdots$$
$$= (0.5)(0.3) + (0.5)(0.7)(0.5)(0.3)$$
$$+ \ (0.5)(0.7)(0.5)(0.7)(0.5)(0.3) + \cdots$$
$$= (0.5)(0.3) \sum_{k=0}^{\infty} (0.35)^k$$
$$= (0.15)\left(\frac{1}{1 - 0.35}\right)$$
$$= \frac{3}{13}$$

Now consider the second scenario. If Andy shoots at Bob and misses, Bob will undoubtedly shoot at (and hit) Charley, since Charley is the more dangerous adversary. Then it will be Andy's turn again. Whether or not he sees another tomorrow will depend on his ability to make that very next shot count. Specifically,

$$P(\text{Andy survives}) = P(\text{Andy hits Bob on second turn})$$
$$= \frac{3}{10}$$

But $\frac{3}{10} > \frac{3}{13}$, so Andy is better off *not* hitting Bob with his first shot. And because we have already argued it would be foolhardy for Andy to shoot at Charley, Andy's optimal strategy is clear—deliberately miss everyone with the first shot.

Example 2.9.2

In the game of craps, the person rolling the dice is called the *shooter*. The game is played at even money, but, as we see in this example, the shooter actually has a slightly less than 50-50 chance of winning.

There are two basic ways the shooter can win: (1) by throwing either a 7 or an 11 on the first roll (this is called a *natural*) or (2) by throwing either a 4, 5, 6, 8, 9, or 10 on the first roll and then throwing that number again *before* rolling a 7 (this is called *making the point*). Let A_1 be the event the shooter throws a natural and let A_4, A_5, A_6, A_8, A_9, and A_{10} be the events that the shooter eventually wins when the point is a 4, 5, 6, 8, 9, or 10, respectively. The A_i's are mutually exclusive, so

$$P(\text{shooter wins}) = P(A_1) + P(A_4) + P(A_5) + P(A_6)$$
$$+ P(A_8) + P(A_9) + P(A_{10})$$

The probability of throwing a natural is easy to compute:

$$P(A_1) = P(7 \text{ or } 11) = P(7) + P(11) = \tfrac{6}{36} + \tfrac{2}{36} = \tfrac{8}{36}$$

To determine the remaining $P(A_i)$'s, we need to think of the game as a series of repeated independent trials. For example, the shooter will win with a point of 4 if he or she rolls a 4 on the first throw and a 4 on the second *or* a 4 on the first, something other than a 4 or a 7 on the second, and a 4 on the third *or* a 4 on the first, something other than a 4 or a 7 on the second and third, and a 4 on the fourth, and so on. Let B be the event that something other than a 4 or a 7 occurs. Then, appealing again to the fact that these probabilities are all mutually exclusive, we can write

$P(A_4) = P(4 \text{ on first} \cap 4 \text{ on second})$

$\quad + P(4 \text{ on first} \cap B \text{ on second} \cap 4 \text{ on third})$

$\quad + P(4 \text{ on first} \cap B \text{ on second} \cap B \text{ on third} \cap 4 \text{ on fourth}) + \cdots$

By inspection, $P(4) = \tfrac{3}{36}$ and $P(B) = \tfrac{27}{36}$. Since each roll is an independent trial,

$$P(A_4) = \left(\frac{3}{36}\right)\left(\frac{3}{36}\right) + \left(\frac{3}{36}\right)\left(\frac{27}{36}\right)\left(\frac{3}{36}\right) + \left(\frac{3}{36}\right)\left(\frac{27}{36}\right)\left(\frac{27}{36}\right)\left(\frac{3}{36}\right) + \cdots$$

$$= \left(\frac{3}{36}\right)^2 \sum_{k=0}^{\infty} \left(\frac{27}{36}\right)^k = \left(\frac{3}{36}\right)^2\left[\frac{1}{1 - \left(\frac{27}{36}\right)}\right]$$

$$= \frac{1}{36}$$

The other $P(A_i)$'s are calculated similarly. Of course, because of symmetry we need only to determine $P(A_4)$, $P(A_5)$, and $P(A_6)$: Since the probability of throwing a 4 is the same as the probability of throwing a 10, $P(A_4) = P(A_{10})$; also, $P(A_5) = P(A_9)$ and $P(A_6) = P(A_8)$. Table 2.9.1 summarizes the results.

TABLE 2.9.1

Winning event, A_i	$P(A_i)$
A_1	$\frac{8}{36}$
A_4	$\frac{1}{36}$
A_5	$\frac{16}{360}$
A_6	$\frac{25}{396}$
A_8	$\frac{25}{396}$
A_9	$\frac{16}{360}$
A_{10}	$\frac{1}{36}$

Adding the seven entries in the second column gives the probability the shooter wins:

$$P(\text{shooter wins}) = \frac{8}{36} + \frac{1}{36} + \cdots + \frac{1}{36}$$

$$= 0.493$$

This confirms what was claimed at the outset, that the odds are against the shooter—although not by much. As games of chance go, craps is relatively fair. (On the other hand, the game is played very rapidly, so the shooter can still manage to lose a lot of money in a short period of time.)

Example 2.9.3

The great French mathematician Pierre Simon Laplace (1749–1827) proposed an independent-trials problem that led to some rather curious interpretations. Suppose we have a coin whose probability of coming up heads is unknown but is equally likely to be $1/N$, $2/N$, . . . , or N/N. That is, if $p = P(\text{heads})$,

$$P\left(p = \frac{i}{N}\right) = \frac{1}{N}, \qquad i = 1, 2, \ldots, N$$

If the first n tosses turn up heads, what is the probability the $(n + 1)$st toss will also be heads?

To cast this question into a workable framework requires that we define $(N + 2)$ events:

1. Let B be the event that the first n tosses come up heads.
2. Let A be the event that the $(n + 1)$st toss comes up heads.
3. Let C_i be the event that $p = i/N$, $i = 1, 2, \ldots, N$.

With the events set up in this fashion, we can state our objective very simply— we want to find $P(A|B)$. As a starting point, note that

$$B = (B \cap C_1) \cup (B \cap C_2) \cup \cdots \cup (B \cap C_N)$$

and, because the $(B \cap C_i)$'s are mutually exclusive,

$$P(B) = \sum_{i=1}^{N} P(B \cap C_i) = \sum_{i=1}^{N} P(B|C_i)P(C_i)$$

But the coin tosses—when conditioned on the C_i's—clearly qualify as independent events (recall Example 2.8.9), so

$$P(B|C_i) = P((\text{1st coin is heads} \cap \cdots \cap n\text{th coin is heads})|C_i)$$

$$= \left(\frac{i}{N}\right)^n$$

Therefore,

$$P(B) = \frac{1}{N} \sum_{i=1}^{N} \left(\frac{i}{N}\right)^n$$

In a similar way, the intersection $A \cap B$ can be written as

$$A \cap B = [(A \cap B) \cap C_1] \cup [(A \cap B) \cap C_2] \cup \cdots \cup [(A \cap B) \cap C_N]$$

and its probability can be derived using Theorem 2.6.1:

$$P(A \cap B) = \sum_{i=1}^{N} P(A \cap B \cap C_i) = \sum_{i=1}^{N} P(A \cap B|C_i)P(C_i)$$

$$= \frac{1}{N} \sum_{i=1}^{N} \left(\frac{i}{N}\right)^{n+1}$$

Dividing $P(A \cap B)$ by $P(B)$ gives $P(A|B)$:

$$P(A|B) = \frac{P(A \cap B)}{P(B)} = \frac{(1/N)\sum_{i=1}^{N} (i/N)^{n+1}}{(1/N) \sum_{i=1}^{N} (i/N)^n} \qquad (2.9.2)$$

A more-convenient expression for $P(A|B)$ can be found by treating the numerator and denominator of Equation 2.9.2 as Riemann sums:

$$\left(\frac{1}{N}\right) \sum_{i=1}^{N} \left(\frac{i}{N}\right)^{n+1} \doteq \int_0^1 x^{n+1}\, dx = \frac{1}{n+2}$$

and

$$\left(\frac{1}{N}\right) \sum_{i=1}^{N} \left(\frac{i}{N}\right)^{n} \doteq \int_0^1 x^{n}\, dx = \frac{1}{n+1}$$

Therefore,

$$P(A|B) \doteq \frac{1/(n+2)}{1/(n+1)} = \frac{n+1}{n+2} \qquad (2.9.3)$$

Equation 2.9.3 is known as *Laplace's rule of succession:* Given the presumed equally likely structure, it says that if some particular event has occurred in n consecutive trials, the probability is $(n + 1)/(n + 2)$ that it will occur on the very next trial.

Comment

Except in the most artificial of situations, Equation 2.9.3—because of all the assumptions that were needed in deriving it—is more of a mathematical curiosity than a practical formula. But this has not prevented people from *trying* to apply it. Laplace himself suggested it could be used to estimate the probability of the sun coming up. Assume, he said, history goes back 5000 years (or 1,826,213 days). Because we know the sun rose on each of those days, the probability of its making an appearance tomorrow is 1,826,214/1,826,215, or 0.999994.

The final two examples in this section are variations of problem types we have seen earlier in simpler settings. The first is basically a two-urn problem but with the distinction that the number of draws from each urn is determined by differentiating a probability. The second is a modification of the binary code transmission problem described in Example 2.7.2—here the signal is fed through three relay stations (each of which can introduce new errors or correct previous ones) before being sent on to the receiver.

Example 2.9.4

A Coast Guard dispatcher receives an SOS from a ship that has run aground off the shore of a small island. Before the captain can relay her exact position, though, her radio goes dead. The dispatcher has n helicopter crews he can send out to conduct a search. He suspects the ship is somewhere either south in area I (with probability p) or north in area II (with probability $1 - p$). Each of the n rescue parties is equally competent and has probability r of locating the ship given it has run aground in the sector being searched. How should the dispatcher deploy the helicopter crews to maximize the probability that one of them will find the missing ship?

Let I and II be the events that the ship is in areas I and II, respectively. Suppose m of the n search teams are sent to area I. Then

$$P(\text{ship is found} \mid \text{I}) = 1 - P(\text{all } m \text{ I-teams fail})$$
$$= 1 - (1 - r)^m$$

Similarly,

$$P(\text{ship is found} \mid \text{II}) = 1 - (1 - r)^{n-m}$$

Combining the two conditional probabilities with $P(\text{I})$ and $P(\text{II})$, we can write

$$P(\text{ship is found}) = p[1 - (1 - r)^m] + (1 - p)[1 - (1 - r)^{n-m}]$$

To find the dispatcher's optimal strategy, we need to think of m as a continuous variable and differentiate P(ship is found):

$$\frac{dP\text{(ship is found)}}{dm} = -p(1-r)^m \ln(1-r) + (1-p)(1-r)^{n-m} \ln(1-r)$$

Setting the derivative equal to 0 and solving for m gives

$$m = \frac{n}{2} + \frac{\ln[(1-p)/p]}{2\ln(1-r)}$$

Suppose a dispatcher has $n = 12$ helicopter crews, each with the same successful search probability of $r = 0.30$. If he feels the odds are 4 to 1 that the ship has run aground somewhere to the south, he should send

$$m = \frac{12}{2} + \frac{\ln(0.20/0.80)}{2\ln(0.70)}$$

$$= 7.9$$

or 8 of the crews to area I and 4 to area II.

Example 2.9.5

A transmitter is sending a binary code ($+$ and $-$ signals) that must pass through three relay stations before being sent on to the receiver (see Figure 2.9.2). At each relay station, there is a 25% chance the signal will be reversed—that is, $P(+$ is sent by relay $i \mid -$ is received by relay $i) = \frac{1}{4} = P(-$ is sent by relay $i \mid +$ is received by relay $i)$, $i = 1, 2, 3$. Suppose $+$ symbols make up 60% of the message being sent. If a $+$ is received, what is the probability a $+$ was sent?

Like Example 2.7.2, this is basically a Bayes' theorem problem, but the three relay stations introduce a more-complex mechanism for transmission error. Let A be the event "$+$ is transmitted from tower" and B be the event "$+$ is received from relay 3." Then, clearly,

$$P(A \mid B) = \frac{P(B \mid A)P(A)}{P(B \mid A)P(A) + P(B \mid A^C)P(A^C)}$$

Notice that a $+$ can be received from relay 3 given that a $+$ was initially sent from the tower if either (1) all the relay stations function properly or (2) any *two* of them make transmission errors. Table 2.9.2 shows the four mutually exclusive ways (1) and (2) can happen. The probabilities associated with the message transmission at each relay station are shown in parentheses. Assuming the relay station outputs are independent events, the probability of an entire transmission

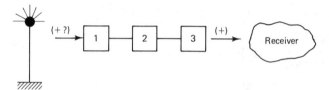

Figure 2.9.2

sequence is simply the product of the probabilities in parentheses in any given row. Those overall probabilities are listed in the last column; their sum, $\frac{36}{64}$, is $P(B|A)$. By a similar analysis (or by a symmetry argument), we can show that $P(B|A^C) = P(+$ is received from relay $3| -$ is transmitted from tower$) = \frac{28}{64}$. Finally, since $P(A)$ and $P(A^C)$ are 0.6 and 0.4, respectively, the conditional probability we are looking for is 0.66:

TABLE 2.9.2

| Tower | Signal transmitted by: | | | Probability |
	Relay 1	Relay 2	Relay 3	
+	$+ \left(\frac{3}{4}\right)$	$- \left(\frac{1}{4}\right)$	$+ \left(\frac{1}{4}\right)$	$\frac{3}{64}$
+	$- \left(\frac{1}{4}\right)$	$- \left(\frac{3}{4}\right)$	$+ \left(\frac{1}{4}\right)$	$\frac{3}{64}$
+	$- \left(\frac{1}{4}\right)$	$+ \left(\frac{1}{4}\right)$	$+ \left(\frac{3}{4}\right)$	$\frac{3}{64}$
+	$+ \left(\frac{3}{4}\right)$	$+ \left(\frac{3}{4}\right)$	$+ \left(\frac{3}{4}\right)$	$\frac{27}{64}$
				$\frac{36}{64}$

$$P(A|B) = \frac{\left(\frac{36}{64}\right)(0.6)}{\left(\frac{36}{64}\right)(0.6) + \left(\frac{28}{64}\right)(0.4)}$$

$$= 0.66$$

(Suppose there were only *two* relay stations. What would be the probability of a + having been transmitted from the tower given that a + was received from relay 2?)

Question 2.9.1 In a certain third-world nation, statistics show that only 2000 out of 10,000 children born in the early 1960s reached the age of 21. If the same mortality rate is operative over the next generation, how many children should a couple plan to have if they want to be at least 75% certain that at least one of their offspring survives to adulthood?

Question 2.9.2 Suppose the probability function describing the lifetime (in years) of an X-ray tube is

$$f(t) = e^{-t}, \qquad t > 0$$

If a dental clinic has three such tubes, what is the probability one of them wears out in its first year, one in its second year, and one in its third year? Assume the tubes wear out independently.

Question 2.9.3 If two fair dice are tossed, what is the smallest number of throws, n, for which the probability of getting at least one double 6 exceeds 0.5? (*Note:* This was one of the first problems that de Mere communicated to Pascal in 1654.)

Question 2.9.4 A string of eight Christmas tree lights is wired in series. If the probability of any particular bulb failing sometime during the holiday season is 0.05 and if the failures are independent events, what is the probability the lights will not remain lit?

Question 2.9.5 Players A, B, and C toss a fair coin in order. The first to throw a head wins. What are their respective chances of winning?

Question 2.9.6 An urn contains w white chips, b black chips, and r red chips. The chips are drawn out at random, one at a time, *with* replacement. What is the probability a white appears before a red?

Question 2.9.7 A penny may be fair or it may have both faces alike. We toss it n times and it comes up heads on each occasion. If our initial judgment was that both options for the coin ("fair" or "both sides alike") were equally probable, what is our revised judgment in light of the data?

Question 2.9.8 Suppose four people (A, B, C, and D) each toss a fair die in order (A first, then B, and so on) until the face showing a 6 appears for the first time. What is the probability C is the one who rolls the first 6?

Question 2.9.9 A fair die is rolled until a 6 shows. What is the probability it will take k rolls for that to happen? What is the probability of the first 6 appearing on an even-numbered roll?

Question 2.9.10 You are playing for the Monopoly Championship of the World. Your opponent is on Go. It is your turn and you have enough money to put a house on either Oriental Avenue, Vermont Avenue, or Connecticut Avenue. The three properties are 6, 8, and 9 spaces away from Go, respectively. Where should you put the house?

Question 2.9.11 A biased coin, $P(\text{heads}) = p$, is tossed repeatedly until a head occurs for the first time.
    a. Write a formula for the probability that the first head appears no later than the jth toss.
    b. Suppose we are able to specify in advance the number of times, j, the coin will be tossed. Show that in order to have a probability of at least $1 - \alpha$ that the first head will occur no later than the jth toss, j has to be greater than or equal to $\log \alpha / \log (1 - p)$.

Question 2.9.12 An incompetent stockbroker elects to invest all her client's money in Ne'er-Do-Well, Inc., currently selling at \$6 a share. Its closing price fluctuates randomly from day to day, going up 1 point with probability $\frac{1}{2}$ and down 1 point with probability $\frac{1}{2}$. If it ever drops to \$5 a share, the client will go bankrupt. What is the client's probability of financial disaster?

APPENDIX 2.A THE BEAUTY CONTEST PROBLEM (OPTIONAL)

We have already encountered a number of problems whose solutions belied our common sense (recall, for instance, Examples 2.4.5, 2.5.7, 2.6.5, 2.7.4, and 2.9.1). More than being just amusing, these problems serve a definite purpose—they help us refine our intuition and explore subtleties we would otherwise overlook. Here we present one of the most "unintuitive" problems in all of probability. While the question it poses can be ever-so-simply stated, its answer, at least at first glance, is totally absurd. It appears throughout the mathematical literature in a number of different formulations; here it will be couched in one of its more familiar guises, as a *beauty contest problem* (see (60)).

> You are to judge a beauty pageant. There are n contestants and you are to select the prettiest. The rules are simple. After seeing the first contestant (and without having seen any of the others), you must decide whether or not you want to declare that girl the winner. If you do, the contest is over. If you elect not to, you get to see the second contestant, and, again, you must decide whether or not *she* is to be your choice. The process continues until you make a positive decision. *You are not allowed to change your*

mind—once a candidate has been rejected, she remains ineligible for the duration of the contest. The question is this: What is your optimal strategy, assuming you wish to maximize your chances of selecting the prettiest contestant—and what is the probability of that best strategy working?

Notice that if we elected to pick the winner *at random,* the probability of our making the correct choice would be $1/n$. Furthermore, our intuition would surely tell us that, *as n increases,* our chances of identifying the prettiest girl—no matter what our strategy might be—would surely decrease to zero. But that proves not to be the case. We will show that *even if n tends to infinity,* there is a strategy that will enable us to single out the prettiest girl almost 37% of the time!

We begin with a simple example that illustrates the sort of approach with which we will eventually come up. Suppose there are only $n = 3$ contestants, and we decide to let the first girl pass by and pick as the winner the next contestant prettier than that first one. Let 3 denote the prettiest contestant (the one who *should* win), 2 the second prettiest, and 1 the least pretty. By direct enumeration, it can easily be seen that the sample space consists of the six possible *orders* in which the contestants can be presented. Let the triple (i, j, k) denote the sample outcome whereby the *i*th-rated contestant goes first, the *j*th-rated contestant second, and the *k*th-rated third. Then the outcomes in S can be written

$$(1, 2, 3)$$
$$(1, 3, 2)*$$
$$(2, 1, 3)*$$
$$(2, 3, 1)*$$
$$(3, 1, 2)$$
$$(3, 2, 1)$$

The sequences marked by asterisks are the ones for which our proposed plan would yield the proper choice. Since all orderings are equally likely, the probability of this particular strategy working is $\frac{3}{6}$, making it clearly superior to the "random" procedure that would have a winning probability of only $\frac{1}{3}$.

Before extending this example into a generalized optimal strategy, we need one additional bit of terminology. A *candidate* will be defined as any contestant prettier than all the others at which we have already looked. For example, if n were 4 and the order of presentation were 3214, then 4 (and 3, trivially) would be the only candidates. If the order were 1234, all four contestants would be candidates.

Theorem 2.A.1. The *form* of the optimal strategy is to pass the first $x - 1$ contestants by and choose as the winner the first *candidate* appearing thereafter.

Proof. If contestant i is a candidate, it makes sense to nominate her as the prettiest if the probability of her being the correct choice exceeds the probability of finding the winner using whatever is the optimal strategy, beginning with the $(i + 1)$st contestant. That is, we should choose contestant i (provided she is a candidate) if

P(we win with contestant/candidate i) $> P$(we win with best strategy from $i + 1$ on)

Clearly, the probability on the right must decrease as i increases (for $i = 0$, its value is precisely what we are seeking; at $i = n - 1$, it achieves its minimum, $1/n$). To complete the proof, we need to know something about the behavior of the probability on the left.

Lemma. Given n contestants, the probability of a *candidate* in the ith position being the eventual winner is directly proportional to i; specifically,

$$P(\text{we win with contestant/candidate } i) = \frac{i}{n}$$

Proof. By definition,

$$P(\text{candidate in } n\text{th position is the overall prettiest}) = 1$$

Also,

P(candidate in $(n - 1)$st position is the overall prettiest)

$$= 1 - P(\text{contestant in } n\text{th position is prettier})$$

$$= 1 - \frac{1}{n} = \frac{n - 1}{n} \tag{2.A.1}$$

Similarly,

P(candidate in $(n - 2)$nd position is the overall prettiest)

$$= 1 - P(\text{prettiest contestant is in either } (n - 1)\text{st or } n\text{th position})$$

$$= 1 - \frac{2}{n} = \frac{n - 2}{n} \tag{2.A.2}$$

Without going into a formal induction proof, it should be clear from Equations 2.A.1 and 2.A.2 that the statement of the lemma is true.

Returning now to the proof of Theorem 2.A.1, we note from the lemma that P(we win with contestant/candidate i) *increases* from $1/n$ to 1. So, since P(we win with best strategy from $i + 1$ on) *decreases* from something greater than $1/n$ to $1/n$, it follows that somewhere along the i-axis there is a point where it is better to pick as the winner a candidate in front of you than to wait for the possibility of someone prettier coming along later. This proves the theorem.

Question 2.A.1 Fill in the details of the induction proof to show that P(we win with contestant/candidate i) $= i/n$.

Having settled on the *form* of the optimal strategy, it remains to determine the value of the cutoff point, x. To do so requires deriving an expression for the probability of such a strategy working, *as a function of* x. Let A_i, $i = 1, 2, \ldots, n$, denote the event "prettiest girl is in position i." Let B be the event "optimal strategy (letting the first $x - 1$ contestants go by) works." Then, from Theorem 2.6.1,

$$P(B) = P(B \cap A_1) + \cdots + P(B \cap A_{x-1}) + P(B \cap A_x)$$
$$+ P(B \cap A_{x+1}) + \cdots + P(B \cap A_n)$$

$$= 0 + \cdots + 0 + P(B|A_x)P(A_x)$$
$$+ P(B|A_{x+1})P(A_{x+1}) + \cdots + P(B|A_n)P(A_n)$$

But $P(A_i) = 1/n$ for all i and, for $i \geq x$,

$$P(B|A_i) = P(\text{prettiest of the first } i - 1 \text{ contestants is somewhere}$$
$$\text{in the first } x - 1 \text{ positions})$$

$$= \frac{x - 1}{i - 1}$$

(If the prettiest of the first $i - 1$ contestants were in some position x through $i - 1$, then that contestant would be a candidate and would be selected by our strategy as the winner. However, the strategy would be in error because, by assumption, the prettiest girl overall is in the ith position (see Figure 2.A.1).) Therefore,

$$P(B) = P(B|A_x)P(A_x) + P(B|A_{x+1})P(A_{x+1}) + \cdots + P(B|A_n)P(A_n)$$
$$= 1\left(\frac{1}{n}\right) + \frac{x-1}{x}\left(\frac{1}{n}\right) + \cdots + \frac{x-1}{n-1}\left(\frac{1}{n}\right)$$
$$= \frac{x-1}{n}\left(\frac{1}{x-1} + \frac{1}{x} + \cdots + \frac{1}{n-1}\right)$$

Now, recall that the optimal x is the smallest integer for which

$P(\text{we win with contestant/candidate } i) > P(\text{we win with best strategy from } i + 1 \text{ on})$

Equivalently, we are seeking the smallest integer x^* such that

$$\frac{x^*}{n} > \frac{x^*}{n}\left(\frac{1}{x^*} + \frac{1}{x^* + 1} + \cdots + \frac{1}{n-1}\right) \qquad \text{(Why?)}$$

or

$$1 > \frac{1}{x^*} + \frac{1}{x^* + 1} + \cdots + \frac{1}{n-1} \qquad (2.\text{A}.3)$$

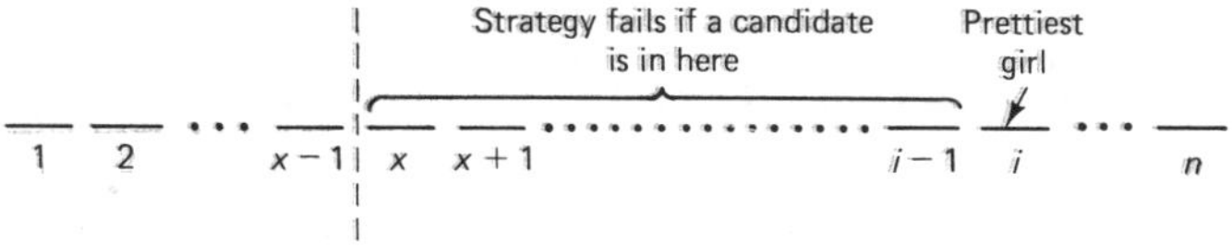

Figure 2.A.1 Diagram showing that $P(B|A_i) = (x - 1)/(i - 1)$.

Table 2.A.1 shows (1) the optimal value x^* for various n and (2) the corresponding probability of the optimal strategy working. Thus, if n were 10, we should let the first $x^* - 1 = 4 - 1 = 3$ contestants pass by and choose as the winner the first girl prettier than the prettiest of those initial three. By following that scheme, we will be guided into making the right selection 39.9% of the time.

TABLE 2.A.1 OPTIMAL VALUES OF x AND WINNING PROBABILITIES

n	x^*	$P(\text{win})$	n	x^*	$P(\text{win})$
1	1	1.000	10	4	0.399
2	1	0.500	20	8	0.384
3	2	0.500	50	19	0.374
4	2	0.458	100	38	0.371
5	3	0.433	∞	$\dfrac{n}{e}$	$\dfrac{1}{e} \cong 0.368$

The solution for large n follows if we approximate the partial sum of the harmonic series given in Inequality 2.A.3. A well-known result in analysis states that

$$\sum_{i=1}^{n} \frac{1}{i} \doteq C + \ln n$$

where C = Euler's constant $(0.5772 \ldots)$. Therefore,

$$\frac{x-1}{n}\left(\frac{1}{x-1} + \frac{1}{x} + \cdots + \frac{1}{n-1}\right) \doteq \frac{x-1}{n}$$

$$\times \{C + \ln(n-1) - [C + \ln(x-2)]\}$$

$$= \frac{x-1}{n} \ln\left(\frac{n-1}{x-2}\right)$$

$$= \frac{x}{n} \ln\left(\frac{n}{x}\right) \tag{2.A.4}$$

$$= P(\text{"pass-by" strategy works})$$

Since $P(B)$ is approximately x/n for large n, it follows from Equation 2.A.4 that $\ln(n/x) \doteq 1$, in which case $x \doteq n/e$. This means that for large n we should let the first $100(1/e)$ percent of the contestants pass by and declare as the winner the first candidate appearing thereafter. If we adopt that approach, our chances of singling out the prettiest girl are an astonishingly high 36.8%:

$$\lim_{n\to\infty} P(B) = \lim_{n\to\infty} P(\text{optimal strategy works}) = \frac{n/e}{n} \ln\left(\frac{n}{n/e}\right) = \frac{1}{e}$$

or approximately 0.368.

Comment

Notice that the rules of the beauty contest problem are not unlike the conditions under which a woman accepts a marriage proposal. Normally, she has to reject a proposal before being "eligible" to receive a second, and, having turned down a suitor once, she seldom has the opportunity to change her mind. If these conditions are met, then Table 2.A.1 offers our bride-to-be some useful guidelines. If, for example, she anticipates being able to attract *five* proposals (if she always said no), it would be in her best romantic interests to decline the first two offers automatically!

3

Combinatorics

They are called wise
who put things in their right order.
St. Thomas Aquinas

How do I love thee?
Let me count the ways.
Elizabeth Barrett Browning

3.1 INTRODUCTION

Combinatorics is a branch of mathematics that deals with various kinds of enumeration problems. Depending on the context, our interest may focus on a single enumeration or on the ratio of two enumerations. For example, a bridge player may wish to know in how many ways he or she can be dealt seven spades and six clubs; more likely, though, the player would primarily be interested in knowing the *probability* of being dealt such a hand, the latter being, as we will see, the ratio of the number of possible 7-spade–6-club hands divided by the total number of *all* 13-card hands.

Having to enumerate outcomes is not a problem we are facing for the first time. Indeed, every exercise in Chapter 2 that involved a finite sample space necessarily required an enumeration. But those enumerations were merely listings, and that kind of approach will not always accomplish our purpose. It is a simple enough task to write down all the outcomes associated with two throws of a die; it is quite another matter to identify all possible seven-spade–six-club bridge hands. To handle problems of this second type—situations where an actual listing of outcomes would be prohibitively long—it is necessary to develop techniques that help us count in a more systematic fashion. Those techniques are what this chapter is all about.

It is not our intention here to present anything resembling a comprehensive survey of combinatorial mathematics. The subject is much too broad for that to be a reasonable objective. Rather, we will confine our attention to some of the basic enumeration results that have direct application to probability problems.

3.2 PERMUTATIONS: ORDERED ARRANGEMENTS

We shall begin our survey of combinatorial mathematics by looking at some theorems that arise in connection with the formation of ordered arrangements. These will be, on the surface, eminently simple results—easy to state and easy to prove—but the principles they embody are of fundamental importance, and their usefulness extends well beyond these initial applications.

By an ordered arrangement (or *permutation of length r*), we will mean a sequence of r objects, $(A_1, A_2, \ldots, A_r)$, where an object's *position* in the sequence matters. Thus, if A_1 is not the same as A_2, the sequences (A_1, A_2) and (A_2, A_1) are, by definition, two different permutations. In counting the number of such arrangements, we need to know something about the nature of the A_i's and the rules for putting them together. Specifically, are the n A_i's from which we are choosing the r that make up our permutation all different? Or, are some of them alike? And are we restricted to using a particular A_i at most once or are repetitions allowed? Figure 3.2.1 previews which of the next several theorems applies to each of the four cases making up the intersection of these two dichotomies.

		Type of Elements	
		All distinct	Grouped
Method of Forming Sequences	No repetitions allowed	Theorem 3.2.1	Theorem 3.2.3
	Repetitions allowed	Theorem 3.2.2	Theorem 3.2.2

Figure 3.2.1

Comment

There are other restrictions that can be imposed on this sort of problem. For example, how many ordered arrangements of length r can be made from n distinct objects ($r \geq n$) if repetitions are allowed and if each of the original n must appear in the permutation at least once? This is a substantially more difficult question, though, than the four options shown in Figure 3.2.1 and is beyond the scope of our intentions. Problems of this type are more properly taken up in texts devoted exclusively to combinatorics. See, for example, (73) or (96).

The obligatory place to start our discussion is with a patently transparent result, yet one so basic to all combinatorial problems that to call it merely a theorem would considerably understate its pervasiveness. It will be referred to here as the *fundamental principle*.

> **Fundamental Principle.** If operation A can be performed in m different ways and operation B in n different ways, the sequence (operation A, operation B) can be performed in $m \cdot n$ different ways.

Proof. At the risk of belaboring the obvious, we can verify the fundamental principle by considering a tree diagram (see Figure 3.2.2). Since each version of A can be followed by any of n versions of B and there are m of the former, the total number of A, B sequences that can be pieced together is obviously the product $m \cdot n$.

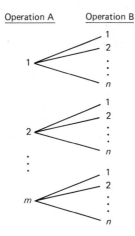

Figure 3.2.2

> **Corollary.** If operation A_i, $i = 1, 2, \ldots, k$, can be performed in n_i ways, $i = 1, 2, \ldots, k$, respectively, then the sequence (operation A_1, operation A_2, $\ldots$, operation A_k) can be performed in $n_1 \cdot n_2 \cdots n_k$ ways.

Comment

We have already encountered a situation where the corollary to the fundamental principle could have been applied. In Section 2.10, the Cartesian product S of the discrete sample spaces $S_1, S_2, \ldots,$ and S_k was defined to be the set of all possible k-tuples, $\{(s_{i_1}^{(1)}, s_{i_2}^{(2)}, \ldots, s_{i_k}^{(k)})\}_{i_1, i_2, \ldots, i_k}$, with $s_{i_j}^{(j)}$ representing an arbitrary element from S_j. If $n_1, n_2, \ldots, n_k$ denote the numbers of outcomes in $S_1, S_2, \ldots, S_k$, respectively, then, by the corollary, the number of k-tuples in S is the product $n_1 \cdot n_2 \cdots n_k$.

Example 3.2.1

In situations where access to a computer must be controlled and monitored, approved users are assigned a *password,* which is a sequence of letters and numbers having a prescribed format. Only if the password is keyed in can a person log on. How many different passwords are possible if each must be in the form

letter	letter	number	number

Couched in the terminology of the corollary to the fundamental principle, a password is an ordered sequence of $k = 4$ operations. Operations A_1 and A_2 can each be "performed" in 26 ways; operations A_3 and A_4 in 10 ways. It follows that the total number of passwords is $26 \cdot 26 \cdot 10 \cdot 10$, or 67,600.

Example 3.2.2

How many integers between 100 and 999 have distinct digits, and how many of those are odd?

Think of the integers as being an arrangement of a hundreds digit, a tens digit, and a units digit (see Figure 3.2.3). The hundreds digit can be filled in any of 9 ways (0s are inadmissible), the tens place in any of 9 ways (anything but what appears in the hundreds place), and the units place in any of 8 ways (the first two digits must not be repeated). Thus, by the corollary to the fundamental principle, the number of integers between 100 and 999 with distinct digits is $9 \cdot 9 \cdot 8$, or 648.

(9)	(9)	(8)	
100s	10s	1s	**Figure 3.2.3**

To compute the number of *odd* integers with distinct digits, we first consider the units place, where any of 5 integers can be positioned (1, 3, 5, 7, or 9). Then, turning to the hundreds place, we have 8 choices (the 0 is inadmissible; so is whatever appeared in the units place). The same number of choices is available for the tens place. Multiplying these numbers together gives $8 \cdot 8 \cdot 5 = 320$ as the number of odd integers in the range 100–999.

Question 3.2.1 A chemical engineer wishes to observe the effect of temperature, pressure, and catalyst concentration on the yield resulting from a certain reaction. If she chooses to include two different temperatures, three pressures, and two levels of catalyst, how many different runs must she make in order to try each temperature-pressure-catalyst combination exactly twice?

Question 3.2.2 A coded message from a CIA operative to his Soviet KGB counterpart is to be sent in the form Q4ET, where the first and last entries must be consonants; the second, an integer between 1 and 9, inclusive; and the third, one of the six vowels. How many different ciphers can be transmitted?

Question 3.2.3 How many terms will be included in the expansion of

$$(a + b + c)(d + e + f)(x + y + u + v + w)$$

Which of the following will be included in that number: *aeu, cdx, bef, xvw*?

Question 3.2.4 A local restaurant offers a choice of 4 appetizers, 14 entrees, 6 desserts, and 5 beverages. In how many ways can a diner "design" his evening meal, assuming he is hungry enough to elect one option from each of the four categories? (Assume that he eats the four courses in the standard order.)

Question 3.2.5 How many ways can a set of four tires be put on a car if all the tires are interchangeable? How many ways are possible if two of the four are snow tires?

Question 3.2.6 In the 1978 World Almanac, enrollment at Virginia Commonwealth University in Richmond, Virginia, was listed at 18,099. Prove that at least two students at VCU had the same set of three initials.

Question 3.2.7 The total number of *even* integers in the range 100–999 with distinct digits must be $648 - 320 = 328$ (recall Example 3.2.2). Find that number directly.

Question 3.2.8 Old-fashioned floor lamps often consisted of three low-wattage bulbs in an outer circle surrounding one large-wattage bulb. Any number of the outer lights could be turned on independently of the center light, but if, for instance, two were on, it would have to be a certain two. If the center light was a three-way bulb, how many different modes of operation did such a lamp have?

Deriving formulas for the counting problems suggested by Figure 3.2.1 is surprisingly easy now that we have the fundamental principle. We begin with the case where all n elements from which the r will be selected are distinct and repetitions are not allowed.

Theorem 3.2.1. Given n distinct objects, we can form ordered arrangements (or *permutations*) of r of those objects in

$$n(n - 1)(n - 2) \cdots (n - r + 1) = \frac{n!}{(n - r)!}$$

ways, repetitions not being allowed.

Proof. Any of the n objects may occupy the first position in the arrangement, any of $n - 1$ the second, and so on; the number of choices available for filling the rth position is $n - r + 1$. The theorem follows, then, from the corollary to the fundamental principle.

Corollary. The number of ways to permute an entire set of n distinct objects is $n!$.

Comment

Computing $n!$ can be quite cumbersome, even for n's that are fairly small—15!, for example, is already in the trillions. Fortunately, there is a fairly easy-to-use approximation: According to *Stirling's formula*,

$$n! \doteq \sqrt{2\pi}\, n^{n+(1/2)} e^{-n}$$

In practice, we apply Stirling's formula by writing

$$\log(n!) \doteq \log(\sqrt{2\pi}) + \left(n + \frac{1}{2}\right) \log(n) - n \log(e)$$

and then finding the antilog of the right-hand side. (For a derivation of Stirling's formula, see (26).)

Example 3.2.3

Years ago, long before Rubik's-cube fever had become epidemic, puzzles were much simpler. One of the more popular combinatorial-related diversions was a 4 × 4 grid consisting of 15 movable squares and one empty space. The object was to maneuver an arbitrary configuration (Figure 3.2.4(a)) into a specific pattern (Figure 3.2.4(b)) as quickly as possible. In how many different ways can the puzzle be arranged?

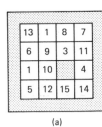
(a)

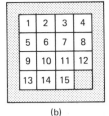
(b)

Figure 3.2.4

Think of the empty space as being square 16 and imagine the four rows of the grid laid end to end to make a 16-number sequence. Each permutation of that sequence corresponds to a different pattern for the grid. By the corollary to Theorem 3.2.1, the number of ways to arrange the puzzle is 16!, or 20,922,789,888,000. (Can all 20 billion arrangements be achieved without physically removing some of the tiles and putting them back somewhere else in the grid?)

Example 3.2.4

Four dogs (A, B, C, D), three hamsters (a, b, c), and three camels (α, β, γ) are lined up at a veterinarian's office waiting to get their rabies shots.

a. How many ways can they position themselves if the dogs are to hold the first four places in line; the hamsters, the next three; and the camels, the last three?

The enumeration here involves an application of the corollary to Theorem 3.2.1, in conjunction with the fundamental principle. Consider a typical admissible arrangement (see Figure 3.2.5). Restricted to the first four positions, the dogs can still be permuted in 4! ways. Similarly, the hamsters and the camels can each be permuted in 3! ways. By the corollary to the fundamental principle, then, there are 864 admissible arrangements:

$$4! \cdot 3! \cdot 3! = 24 \cdot 6 \cdot 6 = 864$$

Figure 3.2.5

b. How many arrangements are possible if members of the same species must stay together?

Part of the required total here has already been counted—the 864 permutations enumerated in part (a) all have the property that members of the same species are lined up together. But now we do not want to restrict our attention to queues where the dogs, for example, are necessarily first. Arrangements like the one shown in Figure 3.2.6 must also be included. By the corollary to Theorem 3.2.1, the three species can be permuted in 3! = 6 ways:

(dogs, hamsters, camels)

(dogs, camels, hamsters)

(hamsters, dogs, camels)

(hamsters, camels, dogs)

(camels, dogs, hamsters)

(camels, hamsters, dogs)

Since each of these six gives rise to 864 distinct arrangements, it follows that the total number of queues having members of the same species grouped together is 6 · 864, or 5184.

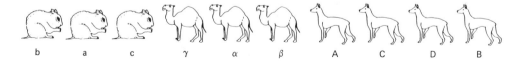

Figure 3.2.6

c. How many different queues can be formed?

For this question, the species are irrelevant: Any arrangement of the ten animals is admissible (see Figure 3.2.7). Thus, the total number of queues is the total number of ways to permute 10 distinct objects—namely, 3,628,800 (10!).

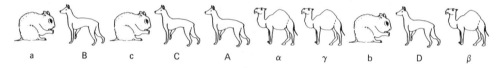

Figure 3.2.7

Example 3.2.5

Consider the set of nine-digit numbers that can be formed by rearranging (without repetition) the integers 1 through 9. In how many of those permutations will the 1 and the 2 precede the 3 and the 4? That is, we want to count sequences like 7 2 5 1 3 6 9 4 8, but not like 6 8 1 5 4 2 7 3 9.

At first glance, the problem of tallying up all the arrangements that satisfy our restriction on 1 and 2 would seem to be enormous—or, at best, decidedly unpleasant. But it turns out the answer can be gotten with surprisingly little effort—provided we approach the question the right way! Imagine just the digits 1 through 4. By the corollary to Theorem 3.2.1, those numbers will give rise to 24 (4!) permutations. Of those 24, exactly 4—(1, 2, 3, 4), (2, 1, 3, 4), (1, 2, 4, 3), (2, 1, 4, 3)—have the property that the 1 and the 2 come before the 3 and the 4. Therefore, since the entire set of digits can be permuted in 9! ways, we can argue that $\frac{4}{24}$ of that figure, or 60,480, is the number of arrangements in which the 1 and the 2 precede the 3 and the 4.

CASE STUDY 3.1

On June 4, 1942, a Japanese naval attack force under the command of Admiral Nagumo and led by four huge carriers, the *Akagi, Kaga, Hiryu,* and *Soryu,* raced toward Midway Island in the central Pacific; their objective—catch the vastly outnumbered American fleet by surprise and finish what the Imperial Air Force had started at Pearl Harbor 6 months earlier. But the battle plan so carefully worked out by Admiral Yamimoto backfired with stunning swiftness. It was the Japanese forces who were caught by surprise: Divebombers launched from two U.S. carriers, the *Yorktown* and the *Enterprise,* quickly crippled the Japanese carriers and, in a matter of hours, all four had sunk. What was left of the Japanese fleet had no choice but to retreat. History would later show that Midway was far more than just a lost battle—it was a turning point in the entire Japanese war effort.

The assault on Midway was repelled because Yamimoto failed to have what he was counting on most—the element of surprise. Only weeks before the invasion, a group of Allied *cryptanalysts* had broken the Japanese code, so when Yamimoto transmitted instructions to his fleet, Nagumo was not the only one listening. The American commanders learned not only when the attack was to be mounted, but the direction from which it was coming and precisely how many ships would be involved.

Sending, intercepting, and deciphering coded battle strategies have been a part of war for almost 2000 years. In the twentieth century, though, methods of enciphering and deciphering have reached totally unprecedented levels of sophistication. Yet behind even the most complex of these systems lie some of the most basic concepts of combinatorics, including the fundamental principle and the notion of a permutation.

A great many enciphering systems are based on the *substitution* of each letter in the original *plaintext* with a (possibly) different letter in the *ciphertext*. The simplest example of this is gotten by sliding the ciphertext alphabet a fixed number of positions along the plaintext alphabet. In Figure 3.2.8 the ciphertext alphabet has been shifted three letters to the left. Using this alphabet, the code for *MALTESE FALCON* is

PDOWHVH IDOFRQ

This is called a *monalphabetic substitution* because the same permutation of the plaintext alphabet is used to encipher every letter in the message. As might be expected, breaking such a code is trivial because all the patterns and characteristics of the alphabet used for the original message are preserved in the ciphertext.

Plaintext:	A	B	C	D	E	F	G	H	I	J	K	L	M	N	O	P	Q	R	S	T	U	V	W	X	Y	Z
Ciphertext:	D	E	F	G	H	I	J	K	L	M	N	O	P	Q	R	S	T	U	V	W	X	Y	Z	A	B	C

Figure 3.2.8

A much safer *polyalphabetic* code was popularized by Vigenère in the sixteenth century. With his method, successive letters in the message are encoded using a series of ciphertext alphabets that have been shifted to different extents, those extents being controlled by a *keyword*. For example, we might want to encode *MALTESE FALCON* using the keyword *MATA HARI*. Think of a series of shifted ciphertext alphabets where the first starts at M, the second at A, the third at T, and so on (see Figure 3.2.9). We then encode the first letter of the message using the first ciphertext alphabet, the second letter with the second alphabet, and so on. Thus *MALTESE FALCON* becomes

YAETLSV NMLVOU

Notice that the two E's in *MALTESE* become different letters in the ciphertext; the first is encoded as an L, the second as a V. Breaking a Vigenère code is obviously much

A	B	C	D	E	F	G	H	I	J	K	L	M	N	O	P	Q	R	S	T	U	V	W	X	Y	Z
M	N	O	P	Q	R	S	T	U	V	W	X	Y	Z	A	B	C	D	E	F	G	H	I	J	K	L
A	B	C	D	E	F	G	H	I	J	K	L	M	N	O	P	Q	R	S	T	U	V	W	X	Y	Z
T	U	V	W	X	Y	Z	A	B	C	D	E	F	G	H	I	J	K	L	M	N	O	P	Q	R	S
A	B	C	D	E	F	G	H	I	J	K	L	M	N	O	P	Q	R	S	T	U	V	W	X	Y	Z
H	I	J	K	L	M	N	O	P	Q	R	S	T	U	V	W	X	Y	Z	A	B	C	D	E	F	G
A	B	C	D	E	F	G	H	I	J	K	L	M	N	O	P	Q	R	S	T	U	V	W	X	Y	Z
R	S	T	U	V	W	X	Y	Z	A	B	C	D	E	F	G	H	I	J	K	L	M	N	O	P	Q
I	J	K	L	M	N	O	P	Q	R	S	T	U	V	W	X	Y	Z	A	B	C	D	E	F	G	H
M	N	O	P	Q	R	S	T	U	V	W	X	Y	Z	A	B	C	D	E	F	G	H	I	J	K	L
A	B	C	D	E	F	G	H	I	J	K	L	M	N	O	P	Q	R	S	T	U	V	W	X	Y	Z
T	U	V	W	X	Y	Z	A	B	C	D	E	F	G	H	I	J	K	L	M	N	O	P	Q	R	S
A	B	C	D	E	F	G	H	I	J	K	L	M	N	O	P	Q	R	S	T	U	V	W	X	Y	Z
H	I	J	K	L	M	N	O	P	Q	R	S	T	U	V	W	X	Y	Z	A	B	C	D	E	F	G
⋮												⋮													

Figure 3.2.9

more difficult than deciphering a message where only a single permutation of the plaintext alphabet has been used.

If the message to be encoded is longer than the keyword (as it was in this case), we just repeat the keyword as many times as necessary. This means that the same set of permuted alphabets will reoccur every k letters, where k is the length of the keyword. We call k the *period* of the code.

During World War II, electric coding machines were developed that produced polyalphabetic substitutions where the period was in the millions! One of the prototypes for these machines was invented by an American, Edward Hebern (see Figure 3.2.10). Attached to what looked like a typewriter were five rotors, each having 26 electrical contacts, labeled A through Z in some order, on each face. Each contact on the left side of a rotor was wired to exactly one contact on the right side (see Figure 3.2.11). If the operator keyed in the letter E, an impulse would be sent through the rotors and the code for that letter—for example, Q—would come out. Then rotor 1 would rotate one position, thus generating a different ciphertext alphabet for the second letter of the message: If the operator keyed in another E, the corresponding contact on rotor 1 would necessarily be something other than G and the final output would be something other than Q. Following each of the first 26 letters in the message, rotor 1 advances one position, each time producing another ciphertext alphabet. After the 26th letter in the message is hit, rotor 1 returns to its original position *and rotor 2 advances one position*. After rotor 1 then advances through a second revolution (for plaintext letters 27 through 52), rotor 2 advances a second position, and so on. If the message is long enough, each rotor will eventually complete a revolution before the entire set of five rotors returns to its original position. By the fundamental principle, a rotor system of this sort is analogous to a Vigenère code where the keyword is $26 \cdot 26 \cdot 26 \cdot 26 \cdot 26$, or $11,881,376$ letters long!

The German coding device, known as the ENIGMA machine, used a more-complicated six-rotor system that generated periods over 100 million letters long. What is absolutely astounding is that even the enormously complex ENIGMA machine proved to be inadequate: Toward the end of the war, both the British and the Russians were able to decipher Hitler's most top-secret messages time and time again.

Question 3.2.9 Theorem 3.2.1 was the first mathematical result known to be proved by induction, that feat being accomplished in 1321 by Levi ben Gerson. Assume that we do not know the fundamental principle. Prove the theorem the way Levi ben Gerson did.

Question 3.2.10 In her sonnet with the famous first line, "How do I love thee? Let me count the ways," Elizabeth Barrett Browning listed eight. Suppose Ms. Browning had decided that writing greeting cards afforded her a better opportunity to express her feelings. Furthermore, suppose that each verse in a card would contain exactly four of the ways mentioned in her sonnet. For how many years could she have corresponded with her favorite beau on a daily basis and never sent the same card twice? (Assume the four verses on a given card are ordered, and no verse can be used more than once.)

Question 3.2.11 The board of a large corporation has six members and needs to elect next year's

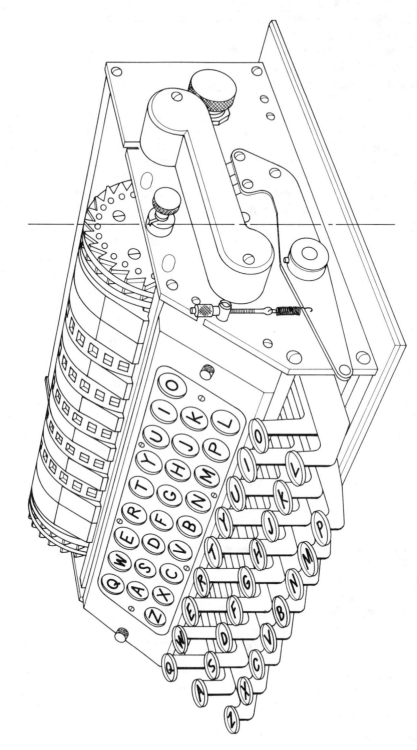

Figure 3.2.10 Edward Hebern's "Electric Code Machine," U.S. Patent 1,683,072. (Adapted with permission from David Kahn, *The Codebreakers* (New York: Macmillan, 1967), p. 416.)

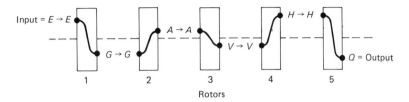

Figure 3.2.11

officers. How many different slates consisting of a president, vice-president, and treasurer could be submitted to the stockholders?

Question 3.2.12 Use Stirling's formula to approximate 30!. (*Note:* The exact answer is 265,252,859, 812,268,935,315,188,480,000,000.)

Question 3.2.13 Suppose a symphony is recorded on all four sides of a two-record album. How many ways can the four sides be played so that at least one is out of sequence?

Question 3.2.14 Four men and four women are to be seated along one side of a table. How many arrangements are possible if the men must sit in alternate chairs? How many arrangements are possible if the two end chairs must be occupied by men?

Question 3.2.15 Consider again the digits 1 through 9 and the associated set of 9! permutations discussed in Example 3.2.5. For how many of those permutations will all the even integers come before any of the odd ones?

Question 3.2.16 Let

$$
A = \begin{pmatrix}
a_{11} & a_{12} & \cdots & a_{1n} \\
a_{21} & a_{22} & \cdots & a_{2n} \\
\vdots & \vdots & \ddots & \vdots \\
a_{n1} & a_{n2} & \cdots & a_{nn}
\end{pmatrix}
$$

denote a square array of numbers. Let $(\lambda_1, \lambda_2, \ldots, \lambda_n)$ be a permutation (without repetition) of the integers $1, 2, \ldots, n$. Define $\epsilon(\lambda_1, \lambda_2, \ldots, \lambda_n)$ to be the sign of the product

$$
\prod_{1 \le r < s \le n} (\lambda_s - \lambda_r)
$$

Then the determinant of A can be written

$$
\det(A) = \sum \epsilon(\lambda_1, \lambda_2, \ldots, \lambda_n) a_{1\lambda_1} \cdots a_{n\lambda_n}
$$

where the summation extends over all possible index sets $\lambda_1, \lambda_2, \ldots, \lambda_n$. How many terms are in the expansion for the determinant of a 5×5 matrix?

Question 3.2.17 In how many ways can a pack of 52 cards be dealt to 13 players, 4 to each, so that every player has one card of each suit? So that one player has a card from each suit but no one else has cards from more than one suit?

Question 3.2.18 How many ways can a 12-member cheerleading squad (6 men and 6 women) pair up to form 6 male-female teams? How many ways can 6 teams line up to do a routine that calls for each woman to stand in front of her partner?

Question 3.2.19 If the definition of $n!$ is to hold for all nonnegative integers n, show that it follows that 0! must equal 1.

If we lift the restriction that no object can be used more than once, the total number of arrangements expands to the figure given in the next theorem. The proof follows immediately from the corollary to the fundamental principle.

Theorem 3.2.2. If repetitions are allowed, the number of ordered arrangements of length r that can be formed from n distinct objects is n^r.

Example 3.2.6

In a famous science fiction story by Arthur C. Clarke, "The Nine Billion Names of God," a computer firm is hired by the lamas in a Tibetan monastery to write a program to generate all possible names of God. For reasons never divulged, the lamas believe that all such names can be written using no more than nine letters. If no letter combinations are ruled inadmissible, is the "nine billion" in the story's title a large enough number to accommodate all possibilities?

No. The lamas are in for a fleecing. The total number of names, N, would be the sum of all one-letter names, two-letter names, and so on. By Theorem 3.2.2, the number of k-letter names is 26^k, so

$$N = 26^1 + 26^2 + \cdots + 26^9 = 5{,}646{,}683{,}826{,}134$$

The proposed list of nine billion, then, would be more than 5.6 trillion names short! (*Note:* The discrepancy between the story's title and the N we just computed is more a language problem than anything else. Clarke is British, and the British have different names for certain numbers than we do. Specifically, our trillion is the English's billion, which means that American editions of Mr. Clarke's story would be more properly entitled "The Nine Trillion Names of God." A more-puzzling question, of course, is why "nine" appears in the title as opposed to "six.")

Example 3.2.7

Proteins are chains of molecules chosen (with repetition) from some 20 different amino acids. In a living cell, proteins are synthesized through the *genetic code*, a mechanism whereby ordered sequences of nucleotides in the messenger RNA dictate the formation of a particular amino acid. The four key nucleotides are adenine, guanine, cytosine, and uracil (A, G, C, and U). Assuming A, G, C, or U can appear any number of times in a nucleotide chain and that all sequences are physically possible, what is the minimum length the nucleotides must have if they are to be able to encode the amino acids?

The answer derives from a trial-and-error application of Theorem 3.2.2. Given a length r, the number of different nucleotide sequences would be 4^r. We are looking, then, for the smallest r such that $4^r \geq 20$. Clearly, $r = 3$.

The entire genetic code is shown in Figure 3.2.12. For a discussion of the duplication and the significance of the three missing triplets, see (92).

Alanine	GCU, GCC, GCA, GCG	Leucine	UUA, UUG, CUU, CUC, CUA, CUG
Arginine	CGU, CGC, CGA, CGG, AGA, AGG	Lysine	AAA, AAG
Asparagine	AAU, AAC	Methionine	AUG
Aspartic acid	GAU, GAC	Phynylalanine	UUU, UUC
Cysteine	UGU, UGC	Proline	CCU, CCC, CCA, CCG
Glutamic acid	GAA, GAG	Serine	UCU, UCC, UCA, UCG, AGU, AGC
Glutamine	CAA, CAG	Threonine	ACU, ACC, ACA, ACG
Glycine	GGU, GGC, GGA, GGG	Tryptophan	UGG
Histidine	CAU, CAC	Tyrosine	UAU, UAC
Isoleucine	AUU, AUC, AUA	Valine	GUU, GUC, GUA, GUG

Figure 3.2.12

Question 3.2.20 In International Morse code, each letter in the alphabet is symbolized by a series of dots and dashes: The letter a, for example, is encoded as $\cdot$ —. What is the minimum number of dots and/or dashes needed to encode the English alphabet?

Question 3.2.21 A word puzzle found in many newspapers has 20 sentences, each requiring that a word be filled in. For each sentence two choices are provided, only one being correct. How many entries must a person submit before one of the solutions is necessarily a winner?

Question 3.2.22 An octave contains 12 distinct notes (on a piano, 5 black keys and 7 white keys). How many different 8-note melodies within a single octave can be written using the white keys only?

Question 3.2.23 Suppose the format for license plates is two letters followed by four numbers.
a. How many different plates can be made?
b. How many different plates can be made if the letters can be repeated but no two numbers can be the same?
c. How many different plates can be made if repetition is allowed but no plate can have four zeros?

The next step in generalizing Theorem 3.2.1 is to consider the case where not all the n objects are distinct. Imagine, for example, permuting the set (A, A, B, C). If the two A's were really an A_1 and an A_2, there would be 24 $(= 4!)$ ordered arrangements; keep the A's identical and that number drops to 12:

$$(A, A, B, C) \quad (B, A, A, C)$$
$$(A, B, A, C) \quad (C, A, A, B)$$
$$(A, B, C, A) \quad (B, A, C, A)$$
$$(A, A, C, B) \quad (B, C, A, A)$$
$$(A, C, A, B) \quad (C, A, B, A)$$
$$(A, C, B, A) \quad (C, B, A, A)$$

In general, we may have r different types of elements, each type being represented n_i times, $i = 1, 2, \ldots, r$. (Here, $r = 3$, $n_1 = 2$, $n_2 = 1$, and $n_3 = 1$.) Theorem 3.2.3 gives a general formula for counting the number of ways those

$n_1 + n_2 + \cdots + n_r$ elements can be permuted. This is an extremely important result, one that we will be seeing over and over again throughout this chapter.

Theorem 3.2.3 The number of ways to arrange n objects, n_1 of one kind, n_2 of a second kind, . . . , and n_r of an rth kind, $\sum_{i=1}^{r} n_i = n$, is

$$\frac{n!}{n_1! n_2! \cdots n_r!}$$

Proof. Let N denote the total number of such arrangements. For any one of those N, the similar objects (if they were actually different) could be arranged in $n_1! n_2! \cdots n_r!$ ways. (Why?) It follows that $N \cdot n_1! n_2! \cdots n_r!$ is the total number of ways to arrange n (distinct) objects. But $n!$ equals that same number. Setting $N \cdot n_1! n_2! \cdots n_r!$ equal to $n!$ gives the result.

Comment

Ratios like $n!/(n_1! n_2! \cdots n_r!)$ are called *multinomial coefficients* because the general term in the expansion of

$$(x_1 + x_2 + \cdots + x_r)^n$$

is

$$\frac{n!}{n_1! n_2! \cdots n_r!} x_1^{n_1} x_2^{n_2} \cdots x_r^{n_r}$$

Example 3.2.8

A sandwich in a vending machine costs 85¢. In how many ways can a customer put in two quarters, three dimes, and a nickel?

Viewed in the context of Theorem 3.2.3, an admissible coin sequence is a permutation of $n = n_1 + n_2 + n_3 = 6$ objects, where

$$n_1 = \text{number of quarters} = 2$$

$$n_2 = \text{number of dimes} = 3$$

$$n_3 = \text{number of nickels} = 1$$

One such sequence is pictured in Figure 3.2.13. It follows that the number of

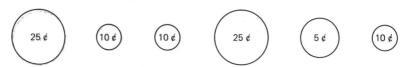

Figure 3.2.13

different ways the coins can be put into the machine is 60:

$$\frac{n!}{n_1! n_2! n_3!} = \frac{6!}{2!3!1!} = 60$$

(Is there another "correct" answer here? Explain.)

Example 3.2.9

What is the coefficient of x^{23} in the expansion of $(1 + x^5 + x^9)^{100}$?
    Note, first, that the exponent 23 can be formed only by combining two
x^9's, one x^5, and 97 1's. It follows that the *coefficient* of x^{23} will be the number
of ways to choose that configuration of factors—in some order—from the 100
$(1 + x^5 + x^9)$'s being multiplied together. By Theorem 3.2.3, that figure is

$$\frac{100!}{2!1!97!}$$

or 485,100.

Example 3.2.10

Prior to the seventeenth century there were no scientific journals, a state of
affairs that made it difficult for researchers to document discoveries. If a scientist
sent a copy of his work to a colleague, there was always a risk that the colleague
might claim it as his own. The obvious alternative—wait to get enough material
to publish a book—invariably resulted in lengthy delays. So, as a sort of interim
documentation, scientists would sometimes send each other anagrams—letter
puzzles that, when properly unscrambled, summarized in a sentence or two what
had been discovered.
    When Christiaan Huygens (1629–1695) looked through his telescope and
saw the ring around Saturn, he composed the following anagram (91):

aaaaaaa, ccccc, d, eeeee, g, h, iiiiiii, llll, mm,

nnnnnnnnn, oooo, pp, q, rr, s, ttttt, uuuuu

How many ways can the 62 letters in Huygens' anagram be arranged?
    Let n_1 (= 7) denote the number of a's, n_2 (= 5) the number of c's, and so
on. Substituting into the appropriate multinomial coefficient, we find

$$N = \frac{62!}{7!5!1!5!1!1!7!4!2!9!4!2!1!2!1!5!5!}$$

as the total number of arrangements. To get a feeling for the magnitude of N,
we need to apply Stirling's formula to the numerator. Write

$$62! \doteq \sqrt{2\pi}\, e^{-62} 62^{62.5}$$

or, equivalently,

$$\log(62!) \doteq \log(\sqrt{2\pi}) - 62 \cdot \log(e) + 62.5 \cdot \log(62)$$

$$\doteq 85.49721$$

The antilog of 85.49721 is 3.142×10^{85}, so

$$N \doteq \frac{3.142 \times 10^{85}}{7!5!1!5!1!1!7!4!2!9!4!2!1!2!1!5!5!}$$

a number on the order of 3.6×10^{60}. Huygens was clearly taking no chances! (*Note:* When appropriately rearranged, the anagram becomes "Annulo cingitur tenui, plano, nusquam cohaerente, ad eclipticam inclinato," which translates to "Surrounded by a thin ring, flat, suspended nowhere, inclined to the ecliptic.")

Example 3.2.11

What is the sum of all the coefficients in the expansion of $(w + x + y + z)^{17}$? From the comment following Theorem 3.2.3,

$$(w + x + y + z)^{17} = \sum_{n_w} \sum_{n_x} \sum_{n_y} \sum_{n_z} \frac{17!}{n_w!n_x!n_y!n_z!} w^{n_w}x^{n_x}y^{n_y}z^{n_z}$$

where the summation is over the integers $n_w \geq 0$, $n_x \geq 0$, $n_y \geq 0$, $n_z \geq 0$, and for which $n_w + n_x + n_y + n_z = 17$. Let $w = x = y = z = 1$. Then

$$(w + x + y + z)^{17} = 4^{17} = \sum_{n_w} \sum_{n_x} \sum_{n_y} \sum_{n_z} \frac{17!}{n_w!n_x!n_y!n_z!}$$

so the sum of the coefficients is 4^{17}, or 17,179,869,184.

Example 3.2.12

In how many ways can the letters of the word

BROBDINGNAGIAN

be arranged without changing the order of the vowels?

The key here is to follow an argument similar to the one we used in Example 3.2.5. Table 3.2.1 shows a breakdown of the word's letter and vowel frequencies. By Theorem 3.2.3, the vowels can be permuted in 30 ways:

$$\frac{5!}{2!2!1!} = 30$$

One of those permutations (O, I, A, I, A) keeps the vowels in their original positions, implying that $\frac{1}{30}$ of *all* the word's permutations should have the vowels in that same order. Our set of admissible arrangements, then, has

$$\frac{1}{30} \cdot \frac{14!}{2!2!1!2!2!3!1!1!} = 30,270,240$$

elements.

TABLE 3.2.1

All letters	Vowels
$A - 2$	$A - 2$
$B - 2$	$I - 2$
$D - 1$	$O - \underline{1}$
$G - 2$	5
$I - 2$	
$N - 3$	
$O - 1$	
$R - \underline{1}$	
14	

Question 3.2.24 Of the 10 stock cars entered in a race, 3 are Fords, 2 are Plymouths, 4 are Mercuries, and 1 is a Chevrolet. In how many ways can they finish, assuming we are interested only in the manufacturers?

Question 3.2.25 Which state name can generate more permutations, *TENNESSEE* or *FLORIDA*?

Question 3.2.26 Imagine six points in a plane, no three being collinear. How many ways can two triangles be drawn using the six points as vertices?

Question 3.2.27 What is the coefficient of x^{13} in the expansion of $(1 + x^3 + x^7)^{18}$?

Question 3.2.28 An irregular n-sided polyhedron (no two faces are congruent) is to be painted using four colors: r of the faces are to be red, w are to be white, b are to be blue, and g are to be green, where $r + w + b + g = n$. How many different coloring schemes are possible?

Question 3.2.29 A tennis tournament has a field of $2n$ entrants, all of whom need to be scheduled to play in the first round. How many different pairings are possible?

Question 3.2.30 In the expansion of $(w + x + y + z)^9$, what is the coefficient of $w^2x^3yz^3$?

Question 3.2.31 How many ways can eight rooks be placed on a chessboard so that no piece can capture any other one? How many ways can they be placed in noncapturing positions if four are white and four are black? If they are numbered 1 through 8?

Question 3.2.32 Vanderbilt plays an 11-game football season.
  a. How many ways can they win 5 games, lose 4, and tie 2?
  b. Let $N(x, y, z)$ denote the number of ways that $x + y + z$ games can end in x wins, y losses, and z ties. Would it make sense to write

$$P(\text{Vanderbilt wins first two games}) = \frac{N(3, 4, 2)}{N(5, 4, 2)}$$

Explain.

Question 3.2.33 How many ways can the letters of the word

$$SLUMGULLION$$

be arranged so that the three L's precede all the other consonants?

Question 3.2.34 How many numbers greater than 4,000,000 can be formed from the digits 2, 3, 4, 4, 5, 5, 5?

Question 3.2.35 In how many ways can the letters of the word

ELEEMOSYNARY

be arranged so that the *S* is always immediately followed by a *Y*?

Question 3.2.36 Consider all possible six-digit numbers formed without repetition from the digits 1 through 6. If the permutations were arranged in increasing order of magnitude, what would be the 275th number on the list?

Question 3.2.37 Let *C* be a "chessboard" of arbitrary shape but containing *m* squares. For each $k \leq m$, let $r_k(C)$ denote the number of ways of placing *k* noncapturing rooks on *C*. Then the *rook polynomial for C, R(x, C)*, is defined by

$$R(x, C) = r_0(C) + r_1(C)x + r_2(C)x^2 + \cdots + r_m(C)x^m$$

Find the rook polynomial for a 4×4 board.

Question 3.2.38 Show that $(k!)!$ is divisible by $k!^{(k-1)!}$. (*Hint:* Think of a related permutation problem.)

Thus far the enumeration results we have seen have dealt with what might be called *linear* permutations—objects being lined up in a row. This is the typical context in which permutation problems arise. Situations sometimes develop, though, where it is necessary to work with nonlinear arrangements of one form or another. The final theorem in this section gives the basic result associated with *circular* permutations. Several other symmetries are examined in the exercises.

Theorem 3.2.4. There are $(n - 1)!$ ways to arrange *n* distinct objects in a circle.

Proof. Fix any object at the "top" of the circle. The remaining $n - 1$ objects can then be permuted in $(n - 1)!$ ways. Since any arrangement with a different object at the top can be reproduced by simply rotating one of the original $(n - 1)!$ permutations, the statement of the theorem holds.

Example 3.2.13

How many different firing orders are theoretically possible in a six-cylinder engine? (If the cylinders are numbered from 1 to 6, a firing order is a list such as 1, 4, 2, 5, 3, 6, giving the sequence in which fuel is ignited in the six cylinders.)

By a direct application of Theorem 3.2.4, the number of distinct firing orders is $(6 - 1)!$, or 120.

Comment

According to legend (38), perhaps the first person for whom the problem of arranging objects in a circle took on life or death significance was Flavius Josephus, an early Jewish scholar and historian. In A.D. 66, Josephus found himself a somewhat reluctant leader of an attempt to overthrow the Roman administration in the town of Judea. But the coup failed, and Josephus and 40 of his comrades ended up being trapped in a cave, surrounded by an angry Roman army. Faced with the prospect of imminent capture, others in the group were intent on committing mass suicide, but Josephus' devotion to the cause did not extend quite that far. Still, he did not want to appear cowardly, so he proposed an alternate plan: All 41 would arrange themselves in a circle; then, one by one, the group would go around the circle and kill every seventh remaining person, starting with whoever was seated at the head of the circle. That way, Josephus argued, only one of them would have to commit suicide, and the entire action would make more of an impact on the Romans. To his relief, the group accepted his suggestion and began forming a circle. Josephus, who was reputed to have had some genuine mathematical ability, quickly made his way to the 25th position. Forty murders later, he was the only person left alive! Is the story true? Maybe yes, maybe no. Any conclusion would be little more than idle speculation. It *is* known, though, that Josephus was the sole survivor of the siege, that he surrendered, and that he eventually rose to a position of considerable influence in the Roman government. And, whether true or not, the legend has given rise to some mathematical terminology: Cyclic cancellations of a fixed set of numbers having the property that a specified number is left at the end are referred to as *Josephus permutations*.

Question 3.2.39 In a V-6 engine, not all firing orders are equally desirable: The engine will run more smoothly if consecutive ignitions are on opposite "arms" of the V. How many of the permutations indicated in Example 3.2.13 meet that criterion?

Question 3.2.40 How many different necklaces can be made using all of seven different colored stones?

Question 3.2.41 How many different dice can be made with faces numbered 1 through 6 if the numbers on each pair of opposite faces must sum to 7?

Question 3.2.42 Consider a square with corners numbered 1 through 4, as shown.

$$\begin{array}{|cc|} \hline 1 & 2 \\ 4 & 3 \\ \hline \end{array}$$

How many different ways can the square be oriented, either by a series of rotations and/or by flipping it any number of times around either its vertical axis or its horizontal axis?

Question 3.2.43 Each of four cubes has its faces painted different colors (making 24 colors in all). How many different ways can the cubes be oriented, end to end?

3.3 COMBINATIONS: UNORDERED SELECTIONS

Order is not always a meaningful characteristic of a collection of n objects, our efforts of the previous section notwithstanding. Consider a poker player being dealt a five-card hand. Whether the player receives a two of hearts, four of clubs, nine of clubs, jack of hearts, and ace of diamonds *in that order* or in any one of the other $5! - 1$ permutations of those particular five cards is irrelevant—the hand is still the same. As the examples in this section will bear out, there are many such situations—problems where our only concern is with the composition of a set of objects, not their precise order.

We will call a collection of r unordered objects a *combination of length r*. For example, given a set of $n = 4$ distinct objects A, B, C, and D, there are *six* ways to form combinations of length 2:

$$(A, B) \qquad (B, C)$$
$$(A, C) \qquad (B, D)$$
$$(A, D) \qquad (C, D)$$

As was true for permutations, counting the number of ways to form combinations is a problem admitting several variations. For our purposes, the most prevalent situation is where we are looking for the number of combinations of length r that can be formed without repetition from a set of n distinct objects. Altogether, though, we will consider *three* variations on the combination problem (see Figure 3.3.1).

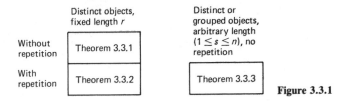

	Distinct objects, fixed length r	Distinct or grouped objects, arbitrary length $(1 \leq s \leq n)$, no repetition
Without repetition	Theorem 3.3.1	Theorem 3.3.3
With repetition	Theorem 3.3.2	

Figure 3.3.1

Finding a physical analogy, or model, to correspond to the statement of a combinatorial theorem often makes it easier to visualize what the theorem is counting. The first result in this section has such a model: The number of ways to form combinations of length r from n distinct objects, chosen without repetition, is equal to the number of ways to reach into an urn containing n distinct objects and draw out, *all at once,* an unordered sample of size r. For example, recall our earlier enumeration with $n = 4$ and $r = 2$: Figure 3.3.2 shows the equivalent urn problem. We will come to see that this is a very useful analogy; there are any number of examples in this section—together with many later in the chapter—that can be conceptually simplified by being put in the context of drawing unordered samples out of an urn.

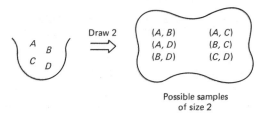

Possible samples of size 2

Figure 3.3.2

Theorem 3.3.1. The number of ways to form combinations of length r from a set of n distinct objects, repetitions not allowed, is denoted by the symbol $\binom{n}{r}$, where

$$\binom{n}{r} = \frac{n!}{r!(n-r)!}$$

Proof. Let the symbol $\binom{n}{r}$ denote the number of combinations satisfying the conditions of the theorem. Because each of those combinations can be ordered in $r!$ ways (why?), the product $r!\binom{n}{r}$ will equal the number of *permutations* of length r that can be formed from n distinct objects (repetitions not allowed). But we already know that n distinct objects can be formed into permutations of length r in $n(n-1) \cdots (n-r+1) = n!/(n-r)!$ ways. Therefore,

$$r!\binom{n}{r} = \frac{n!}{(n-r)!}$$

Solving for $\binom{n}{r}$ gives the result.

Comment

The symbol $\binom{n}{r}$ is read, "n things taken r at a time" or "n choose r." Both expressions are in deference to the sampling-from-an-urn model associated with the problem of forming combinations.

Comment

An equivalent way of deriving Theorem 3.3.1 is to establish a one-to-one relationship between a combination of length r and an ordered arrangement of n objects, where r of the latter belong to one type and $n - r$ belong to another. Once that correspondence is set up, Theorem 3.3.1 is just a special case of Theorem 3.2.3. At this point, though, it is best to emphasize the distinctions in these two problems rather than their similarities. For that reason, we elected to re-prove Theorem 3.3.1.

Comment

Like their Theorem 3.2.3 counterparts, terms such as $\binom{n}{r}$ appear in the expansion of a sum raised to a power—specifically,

$$(x+y)^n = \sum_{r=0}^{n} \binom{n}{r} x^r y^{n-r}$$

Since the base here involves *two* summands, x and y, the constants $\binom{n}{r}$ are called *binomial coefficients*.

Example 3.3.1

Eight politicians meet at a fund-raising dinner. How many greetings can be exchanged if each politician shakes hands with every other politician once and only once?

Imagine the politicians to be eight chips—A through H—in an urn. A handshake corresponds to an unordered sample of size 2 chosen from that urn. Because repetitions are not allowed (even the most overzealous of campaigners would not shake hands with himself!), Theorem 3.3.1 applies, and the total number of handshakes is

$$\binom{8}{2} = \frac{8!}{2!6!}$$

or 28.

Example 3.3.2

The basketball recruiter for Swampwater Tech has scouted 16 former NBA starters that he thinks he can pass off as JUCO transfers—6 are guards, 7 are forwards, and 3 are centers. Unfortunately, his slush fund of illegal alumni donations is at an all-time low and he can afford to buy new Trans-Ams for only 9 of the players. If he wants to keep 3 guards, 4 forwards, and 2 centers, in how many ways can he parcel out the cars?

This is a combination problem that also requires an application of the fundamental principle. First, note there are $\binom{6}{3}$ *sets* of three guards that can be chosen to receive Trans-Ams (think of drawing a set of 3 names out of an urn containing 6 names). Similarly, the forwards and the centers can be bribed in $\binom{7}{4}$ and $\binom{3}{2}$ ways, respectively. It follows from the fundamental principle, then, that the total number of ways to divvy up the cars is the product

$$\binom{6}{3} \cdot \binom{7}{4} \cdot \binom{3}{2}$$

which reduces to $20 \cdot 35 \cdot 3$, or 2100.

Example 3.3.3

A chemist is trying to synthesize part of a straight-chain aliphatic hydrocarbon polymer that consists of 21 radicals—10 ethyls (E), 6 methyls (M), and 5 propyls (P). Assuming all arrangements of radicals are physically possible, how many different polymers can be formed if no two of the methyl radicals are to be adjacent?

Imagine arranging the E's and the P's without the M's: Figure 3.3.3 shows one such possibility. Consider the 16 "spaces" between and outside the E's and the P's, as indicated by the arrows in Figure 3.3.3. In order for the M's to be nonadjacent, they must occupy any 6 of those locations. But those 6 spaces can be chosen in $\binom{16}{6}$ ways. And for each of the $\binom{16}{6}$ positionings of the M's, the E's and P's can be permuted in $\frac{15!}{10!5!}$ ways (Theorem 3.2.3). So, by the fundamental

$$E \quad E \quad P \quad P \quad E \quad E \quad E \quad P \quad E \quad P \quad E \quad P \quad E \quad E \quad E$$

Figure 3.3.3

principle, the total number of polymers having nonadjacent methyl radicals is 24,048,024:

$$\binom{16}{6} \cdot \frac{15!}{10!5!} = \frac{16!}{10!6!} \cdot \frac{15!}{10!5!} = (8008)(3003) = 24,048,024$$

Question 3.3.1 The crew of *Apollo 17* consisted of two pilots and one geologist. Suppose that NASA had actually trained a total of nine pilots and four geologists. How many possible *Apollo 17* crews could have been formed?
a. Assume the two pilot positions have identical duties.
b. Assume the two pilot positions are really a pilot and a copilot.

Question 3.3.2 How many straight lines can be drawn between five points ($A, B, C, D,$ and E), no three of which are collinear?

Question 3.3.3 Show that

$$r\binom{n}{r} = n\binom{n-1}{r-1}$$

Question 3.3.4 A woman's purse is snatched by two men. A day later the police show her a lineup of seven suspects. How many different ways can the woman pick two people out of the lineup? If the two perpetrators are among the seven, how many ways can she pick out two people so that exactly one of the two is guilty?

Question 3.3.5 Consider the following sequence of figures formed by arranging circles in a triangular pattern:

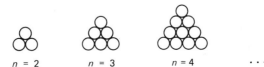

$n = 2 \qquad\qquad n = 3 \qquad\qquad n = 4 \qquad\qquad \cdots$

Show that the number of circles in the nth such figure is $\binom{n+1}{2}$.

Question 3.3.6 The Alpha Beta Zeta sorority is trying to fill a pledge class of 9 new members during fall rush. Among the 25 available candidates, 15 have been judged marginally acceptable and 10 are highly desirable. How many ways can the pledge class be chosen to give a 2-to-1 ratio of highly desirable to marginally acceptable candidates?

Question 3.3.7 A boat's crew consists of 8 people, of whom two can row only on the stroke side of the boat and three only on the bow side. In how many ways can the crew be arranged?

Question 3.3.8 See the figure at the top of the next page. A person wants to walk from X to Y and will consider only routes that pass through intersection O. How many different paths can the person take, assuming he or she never wants to go out of the way?

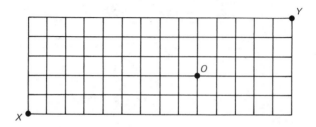

Question 3.3.9 Use Stirling's formula to show that

$$\binom{2n}{n} \doteq (\pi n)^{-1/2} 2^{2n}$$

Question 3.3.10 Ten basketball players meet in the school gym for a pickup game. How many ways can they be split into two teams of five each?

Question 3.3.11 Four men and four women are to be seated in a row.
  a. How many arrangements are possible?
  b. How many arrangements are possible if the men are required to sit in alternate chairs?
  c. How many arrangements are possible if the four men are considered indistinguishable and the four women are considered indistinguishable?
  d. How many arrangements are possible if the four men are considered indistinguishable but the four women are considered distinguishable?

Question 3.3.12 Suppose the license-plate format of Question 3.2.23 were modified so that a tag had to include two letters and four numbers (repetitions allowed), but they could be in any order. How many plates could be made?

Question 3.3.13 Vanessa is a cabaret singer who always opens her act by telling four jokes. Her current engagement is scheduled to run for 4 months. If she gives one performance a night and never wants to repeat the same set of jokes on any two nights, what is the minimum number of jokes Vanessa must have in her repertoire?

Question 3.3.14 How many ways can the letters of the word *MISSISSIPPI* be arranged so that no two *I*'s are adjacent?

Question 3.3.15 Under a certain condition, a good approximation to $(1 + x)^n$ can be gotten by using only the first several terms in its binomial expansion. What is that condition? Explain.

A formula for the number of combinations that can be formed *with* repetition is simple to state but a little more difficult to prove than the other results we have seen thus far. The trick is to reduce the problem to one we have already solved.

> **Theorem 3.3.2.** The number of ways to form combinations of length r from a set of n distinct objects, repetitions allowed, is equal to
>
> $$\binom{n + r - 1}{r}$$

Proof. Think of the n objects as being numbered 1 to n. Then a combination of length r, with repetitions, can be written $(i_1, i_2, \ldots, i_r)$, where, without loss of generality, we can arrange the objects from smallest to largest: $i_1 \leq i_2 \leq i_3 \leq \cdots \leq i_r$. Now, suppose we add 0 to i_1, 1 to i_2, 2 to i_3, and so on. Consider the set $(i_1 + 0, i_2 + 1, i_3 + 2, \ldots, i_r + r - 1)$. It will necessarily be true that no two entries in this transformed set are equal—$i_1 + 0 < i_2 + 1 < i_3 + 2 < \cdots < i_r + r - 1$. For example, if n were 6 and r were 5, the combination $(i_1, i_2, \ldots, i_5) = (1, 2, 2, 4, 4)$ would become $(i_1 + 0, i_2 + 1, \ldots, i_5 + 4) = (1, 3, 4, 7, 8)$.

Note that counting sets of the form $(i_1 + 0, i_2 + 1, \ldots, i_r + r - 1)$ is a familiar problem; since $i_1 + 0 \geq 1$ and $i_r + r - 1 \leq n + r - 1$, what we are forming are subsets of length r, *without repetition*, from the integers 1 through $n + r - 1$. By Theorem 3.3.1, that can be done in $\binom{n+r-1}{r}$ ways. The statement of the theorem follows, then, from the obvious one-to-one correspondence between combinations of the form $(i_1, i_2, \ldots, i_r)$ and those of the form $(i_1 + 0, i_2 + 1, \ldots, i_r + r - 1)$.

Example 3.3.4

Suppose f is an analytic function of three variables: $f = f(x, y, z)$. How many potentially different sixth-order partial derivatives of f can be formed?

Since the order in which a partial derivative is taken is irrelevant—that is,

$$\frac{\partial^2}{\partial x \partial y} f(x, y, z) = \frac{\partial^2}{\partial y \partial x} f(x, y, z)$$

—all that distinguishes one sixth-order partial from another are the numbers of times the differentiation was done with respect to x, with respect to y, and with respect to z. Each sixth-order partial, then, can be represented by a combination of length $r = 6$ from the $n = 3$ "objects," ∂x, ∂y, and ∂z. Thus, the combination $(\partial x, \partial x, \partial y, \partial y, \partial y, \partial z)$ indicates a sixth-order partial where the differentiation is done with respect to x twice, y three times, and z once. It follows that the total number of potentially different sixth-order partials is

$$\binom{3 + 6 - 1}{6} = \binom{8}{6}$$

or 28. (Are all 28 of the combinations necessarily different functions?)

Example 3.3.5

If two distinguishable dice are tossed—one red and one green—we know from the fundamental principle that the total number of outcomes is 36 $(= 6 \cdot 6)$. How many outcomes are possible if the dice are *indistinguishable*?

Note that rolling two indistinguishable dice yields outcomes that can be thought of as combinations of length two, where each component of the combination is an integer from 1 to 6, inclusive, and repetitions are allowed. Thus, $(3, 4)$ corresponds to the outcome where one die comes up 3 and the other, 4.

From Theorem 3.3.2, the number of such outcomes is 21:

$$\binom{n + r - 1}{r} = \binom{6 + 2 - 1}{2} = \binom{7}{2} = 21$$

Figure 3.3.4 shows a listing of all the possibilities. (How many ways can *three* indistinguishable dice be rolled?)

(1, 1)	(1, 4)	(2, 2)	(2, 5)	(3, 4)	(4, 4)	(5, 5)
(1, 2)	(1, 5)	(2, 3)	(2, 6)	(3, 5)	(4, 5)	(5, 6)
(1, 3)	(1, 6)	(2, 4)	(3, 3)	(3, 6)	(4, 6)	(6, 6)

Figure 3.3.4 Ways to roll two indistinguishable dice.

Example 3.3.6

How many numbers are there between 00000 and 99999 whose digits form a nondecreasing sequence? (The number 22389 is a nondecreasing sequence; 12648 is not.)

The solution here is interesting because of the way order is first ignored and then reintroduced. We will see other problems later on that require a similar approach. Think of choosing an unordered sample of size 5, with repetition, from the digits 0 through 9. Note that any such sample—say, 06242—can be rearranged in a unique way to form a nondecreasing sequence: 06242, for example, can be permuted to form 02246. From Theorem 3.3.2, then, the number of five-digit, nondecreasing sequences is $\binom{10+5-1}{5} = \binom{14}{5}$, or 2002.

Question 3.3.16 A tropical-fish enthusiast wants to add three fish to her collection. At the pet store she has her choice of tiger barbs, neon tetras, swordtails, and angel fish. How many different additions can she make to her aquarium?

Question 3.3.17 Show that the number of different throws that can be made with n indistinguishable dice is

$$(1 + n)\left(1 + \frac{n}{2}\right)\left(1 + \frac{n}{3}\right)\left(1 + \frac{n}{4}\right)\left(1 + \frac{n}{5}\right)$$

Question 3.3.18 How many numbers are there between 0000 and 9999 whose digits form a nonincreasing sequence?

Question 3.3.19 Show that the number of ways to form combinations of length r (without repetition) from the n numbers 1, 2, . . . , n such that no two numbers are consecutive is

$$\binom{n + r + 1}{r}$$

(*Hint:* Follow an approach similar to the one used in the proof of Theorem 3.3.2.)

Question 3.3.20 The Smith family has set aside $20 (in $1 bills) as Christmas bonuses to be divided up among the paper carrier, the mail carrier, and the baby sitter. How many ways can the money be given away?

Question 3.3.21 Use Newton's binomial theorem to show that

$$\binom{n + r - 1}{n} = (-1)^n \binom{-r}{n}$$

Question 3.3.22 How many nonnegative integer solutions does the equation

$$x_1 + x_2 + x_3 + x_4 + x_5 = 63$$

have?

We conclude this section with two results concerned with the formation of combinations without repetition where the length of the combination is unspecified and can be anything from 1 to n. Neither of these results finds as many applications as the first two theorems in this section, but the corollary to Theorem 3.3.3 will turn up as an intermediate step in some of the derivations at which we look later in this chapter.

Theorem 3.3.3. Out of n objects—n_1 of a first kind, n_2 of a second kind, . . . , and n_r of an rth kind—we can form combinations (without repetition) in a total of

$$(n_1 + 1)(n_2 + 1) \cdots (n_r + 1) - 1$$

ways. (The length of a combination can be anything from 1 to n, inclusive.)

Proof. In forming a combination, we may include 0, 1, 2, . . . , or all n_1 of the objects from the first group; 0, 1, . . . , or all n_2 from the second, and so on. Thus, each group offers $(n_i + 1)$ choices, $i = 1, 2, \ldots, r$. By the fundamental principle, then, the total number of selections should be the product $(n_1 + 1)(n_2 + 1) \cdots (n_r + 1)$. Notice, however, that one of the combinations formed in this way contains 0 objects from each group. If the null selection is considered inadmissible—and subtracted off—the result follows.

Corollary. Excluding the null selection, the total number of ways to form combinations (without repetition) from a set of n objects, all distinct, is

$$2^n - 1$$

Example 3.3.7

A Braille letter is formed by raising at least one dot in the six-dot matrix

· ·

· ·

· ·

For example, the letter e is written

Punctuation marks also have specified dot patterns, as do certain common words, suffixes, letter combinations, and so on. In all, how many different characters can be transcribed using Braille?

Think of the dots here as being six distinct objects—numbered, say, from 1 to 6. In forming a Braille letter, we have two options for each dot: We can (1) raise it or (2) not raise it. The *six* dots, then, generate 2^6 distinct patterns, but one of those has no raised dots, a configuration that would obviously be of no use to a blind person. Thus, the number of admissible patterns is 63 $(= 2^6 - 1)$. Figure 3.3.5 shows the entire Braille alphabet.

Example 3.3.8

Let N be a positive integer. How many divisors does it have?

The factorization theorem, a fundamental result in the theory of numbers, says that any positive integer can be uniquely expressed as a product of one or more positive primes. Applying that characterization here, we can write

$$N = p_1^{n_1} \cdot p_2^{n_2} \cdots p_k^{n_k}$$

where the p_i's are distinct primes. It follows from Theorem 3.3.3 that

Number of divisors of $N = (n_1 + 1)(n_2 + 1) \cdots (n_k + 1)$

(Why have we not subtracted the 1 that appears in the statement of Theorem 3.3.3?)

Question 3.3.23 A chain of fast-food restaurants offers customers a choice of eight extras to put on their hamburgers. How many different hamburgers can be made? (Assume the order in which the extras are added is irrelevant.) Should the null selection be considered inadmissible?

Question 3.3.24 How many hamburgers of the kind described in Question 3.3.23 could be made if the order in which the extras were added *did* matter?

Question 3.3.25 How many divisors does the number 600 have?

Question 3.3.26 In baseball there are 24 different base-out configurations (runner on first, two outs; bases loaded, none out; and so on). Suppose a new game, sleazeball, is played where there are seven bases (excluding home plate) and each team gets five outs an inning. How many base-out configurations would be possible in sleazeball?

Question 3.3.27 Prove the statement of the corollary without using Theorem 3.3.3.

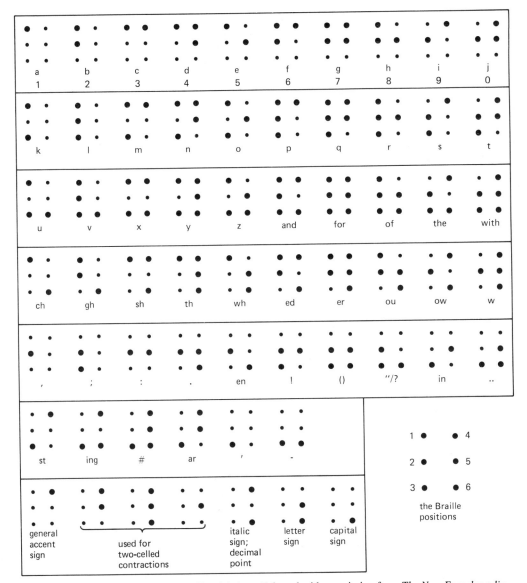

Figure 3.3.5 The Braille alphabet. (Adapted with permission from *The New Encyclopædia Britannica,* 15th ed., Vol. 3 (Chicago: Encyclopædia Britannica, 1982), p. 111.)

3.4 PROPERTIES OF BINOMIAL COEFFICIENTS

The symbol $\binom{n}{r}$, introduced in Theorem 3.3.1, is a familiar one in many branches of mathematics. Sometimes it appears by virtue of its relationship to the binomial expansion; other times, it is used because of its combinatorial significance. In this section we look at some of the basic mathematical properties of $\binom{n}{r}$, properties not necessarily related to probability. In the process, we will encounter some techniques for verifying identities that will prove quite useful later on.

Among the many relationships that involve binomial coefficients, perhaps the most familiar is *Pascal's triangle* (Figure 3.4.1). Numbers in the nth row are assigned the values $\binom{n}{j}$, $j = 0, 1, \ldots, n$, with each entry being the sum of the two numbers appearing diagonally above it. Example 3.4.1 gives a combinatorial argument that justifies Pascal's triangle. The same result can be proved directly by simply expanding the various factorials involved (see Question 3.4.1). (Despite its name, Pascal's triangle was not discovered by Pascal. Its basic structure was known hundreds of years before the French mathematician was born. It was Pascal, though, who first made extensive use of its properties.)

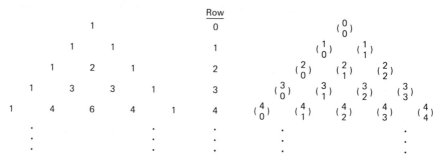

Figure 3.4.1 Pascal's triangle.

Example 3.4.1

Prove the identity pictured in Pascal's triangle, that

$$\binom{n + 1}{k} = \binom{n}{k} + \binom{n}{k - 1}$$

Consider a set of $n + 1$ distinct objects $A_1, A_2, \ldots, A_{n+1}$. We can obviously draw samples of size k from that set in $\binom{n+1}{k}$ different ways. Now, consider any particular object—for example, A_1. Relative to A_1, each of those $\binom{n+1}{k}$ samples belongs to one of two categories: those containing A_1 and those not containing A_1. To form samples containing A_1, we need to select $k - 1$ additional objects from the remaining n. This can be done in $\binom{n}{k-1}$ ways. Similarly, there are $\binom{n}{k}$ ways to form samples not containing A_1. Therefore, $\binom{n+1}{k}$ must equal $\binom{n}{k} + \binom{n}{k-1}$.

Comment

Pascal's triangle has been the subject of considerable mathematical interest for centuries, and many of its properties (some of which we will prove later in this section) are well known. For example, the sum of the entries in the ith row is 2^i. Also, it has an obvious symmetry,

$$\binom{n}{r} = \binom{n}{n - r}$$

and a not-so-obvious one, that for fixed n the sum of the entries for j even equals the sum for j odd. A more curious relationship emerges if we write the triangle in the format shown in Figure 3.4.2. Notice that the sum of the entries along successive diagonals are the terms in the famous Fibonacci series, 1, 1, 2, 3, 5, 8, 13, and so on.

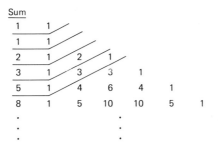

Figure 3.4.2

Among the more recently discovered relationships holding for Pascal triangles is an interesting multiplication property. Suppose we consider any entry in the interior of the triangle and the hexagon of numbers surrounding that entry— for example, the numbers 4, 6, 10, 20, 15, and 5 encircling 10 (see Figure 3.4.3). Now, imagine two triangles inscribed in the hexagon, each having every alternate number as a vertex (how many ways can that be done?). It can be shown that the product of the vertices for the two inscribed triangles will be the same, regardless of which number in the triangle we select initially. Here the two products are $5 \cdot 6 \cdot 20$ and $4 \cdot 15 \cdot 10$, both of which equal 600.

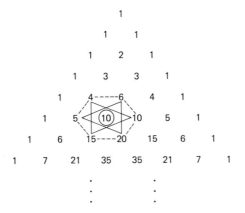

Figure 3.4.3

Question 3.4.1 Prove that

$$\binom{n + 1}{k} = \binom{n}{k} + \binom{n}{k - 1}$$

directly, without appealing to any combinatorial arguments.

Question 3.4.2 How many ways can you form the word *ABRACADABRA* in the array shown at the top of the next page (68)? (Assume you must start at the top *A* and move diagonally downward to the bottom *A*.)

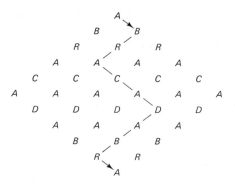

Question 3.4.3 Find a recursion formula for $\binom{n}{k+1}$ in terms of $\binom{n}{k}$.

Question 3.4.4 Use Pascal's triangle to show that

$$\binom{n}{0} + \binom{n+1}{1} + \binom{n+2}{2} + \cdots + \binom{n+r}{r} = \binom{n+r+1}{r}$$

What would be a combinatorial problem having the left-hand side of this identity as its solution? (*Hint:* Consider forming permutations out of $n + r$ objects, n of one kind and r of another.)

Question 3.4.5 Show that $\binom{2n}{n}$ is always an even number.

Question 3.4.6 Show that

$$\binom{n}{r}\binom{r}{k} = \binom{n}{k}\binom{n-k}{r-k}$$

Question 3.4.7 Generalize the proof of the Pascal relationship given in Example 3.4.1 to show that

$$\binom{n+k}{r+k} = \binom{n}{r} + \binom{k}{1}\binom{n}{r+1} + \binom{k}{2}\binom{n}{r+2} + \cdots + \binom{k}{r}\binom{n}{r+k}$$

(*Hint:* Consider the disposition of k objects, $A_1, A_2, \ldots, A_k$.)

Question 3.4.8 Use the recurrence relation of Example 3.4.1 to get a formula for the sum of the first n integers. (*Hint:* Write the relationship in the form

$$\binom{n}{k-1} = \binom{n+1}{k} - \binom{n}{k}$$

and evaluate both sides for $k = 2$ and various values for n.)

The final examples in this section present two other well-known combinatorial identities. Of particular interest is the way in which the two are proved, by manipulating expansions of polynomials. Variations of this approach are widely used in more advanced combinatorial work (see, for example, (51) or (89)).

Example 3.4.2

Show that

$$\binom{n}{1} + \binom{n}{3} + \cdots = \binom{n}{0} + \binom{n}{2} + \cdots$$

for any n (recall the comment following Example 3.4.1).

Consider the expansion of $(x - y)^n$:

$$(x - y)^n = \sum_{k=0}^{n} \binom{n}{k} x^k (-y)^{n-k} \qquad (3.4.1)$$

Let $x = y = 1$. Then $x - y = 0$, and Equation 3.4.1 reduces to

$$0 = \sum_{k=0}^{n} \binom{n}{k} (-1)^{n-k}$$

which can be written

$$\binom{n}{0} + \binom{n}{2} + \cdots = \binom{n}{1} + \binom{n}{3} + \cdots$$

Example 3.4.3

Prove that

$$\binom{n}{1} + 2\binom{n}{2} + \cdots + n\binom{n}{n} = n2^{n-1}$$

This time we begin with the expansion of $(1 + x)^n$:

$$(1 + x)^n = \sum_{k=0}^{n} \binom{n}{k} x^k (1)^{n-k} \qquad (3.4.2)$$

Differentiating both sides of Equation 3.4.2 with respect to x gives

$$n(1 + x)^{n-1} = \sum_{k=0}^{n} \binom{n}{k} k x^{k-1} \qquad (3.4.3)$$

Now, let $x = 1$. This simplifies the left-hand side of Equation 3.4.3 to $n2^{n-1}$, while the right-hand side reduces to

$$\sum_{k=0}^{n} k\binom{n}{k} = \binom{n}{1} + 2\binom{n}{2} + \cdots + n\binom{n}{n}$$

Question 3.4.9 Prove that

$$\binom{n}{0} + \binom{n}{1} + \cdots + \binom{n}{n} = 2^n$$

(*Hint:* Consider the expansion of $(x + y)^n$.)

Question 3.4.10 Give a combinatorial argument to justify the identity in Question 3.4.9. (*Hint:* Consider the problem of forming all possible samples from a set of n distinct objects.)

Question 3.4.11 Prove that

$$\binom{n}{0}^2 + \binom{n}{1}^2 + \cdots + \binom{n}{n}^2 = \binom{2n}{n}$$

(*Hint:* Rewrite the left-hand side as

$$\binom{n}{0}\binom{n}{n} + \binom{n}{1}\binom{n}{n-1} + \binom{n}{2}\binom{n}{n-2} + \cdots$$

and consider the problem of selecting a sample of n objects from an original set of $2n$ objects.)

Question 3.4.12 Prove that successive terms in the sequence $\binom{n}{0}$, $\binom{n}{1}$, . . . , $\binom{n}{n}$, first increase and then decrease. (*Hint:* Examine the ratio of two successive terms, $\binom{n}{j+1}/\binom{n}{j}$.)

Question 3.4.13 Show that

$$\sum_{r=0}^{n}\binom{n}{r}(x-1)^r = x^n$$

Question 3.4.14 Show that

$$n(n-1)2^{n-2} = \sum_{k=2}^{n} k(k-1)\binom{n}{k}$$

Question 3.4.15 Show that

$$1!\, 1 + 2!\, 2 + 3!\, 3 + \cdots + n!\, n = (n+1)! - 1$$

(*Hint:* Use the fact that $(n+1)! - n! = n!\, n$.)

Question 3.4.16 Show that

$$\sum_{k=0}^{n}\frac{1}{k+1}\binom{n}{k} = \frac{2^{n+1}-1}{n+1}$$

Question 3.4.17 Imagine n molecules of a gas confined to a rigid container divided into two chambers by a semipermeable membrane. If i molecules are in the left chamber, the *entropy* of the system is defined by the equation

$$\text{Entropy} = \log\binom{n}{i}$$

If n is even, for what configuration of molecules will the entropy be maximized? (Entropy is a concept physicists find useful in characterizing heat exchanges, particularly those involving gases. In general terms, the entropy of a system is a measure of its disorder: As the "randomness" of the position and velocity vectors of a system of particles increases, so does its entropy.)

3.5 COMBINATORIAL PROBABILITY

The notion that the counting principles we have been discussing can be applied to the solution of probability problems was alluded to in several exercises appearing in previous sections but never actually formalized. Here we will look at a variety of situations where the *probability* of something happening is a more relevant piece of information than the number of *ways* it can happen. Examples of this occur all the time in gambling. A poker player, for instance, would certainly be interested in knowing the probability of being dealt a full house. Would the player want to know *how many* full houses he or she could be dealt? No, not really. By itself, that information would not be of much help in planning a strategy. Fortunately, converting enumerations into probabilities is usually a relatively simple task.

The relationship between combinatorics and probability derives immediately from the basic principles introduced at the beginning of Chapter 2. Suppose an experiment generates a sample space S of equally likely outcomes. Let A be any subset of those outcomes. By the classical definition of probability (recall Section 2.3),

$$P(A) = \frac{\text{number of outcomes in } A}{\text{number of outcomes in } S} \tag{3.5.1}$$

Any combinatorial probability problem, then, reduces to a quotient of two enumerations.

The first example of Equation 3.5.1 that we consider is an often-quoted one. Known as the *birthday problem,* it has a simple statement and a simple solution, but the answer is strongly contrary to our intuition.

Example 3.5.1

Suppose a group of k randomly selected individuals is assembled. What is the probability that at least two of them will have the same birthday?

Most people would guess that for k relatively small—for example, less than 50—a match would be very unlikely. It can be easily shown, though, that the odds of at least one match are better than 50-50 if as few as 23 people are present. And when the group does number 50, there is a 97% chance of at least one match!

The counting results to which we need to appeal are Theorems 3.2.1 and 3.2.2. Omitting leap year, there are 365 possible birthdays for each of the k people. If we imagine the people to be ordered, there are 365^k corresponding permutations of their k birthdays, repetitions allowed. If repetitions are not allowed, the number of permutations of length k reduces to $365!/(365 - k)!$. Assume each person has an equal chance of being born on any particular day. Then

$$P(\text{all } k \text{ birthdays are different}) = \frac{(365)!/(365 - k)!}{(365)^k}$$

and

$$P(\text{at least two have same birthday}) = 1 - P(\text{all birthdays are different})$$

$$= 1 - \frac{(365)!/(365 - k)!}{(365)^k} = P_k$$

Table 3.5.1 gives the value of P_k for $k = 15, 22, 23, 40, 50,$ and 70.

TABLE 3.5.1

k	$P_k = P(\text{at least one match})$
15	0.253
22	0.475
23	0.507
40	0.891
50	0.970
70	0.999

Comment

To facilitate the computation of the P_k's, it was assumed that the equally likely model would describe the distribution of birthdays in the general population. That, of course, is not entirely true, since births are more common during the summer than during the winter. It has been shown, though, that any such nonuniformity serves only to *increase* the value of P_k (61). Thus, with 23 people, the smallest possible probability of at least one match is 0.507, and that occurs when each birthday is equally likely.

Comment

Presidential biographies offer one opportunity to confirm the unexpectedly large values Table 3.5.1 gives for P_k. And they do. Among the $k = 38$ presidents before Carter, two did have the same birthday—Harding and Polk were both born on November 2. More surprising, though, are the dates of death of the presidents, where there were four matches: Adams, Jefferson, and Monroe all died on July 4 and Fillmore and Taft both died on March 8.

Example 3.5.2

A somewhat inebriated conventioneer finds himself in the embarrassing position of being unable to discern whether he is walking forward or backward—or, what is worse, to predict in which of those directions his next step will be. If he is equally likely to walk forward or backward, what is the probability that after hazarding n such maneuvers, he will have moved forward a distance of r steps?

Let x denote the number of steps he takes forward and y denote the number backward. Then

$$x + y = n$$

and

$$x - y = r$$

Solving these equations simultaneously, we get $x = (n + r)/2$ and $y = (n - r)/2$. Thus, out of the 2^n total ways he can take n steps, the number of permutations for which he ends up r steps forward is $n! \left/ \left[\left(\frac{n + r}{2} \right)! \left(\frac{n - r}{2} \right)! \right] \right.$ (recall Theorem 3.2.3). The probability that he shows a net advance of r steps is the quotient

$$\frac{\dbinom{n}{\frac{n + r}{2}}}{2^n}$$

Example 3.5.3

Two monkeys, Mickey and Marian, are strolling along a moonlit beach when

Mickey sees an abandoned Scrabble set. Investigating, he notices that some of the letters are missing and what remain are the following 59:

A	B	C	D	E	F	G	H	I	J	K	L	M
4	1	2	2	7	1	1	3	5	0	3	5	1

N	O	P	Q	R	S	T	U	V	W	X	Y	Z
3	2	0	0	2	8	4	2	0	1	0	2	0

Mickey, being of a romantic bent, would like to impress Marian, so he rearranges the letters in hopes of spelling out something clever. (*Note:* The rearranging is random because Mickey cannot spell; fortunately, Marian cannot read, so it really doesn't matter.) What is the probability that Mickey gets lucky and spells out

> She walks in beauty, like the night
> Of cloudless climes and starry skies

As we might imagine, Mickey would have to get *very* lucky! The total number of ways to permute 59 letters—four A's, one B, two C's, and so on—is a direct application of Theorem 3.2.3:

$$\frac{59!}{4! \ 1! \ 2! \cdots 2! \ 0!}$$

But of that number, only one is the couplet he is hoping for. So, since he is arranging the letters randomly, making all permutations equally likely, the probability of his spelling out Byron's lines is

$$\frac{1}{\dfrac{59!}{4! \ 1! \ 2! \cdots 2! \ 0!}}$$

or, using Stirling's formula, about 1.7×10^{-61}. Love may conquer all, but it won't beat those odds: Mickey would be well advised to direct his libido elsewhere.

Question 3.5.1 Premack (69) describes an experiment set up to see whether language skills can be taught to anthropoid apes. Among the subjects was an African-born, 5-year-old chimpanzee named Sarah. For one of the tests designed to measure comprehension, Sarah was asked to "read" a sentence and respond appropriately. She was to answer by first selecting a set of four plastic symbols from among a set of eight and then arranging the four in descending order on a board. For each sentence there was only one correct answer. What is the probability of Sarah's getting the right answer by chance?

Question 3.5.2 A test tube at the top of the next page contains $2n$ grains of sand; n are white and n are black. The test tube is thoroughly shaken. What is the probability that the two kinds of sand segregate perfectly—that is, all one color are on the bottom and all the other color are on top? (*Hint:* Imagine the test tube turned on its side and the $2n$ grains of sand arranged in a row.)

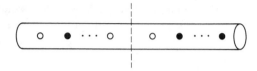

Question 3.5.3 What is the probability that the four aces in a well-shuffled poker deck will all be adjacent? (*Hint:* As a "model," consider the aces to be joined together and shuffled as a single card.)

Question 3.5.4 A committee of 50 politicians is to be chosen from among the 100 U.S. senators. If the selection is done at random, what is the probability that each state will be represented?

Question 3.5.5 If three fair dice are tossed—one red, one white, and one green—what is the probability that the sum of the faces showing will be less than or equal to 5? Greater than or equal to 5?

Question 3.5.6 Bordering a large estate is a stand of 15 European white birches, all in a row. Three of the trees have leaves damaged by some sort of parasite, and those 3 trees are adjacent to one another. What probability would you associate with that event? Does it seem reasonable to argue that the infestation is contagious? Explain.

Question 3.5.7 Suppose 15 politicians, including Ronald Reagan and Edward Kennedy, line up randomly for a picture. What is the probability that exactly three people will be standing between them? What is the probability that Reagan will stand to the left of Kennedy?

Question 3.5.8 An apartment building has eight floors. If seven people get on the elevator on the first floor, what is the probability they all want to get off on different floors? On the same floor? What assumption are you making? Does it seem reasonable? Explain.

Question 3.5.9 A certain variety of laboratory mouse exhibits two distinct fur patterns: agouti and nonagouti. Suppose a medical researcher reaches into a large cage containing equal numbers of each kind and picks out at random, one by one, seven of the animals. How likely is it that all seven will have the same fur type? Is your answer exact or just an approximation? If the latter, what does the closeness of the approximation depend on?

Question 3.5.10 How long must a series of random digits be in order to have an a priori probability of at least 0.8 that it contains at least one 3?

Question 3.5.11 Four married couples are invited to a banquet and are seated at random around a circular table. What is the probability that each husband will be sitting next to his wife? What is the probability that each husband will be sitting immediately to the right of his wife?

Question 3.5.12 Consider a set of 10 urns, 9 of which each contains three white chips and three red chips. The tenth contains five white chips and one red chip. An urn is picked at random. Then a sample of size 3 is drawn without replacement from that urn. If all three chips drawn are white, what is the probability the urn being sampled is the one with five white chips?

Question 3.5.13 Thirteen tombstones in a country churchyard are arranged in three rows, four in the first row, five in the second, and four in the third. Suppose that two women are buried in each row. Assuming each arrangement to be equally likely, what is the probability that in each row the women occupy the two leftmost positions? Is it reasonable to assume that the arrangements will be equally likely? Explain.

Question 3.5.14 Each of k urns contains n chips numbered 1 through n. One chip is drawn from each urn. What is the probability that r is the largest number drawn?

Question 3.5.15 The letters of the phrase *witches and ghosts* are written on a deck of 16 cards, one letter to a card. The cards are shuffled and the top five are dealt face up in a row. What is the probability that those five, *in the order drawn*, spell out the word *twits*?

Question 3.5.16 An urn contains w white chips and $N - w$ black chips. Chips are drawn successively (with replacement) until a white appears for the first time. Let X denote the number of selections required to get the first white chip. Show that

$$P(X > k) = \frac{(N - w)^k}{N^k}$$

Question 3.5.17 Suppose a randomly selected group of k people are brought together. What is the probability that exactly one pair has the same birthday?

Question 3.5.18 A three-digit number is formed at random from the digits 1 through 7, without repetition. What is the probability the number will be less than 289?

Question 3.5.19 Six dice are rolled one time. What is the probability that each of the six faces appears?

Question 3.5.20 An urn contains w white chips and b black chips. Chips are drawn out one at a time (without replacement) until those of only one color remain. What is the probability that the ones remaining are all white?

Question 3.5.21 Rufus the Clown always hands out balloons to children during the annual Memorial Day parade. When the festivities begin, Rufus has $2n$ balloons, which he evenly divides between the two pockets in his costume. Once the parade gets underway, he begins distributing the balloons, sometimes drawing from his right pocket, other times from his left. Assume he hands out the balloons one at a time and that he is equally likely to draw from either pocket. If at some point during the parade he finds one pocket is empty, what is the probability that exactly k balloons are left in the other pocket?

Question 3.5.22 Suppose each of ten sticks is broken into a long part and a short part. The 20 parts are arranged into 10 pairs and glued back together, so that again there are 10 sticks. What is the probability that each long part will be paired with a short part? (*Note:* This problem is a model for the effects of radiation on a living cell. Each chromosome, as a result of being struck by ionizing radiation, breaks into two parts, one part containing the centromere. The cell will die unless the fragment containing the centromere recombines with one not containing a centromere.)

One of the more instructive—and to some, one of the more useful—applications of combinatorics is the calculation of probabilities associated with various poker hands. It will be assumed in what follows that five cards are dealt from a poker deck and that no other cards are showing, although some may already have been dealt. The sample space is the set of

$$\binom{52}{5} = \frac{52!}{5! \, 47!} = 2{,}598{,}960$$

different hands, each having probability $1/2{,}598{,}960$. Example 3.5.4 derives the probabilities of being dealt a *full house*, *one pair*, *two pairs*, a *straight*, and a *flush*. Probabilities for other kinds of poker hands are gotten in much the same way. Example 3.5.5 shows some similar computations for a bridge hand.

Example 3.5.4

a. *Full house*. A full house consists of three cards of one denomination and two of another. Denominations for the three of one kind can be chosen in

$\binom{13}{1}$ ways. Then, given that a denomination has been decided on, the three requisite suits can be selected in $\binom{4}{3}$ ways. Applying the same reasoning to the pair gives $\binom{12}{1}$ available denominations, each having $\binom{4}{2}$ possible choices for suits. Thus, by the fundamental principle,

$$P(\text{full house}) = \frac{\binom{13}{1}\binom{4}{3}\binom{12}{1}\binom{4}{2}}{\binom{52}{5}} = 0.00144$$

b. *One pair*. To qualify as a one-pair hand, the five cards must include two of the same denomination and three "single" cards—cards whose denominations match neither the pair nor each other. For the pair, there are $\binom{13}{1}$ possible denominations and, once selected, $\binom{4}{2}$ possible suits. Denominations for the three single cards can be chosen in $\binom{12}{3}$ ways (see Question 3.5.23), and each card can have any of $\binom{4}{1}$ suits. Multiplying all these factors together and dividing by $\binom{52}{5}$ gives a probability of 0.42:

$$P(\text{one pair}) = \frac{\binom{13}{1}\binom{4}{2}\binom{12}{3}\binom{4}{1}\binom{4}{1}\binom{4}{1}}{\binom{52}{5}} = 0.42$$

c. *Two pairs*. A typical two-pair hand is, for instance, a king of diamonds and a king of clubs, a 7 of hearts and a 7 of clubs, and a 5 of diamonds—that is, the two pairs must not have the same denomination (if they did, the hand would be labeled *four of a kind*), and the denomination of the fifth card must not match that of either of the pairs (otherwise, the hand would be a full house). For the two pairs, there are $\binom{13}{2}$ combinations of denominations and $\binom{4}{2}\binom{4}{2}$ suits; for the fifth card, $\binom{11}{1}$ denominations and $\binom{4}{1}$ suits. Thus,

$$P(\text{two pairs}) = \frac{\binom{13}{2}\binom{4}{2}\binom{4}{2}\binom{11}{1}\binom{4}{1}}{\binom{52}{5}} = 0.048$$

d. *Straight*. A straight is five cards having consecutive denominations—for example, a 4 of diamonds, 5 of hearts, 6 of hearts, 7 of clubs, and 8 of diamonds. An ace may be counted high or low, meaning that (10, jack, queen, king, ace) is a straight and so is (ace, 2, 3, 4, 5). Altogether, there are 10 sets of consecutive denominations of length 5: (ace, 2, 3, 4, 5), (2, 3, 4, 5, 6), . . . , (10, jack, queen, king, ace). No restrictions are put on the suits, so each card can be either a diamond, heart, club, or spade. Therefore,

$$P(\text{straight}) = \frac{10\binom{4}{1}^5}{\binom{52}{5}} = 0.00394$$

What we have just calculated, though, is not entirely correct. While it *is* true that the proportion of hands that qualify as a straight is 0.00394, included in that number are hands that are straights in the same suit. These are called *straight flushes* and constitute a type themselves. Obviously, a straight flush, being rarer, would beat a straight. To calculate the probability that a straight would be "called" a straight, we need to subtract from the numerator all the hands that are straight flushes (see Question 3.5.24).

e. *Flush*. A flush is five cards in the same suit, denominations being of no importance. Thus, for a flush in, say, hearts, five cards must be selected from that suit's 13 denominations. This can be done in $\binom{13}{5}$ ways. A similar number of hands qualify as flushes in each of the other three suits, implying that

$$P(\text{flush}) = \frac{4\binom{13}{5}}{\binom{52}{5}} = 0.0020$$

(Do we need to qualify $P(\text{flush})$ like we did $P(\text{straight})$ in part (d)?)

Example 3.5.5

In a bridge hand, what is the probability each player will be dealt exactly one ace?

Unlike the computations required for poker hands, here we need to consider simultaneously *four* 13-card hands. To visualize the basic counting technique, think of the 52 cards laid out in a row. Under 13 of the cards imagine an N; under 13 others, an S; under 13 others, an E; and under the remaining 13, a W. That permutation of N's, S's, E's, and W's would determine the hands received by North, South, East, and West, respectively.

Clearly, the total number of ways to deal the four hands will equal the total number of ways to permute the four sets of letters—namely,

$$\frac{52!}{13!\ 13!\ 13!\ 13!}$$

By a similar argument, the aces can be distributed—one to a player—in $4!/1!\ 1!\ 1!\ 1!$ ways, and for each of those distributions, the remaining 48 cards can be dealt in $48!/12!\ 12!\ 12!\ 12!$ ways. Thus, by the fundamental principle, the probability of each player receiving exactly one ace is

$$\frac{\dfrac{4!}{(1!)^4} \cdot \dfrac{48!}{(12!)^4}}{\dfrac{52!}{(13!)^4}} = \frac{4!(13)^4 48!}{52!}$$

or 0.105.

Question 3.5.23 For one-pair hands, why is the number of denominations for the three single cards equal to $\binom{12}{3}$ rather than $\binom{12}{1}\binom{11}{1}\binom{10}{1}$?

Question 3.5.24 What is the probability of being dealt a *straight flush*? (*Note:* A straight flush is five cards in the same suit having denominations that are consecutive.) What is the probability that a poker player is dealt a hand that is called a straight—one that qualifies as a straight but is not a straight flush?

Question 3.5.25 Five cards are dealt from a poker deck. What is the probability that four have the same denomination?

Question 3.5.26 A poker player is dealt a 3 of diamonds, an 8 of clubs, a 9 of clubs, a 10 of spades, and an ace of hearts. If she discards the 3 and the ace and draws two others, what are the chances she ends up with a straight?

Question 3.5.27 A five-card hand is dealt from a standard poker deck. In the game of Night Whammy, a *blue turtle* is defined to be a hand consisting of two face cards (jack, queen, or king) and three numerical cards (2 through 10) of different but even denominations. What is the probability of a blue turtle?

Question 3.5.28 A coke hand in bridge is one where none of the 13 cards is an ace or is higher than a 9. What is the probability of being dealt such a hand?

Question 3.5.29 In a bridge hand, what is the probability that East, West, North, and South receive three, six, four, and no spades, respectively?

Question 3.5.30 A pinochle deck has 48 cards, two of each of six denominations (9, jack, queen, king, 10, ace), and the usual four suits. Among the many hands that count for meld is a *roundhouse*, which occurs when a player has a king and queen of each suit. In a hand of 12 cards, what is the probability of getting a *bare roundhouse* (a king and queen of each suit and no other kings or queens)?

Question 3.5.31 If five cards are dealt from the top of a poker deck we know that P(full house) $= 0.0014$. Suppose, though, one card is dealt from the deck face down and *then* a five-card hand is dealt. Show that the probability of this "second" five-card hand being a full house is still 0.0014. Give both an intuitive argument and a formal argument.

The final example in this section is a special case of the *ballot theorem* first proved by the French mathematician Bertrand in the latter part of the nineteenth century. The enumeration it requires is accomplished using an interesting reflection principle that we are seeing here for the first time. Generalizations of the problem are discussed at considerable length in (26).

Example 3.5.6

A line of $2n$ surly Pittsburgh Steeler fans are waiting in front of a ticket window trying to buy \$10 Super Bowl tickets. Half of them have \$10 bills; the other half have \$20 bills. The cashier has no change to start with. What is the probability that those with \$10 bills will be positioned in the queue in such a way that the cashier always has enough change, and all $2n$ fans are able to buy their tickets without being hassled?

To begin, we define a set of indicator variables, X_i, $i = 1, 2, \ldots, 2n$, where

$$X_i = \begin{cases} +1 & \text{if } i\text{th fan has a \$20 bill} \\ -1 & \text{if } i\text{th fan has a \$10 bill} \end{cases}$$

Then $S_k = \sum_{i=1}^{k} X_i$, $k = 1, 2, \ldots, 2n$, denotes the partial sums of the X_i's. Clearly, all $2n$ patrons will be able to buy tickets only if $S_k \leq 0$, for all k.

Notice that the S_k's can be represented by a segmented line having $2n$

"sides." Because the entire queue contains n +1's and n −1's, each S_k line starts at $(k, S_k) = (0, 0)$ and ends at $(2n, 0)$. For example, if $2n = 6$, the queue ($10, $20, $20, $10, $10, $20) would generate the graph pictured in Figure 3.5.1.

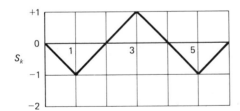

Figure 3.5.1

There is a unique correspondence between a path and the queue positions of the $20 bills. But the latter can be chosen in $\binom{2n}{n}$ ways, so the total number of paths from $(0, 0)$ to $(0, 2n)$ is that same number. Furthermore, since each path is equally likely, $\binom{2n}{n}$ will be the denominator in our expression for the probability that all $2n$ make it through the line.

To get an expression for the "favorable" paths, notice that every S_k series either (1) never goes above 0 or (2) equals +1 at least once. If a path *does* touch +1, we will consider its reflection (about the line $S_k = +1$) from that point on. The dotted line in Figure 3.5.2 shows the reflected path for the queue graphed in Figure 3.5.1. The remainder of the derivation hinges on the observation that each different path that touches +1 at least once will be associated with a different reflection. So, we can count the former by counting the latter.

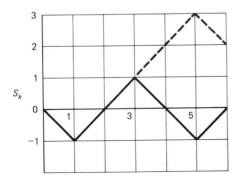

Figure 3.5.2

Starting at the point $S_k = +1$, the actual path must have one more −1 side than +1 side if it is to return to the point $(2n, 0)$. Therefore, the reflected path must have one more +1 side than −1 side. It follows that all reflected paths must end at the point $(2n, +2)$. Of course, the number of paths from $(0, 0)$ to $(2n, +2)$ will be the number of ways to arrange $(n + 1)$ +1's and $(n - 1)$ −1's—namely, $(2n)!/[(n - 1)! \, (n + 1)!]$, or $\binom{2n}{n+1}$.

Thus, if there are $\binom{2n}{n}$ paths from $(0, 0)$ to $(0, 2n)$ and if $\binom{2n}{n+1}$ of those touch the line $S_k = +1$ at least once, it must be true that $\binom{2n}{n} - \binom{2n}{n+1}$ will be the number of paths for which $S_k \leq 0$, for all k. Therefore,

$$P(\text{all } 2n \text{ can buy tickets}) = \frac{\binom{2n}{n} - \binom{2n}{n+1}}{\binom{2n}{n}}$$

$$= \frac{1}{n+1}$$

Question 3.5.32 Generalize the solution to the problem posed in Example 3.5.6 to include the case where the cashier has an initial $2k$ \$10 bills to use as change.

Question 3.5.33 Suppose a fair coin is tossed $2n$ times. Define

$$X_i = \begin{cases} +1 & \text{if a head appears on } i\text{th toss} \\ -1 & \text{if a tail appears on } i\text{th toss} \end{cases}$$

and let $S_k = \Sigma_{i=1}^{k} X_i$. What is the probability that $S_{2n} = 0$?

Question 3.5.34 Suppose a fair coin is tossed $2n$ times. Let S_k be the same cumulative sum as defined in Question 3.5.33. It can be shown (see (26)) that the probability of $2j$ of the $2n$ sides of the resulting polygonal path being below the line $S_k = 0$ is given by

$$p_{2j, 2n} = \frac{\binom{2j}{j}\binom{2n-2j}{n-j}}{2^{2n}}$$

Evaluate $p_{2j, 2n}$ for $2n = 20$ and all admissible values of j. Does the resulting distribution surprise you? Explain.

3.6 THE BINOMIAL, MULTINOMIAL, AND HYPERGEOMETRIC DISTRIBUTIONS

The problems discussed in the previous section were all combinatorial in nature, but beyond that they shared no special similarity. The birthday problem was an application of Theorems 3.2.1 and 3.2.2, computing the probabilities of poker hands was an exercise in the $\binom{n}{k}$ symbol coupled with the fundamental principle, and all the others appealed—at least in part—to Theorem 3.2.3. It is also true that none of the examples just mentioned can be considered a problem "type" in the sense that it serves as a model for a wide variety of other situations. In this section we take up three combinatorial probability problems that *do* have much in common and *are* widely used as data models. The *binomial, multinomial,* and *hypergeometric* distributions have all made cameo appearances in several Chapter 2 and Chapter 3 examples and exercises but have never been formally defined or explained at any length. Pay particular attention here to the broad range of phenomena to which these distributions can effectively be applied.

First, the binomial. The mathematical background for the binomial derives from the repeated independent trials discussion in Section 2.9. As a data model, it applies to situations where each outcome in the sample space is a series of n independent trials, each having one of only two possible outcomes, *success* or *failure*. If a coin, for example, is tossed $n = 2$ times with heads being designated as a success, the sample space consists of $2^n = 2^2 = 4$ possible outcomes: ({H, H}, {T, T}, {H, T}, and {T, H}). Let p denote the likelihood of the coin coming up heads on any given toss. Then the probabilities associated with the four possible sequences of Bernoulli trials are p^2, $(1 - p)^2$, $p(1 - p)$, and $(1 - p)p$, respectively (see Table 3.6.1).

TABLE 3.6.1

Outcome	Probability
(H, H)	p^2
(T, T)	$(1 - p)^2$
(H, T)	$p(1 - p)$
(T, H)	$(1 - p)p$

It was mentioned in Section 2.9 (and we will see this coming up in the examples again and again) that often the most-relevant information to be gleaned from a series of n Bernoulli trials is not which trials ended in success and which in failure but, rather, *how many* ended in success. Let Y denote that number. What we would like to find is the probability that Y takes on the value k, where k is some integer between 0 and n, inclusive. At this point, in terms of what we already know, deriving the probability function for Y is very easy: By the fundamental principle, $P(Y = k)$ will equal the number of ways to position k successes (and $n - k$ failures) in n trials times the probability associated with any particular ordering. The latter figure is $p^k q^{n-k}$ (recall the comment following Definition 2.9.1); the former, by virtue of Theorem 3.2.3, is $\binom{n}{k}$.

Theorem 3.6.1. Let Y denote the number of successes in n independent Bernoulli trials, where the probability of success at any particular trial is p (and the probability of failure is $q = 1 - p$). Then Y is said to have a *binomial distribution* and

$$P(Y = k) = \binom{n}{k} p^k q^{n-k}, \qquad k = 0, 1, \ldots, n$$

The next several examples show some applications of the binomial distribution. Notice that, unlike the situation we encountered in Section 3.5, all these problems have essentially the same solution: Each appeals directly to the statement of Theorem 3.6.1. The only variation of note is whether a *single* binomial term or a *sum* of binomial terms is needed.

Example 3.6.1

In a nuclear reactor, the fission process is controlled by inserting a number of special rods into the radioactive core; the purpose of these rods is to absorb the neutrons emitted by the critical mass and thereby slow down the nuclear chain reaction. When functioning properly, these rods serve as the first-line defense against a disastrous core meltdown.

Suppose that a particular reactor has 10 of these control rods (actually, there would be more than 100), each operating independently and each having a 0.80 probability of being properly inserted in the event of an incident. Furthermore, suppose that a meltdown will be prevented if at least half the rods perform satisfactorily. What is the probability that, upon demand, the system will fail?

If Y denotes the number of control rods that function as they should, a system failure occurs if $Y \le 4$. By Theorem 3.6.1, the probability of that happening is 0.007:

$$P(\text{system will fail}) = P(Y \le 4) = \sum_{k=0}^{4} \binom{10}{k}(0.80)^k(0.20)^{10-k}$$

$$= \binom{10}{0}(0.80)^0(0.20)^{10} + \cdots + \binom{10}{4}(0.80)^4(0.20)^6$$

$$= 0.000 + 0.000 + 0.000 + 0.001 + 0.006$$

$$= 0.007$$

Example 3.6.2

In planning the rescue attempt of the U.S. hostages held in Iran in the spring of 1980, Pentagon officials determined that a minimum of 6 Sea Stallions would have to get through in order for the mission to be a tactical success (63). Furthermore, they estimated that if 8 of the helicopters were launched from the U.S.S. *Nimitz,* the probability of at least 6 making it back was 0.965. After-the-fact criticism of the venture was often directed at the seemingly small number of aircraft initially deployed. What would have been the mission's success probability had the plan called for sending in 10 Sea Stallions?

Let the variable Y denote the number of helicopters completing the mission, with n being the number launched and p being the probability that any given one successfully returns. What we are given is one conditional probability—namely, $P(Y \ge 6 \mid n = 8) = 0.965$—and we are asked to find another, $P(Y \ge 6 \mid n = 10)$. If we assume that Y can be described by a binomial model (see Question 3.6.4), then

$$P(Y \ge 6 \mid n = 8) = 0.965 = \sum_{k=6}^{8} \binom{8}{k}p^k(1-p)^{8-k}$$

$$= \binom{8}{6}p^6(1-p)^2 + \binom{8}{7}p^7(1-p)^1 + \binom{8}{8}p^8(1-p)^0$$

$$(3.6.1)$$

By trial and error, we can find that $p = 0.904$ is the solution to Equation 3.6.1. Substituting that value back into a binomial model with $n = 10$ gives

$$P(Y \geq 6 | n = 10) = \sum_{k=6}^{10} \binom{10}{k} (0.904)^k (0.096)^{10-k}$$

$$= 0.998$$

Thus, deploying an additional two helicopters would have increased the mission's success chances by 3.3% (assuming every other aspect of the situation remained the same).

Example 3.6.3

During the 1978 baseball season, Pete Rose of the Cincinnati Reds set a National League record by hitting safely in 44 consecutive games. Assume that Rose is a .300 hitter and that he comes to bat (officially) four times each game. If each at bat is assumed to be an independent and identically distributed Bernoulli trial, what probability would be associated with Rose's streak?

Let

$$Y = \text{number of hits in a given game}$$

$$n = \text{number of at bats each game (4)}$$

$$p = P(\text{Rose hits safely on a given at bat}) = 0.300$$

Note that the streak will be continued if, on a given day, $Y \geq 1$. But,

$$P(Y \geq 1) = 1 - P(Y = 0) = 1 - \binom{4}{0} (0.300)^0 (0.700)^4$$

$$= 0.76$$

Since the streak is an intersection of 44 (independent) events, each of the form $Y \geq 1$,

$$P(\text{Rose hits safely in 44 consecutive games}) = (0.76)^{44}$$

$$= 0.0000057$$

That is, the chances are less than 6 in a million that a .300 hitter would bat safely in a given series of 44 consecutive games. (Is this the only way to measure the unlikelihood of a 44-game hitting streak? See Question 3.6.13.)

Example 3.6.4

For reasons not entirely clear, Doomsday Airlines books a daily shuttle service from Altoona to Hoboken. They offer two round-trip flights, one on a two-engine prop plane, the other on a four-engine prop plane. Suppose that each engine on each plane will fail independently with the same probability p and that each plane will arrive safely at its destination only if at least half its engines remain in working order. Assuming you want to live, for what values of p would you prefer to fly on the two-engine plane?

Let Y denote the number of engines on each plane that remain operable.

For the two-engine plane,

$$P(\text{flight lands safely}) = P(Y \geq 1) = \sum_{k=1}^{2} \binom{2}{k}(1 - p)^k p^{2-k} \qquad (3.6.2)$$

For the four-engine plane,

$$P(\text{flight lands safely}) = P(Y \geq 2) = \sum_{k=2}^{4} \binom{4}{k}(1 - p)^k p^{4-k} \qquad (3.6.3)$$

When to opt for the two-engine plane reduces to an algebra problem: We look for the values of p for which

$$\sum_{k=1}^{2} \binom{2}{k}(1 - p)^k p^{2-k} > \sum_{k=2}^{4} \binom{4}{k}(1 - p)^k p^{4-k} \qquad (3.6.4)$$

To minimize the number of terms that need to be manipulated, it proves expedient to rephrase the problem using the complements of Equations 3.6.2 and 3.6.3. The set of p values for which Inequality 3.6.4 is true is equivalent to the set for which

$$\sum_{k=0}^{1} \binom{4}{k}(1 - p)^k p^{4-k} > \binom{2}{0}(1 - p)^0 p^2$$

Simplifying the inequality

$$\binom{4}{0}(1 - p)^0 p^4 + \binom{4}{1}(1 - p)^1 p^3 > \binom{2}{0}(1 - p)^0 p^2$$

gives

$$(3p - 1)(p - 1) < 0 \qquad (3.6.5)$$

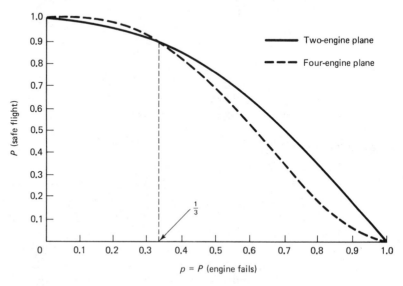

Figure 3.6.1

But $p - 1$ is never positive, so Inequality 3.6.5 will be true only when $3p - 1 > 0$, which gives $p > \frac{1}{3}$ as the desired solution set. Figure 3.6.1 is a graph of the two safe-return probabilities, $P(Y \geq 1 \mid n = 2)$ and $P(Y \geq 2 \mid n = 4)$, as a function of p.

Question 3.6.1 Samuel Pepys was the greatest diarist of the English language. He was also a friend of Sir Isaac Newton and, in 1693, sought the latter's advice in a matter related to gambling. Phrased in modern terminology, Pepys' question was a binomial problem: Is it more likely to get at least one 6 when 6 dice are rolled, at least two 6's when 12 dice are rolled, or at least three 6's when 18 dice are rolled? After considerable correspondence (see (80)), Newton was able to convince a skeptical Pepys that the former has the greatest likelihood. Compute these three probabilities.

Question 3.6.2 Each day a stock price moves up one point or down one point with probabilities $\frac{1}{4}$ and $\frac{3}{4}$, respectively. What is the probability that after 4 days the stock will have returned to its original price? Assume the daily price fluctuations are independent events.

Question 3.6.3 Experience has shown that only $\frac{1}{3}$ of all patients having a certain disease will recover if given the standard treatment. A new drug is to be tested on a group of 12 volunteers. If the FDA requires that at least 7 of these patients recover before it will license the new drug, what is the probability the treatment will be discredited even if it has the potential to increase an individual's recovery rate to $\frac{1}{2}$?

Question 3.6.4 Recall the hostage-rescue question posed in Example 3.6.2. What critical assumption are we making when we presume that Y, the number of helicopters returning, has a binomial distribution? Does the assumption seem reasonable in this particular context? Explain.

Question 3.6.5 Suppose that since the early 1950s some 10,000 independent UFO sightings have been reported to civil authorities. If the probability that any particular sighting is genuine is on the order of 1 in 100,000, what is the probability that at least one sighting was genuine?

Question 3.6.6 The captain of a Navy gun boat orders a volley of n missiles to be fired at random along a 500-ft stretch of shoreline that the captain hopes to establish as a beachhead. Dug into the beach is a 30-ft long bunker serving as the enemy's first line of defense. If n is 25, what is the probability that exactly 3 shells will hit the bunker?

Question 3.6.7 Let $b(k; n, p)$ denote the probability that the variable in a binomial model equals k—that is,

$$b(k; n, p) = \binom{n}{k} p^k (1 - p)^{n-k}$$

Prove the recursion formula,

$$b(k; n, p) = \frac{(n - k + 1)p}{k(1 - p)} \cdot b(k - 1; n, p)$$

Question 3.6.8 In his 1889 publication *Natural Inheritance,* the renowned British scientist Sir Francis Galton described a pinball-type board that he called a *quincunx.* As pictured, the quincunx has five rows of pegs, the pegs in each row being the same distance apart. At the bottom of the board are five cells, numbered 0 through 4. A ball is introduced at the top of the board between the pegs in the first row. After wedging past those two pegs, it will hit the middle peg in the second row and veer either to the right or to the left, then strike a peg in the third row, again

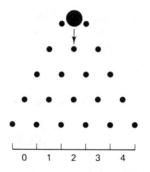

veering to the right or to the left, and so on. If the ball has a 50-50 chance of going either direction each time it hits a peg, what is the probability it ends up in cell 3?

Question 3.6.9 Let $B(k; n, p)$ denote the lower tail of the binomial distribution:

$$B(k; n, p) = \sum_{r=0}^{k} \binom{n}{r} p^r (1 - p)^{n-r}$$

Show that

$$B(k; n, p) = (n - k)\binom{n}{k} \int_0^{1-p} t^{n-k-1}(1 - t)^k \, dt$$

(*Hint:* Differentiate both sides of the equation with respect to p.)

Question 3.6.10 Suppose n points are distributed at random along the interval $(0, T)$. What is the probability that exactly k are in the subinterval (a, b)?

Question 3.6.11 Two baseball teams are negotiating a format for deciding how to determine the league champion. Two possibilities have been mentioned: a best-two-of-three playoff series or a best-three-of-five series. Suppose the members of one of the teams estimate that they have a 55% chance of defeating their opponent on any given day. Which of the two playoff schemes should they support? Compute the two relevant probabilities. Assume each game can be considered an independent Bernoulli trial. Does your answer make sense intuitively?

Question 3.6.12 For persons in a certain state convicted of grand theft auto and given 3-year sentences, the length of time actually served, x, is described by the probability function $f(x) = (\frac{1}{9})x^2$, $0 < x < 3$. Suppose the records of three former prisoners, chosen at random, are examined. What is the probability that exactly two of the three were released in less than a year?

Question 3.6.13 Recall Example 3.6.3. Suppose a .300 hitter plays a full 162-game schedule. What is the probability that *sometime during the season* the player has exactly one 44-game hitting streak and nothing longer?

Question 3.6.14 A friend has a penny that may be a normal fair coin or both sides may be heads or both may be tails. Suppose the friend tosses the coin n times and it comes up heads on each occasion. For what value of n does the probability of the coin's being two-sided fall below $\frac{1}{100}$ for the first time?

Question 3.6.15 One of the printed circuits in a color TV chassis has 12 connections. The probability density function for the lifetime (in operating hours) of each of those connections is

$$f(x) = (\tfrac{1}{14,000})e^{-x/14,000}, \qquad x > 0$$

If three or more of the connections break, the picture will develop a vertical roll. Suppose a set has a 2-year warranty and that it is used an average of 5 hours a day. What is the probability the picture will develop a vertical roll before the warranty expires?

Question 3.6.16 If a couple intend to have four children, is it more likely they will have two boys and two girls or three of one sex and one of the other? Assume the probability of any child being a boy is $\frac{1}{2}$.

Question 3.6.17 The gunner on a small assault boat fires 6 missiles at an attacking plane. Each has a 20% chance of being on target. If two or more of the shells find their mark, the plane will crash. At the same time, the pilot of the plane fires 10 air-to-surface rockets, each of which has a 0.05 chance of critically disabling the boat. Would you rather be on the plane or on the boat?

Question 3.6.18 A fair coin is tossed $2n$ times. Show that the probability of getting exactly n heads is less than $1/\sqrt{n}$, meaning it goes to 0 as n goes to infinity. (*Hint:* Show that

$$\frac{(2n)!}{2^{2n}(n!)^2} = \left(1 - \frac{1}{2}\right)\left(1 - \frac{1}{4}\right) \cdots \left(1 - \frac{1}{2n}\right)$$

and make use of the fact that for $x > 0$, $1 - x < e^{-x}$.)

Question 3.6.19 Among the early probability questions that piqued the curiosity of Pascal and Fermat was the *problem of points*. Suppose A and B are playing a series of games where the winner of the overall match is the first player to win a total of k games. However, the contest is interrupted before either competitor has achieved that number. If A needs to take m more games to win the match, while B is n short of the necessary k, how should the stakes be divided? Assume (1) the stakes should be divided in proportion to A's and B's relative chances of winning, (2) A has a probability p of winning any particular game, and (3) the games are independent. (*Hint:* Consider the consequences if A and B play an additional $m + n - 1$ games.)

One of the obvious ways to generalize the statement of Theorem 3.6.1 is to consider situations in which one of k outcomes can occur at each trial, rather than just a success or a failure. Let $x_1, x_2, \ldots, x_k$ denote those possible outcomes, and suppose they occur with probabilities $p_1, p_2, \ldots, p_k$, respectively. Of course,

$$\sum_{i=1}^{k} p_i = 1$$

Note that if n such trials are observed, the resulting distribution of x-values can be summarized by a new set of variables, $Y_1, Y_2, \ldots, Y_k$, where

Y_i = the number of trials for which the outcome was x_i, $\quad i = 1, 2, \ldots, k$

(What must $\sum_{i=1}^{k} Y_i$ equal?)

To parallel our development of the binomial, we want to find an expression giving the simultaneous probability that $Y_1 = y_1$, $Y_2 = y_2, \ldots, Y_k = y_k$—that is, the probability that (*irrespective of order*) y_1 of the trials end in outcome x_1, y_2 show outcome x_2, and so on. Theorem 3.6.2 states the result. The proof is based on Theorem 3.2.3; it will be left as an exercise.

Theorem 3.6.2. Consider a series of n independent trials, each resulting in one of the k outcomes $x_1, x_2, \ldots, x_k$, with probabilities $p_1, p_2, \ldots, p_k$, respectively. Let Y_i denote the number of trials for which x_i occurs, $i = 1, 2, \ldots, k$. The probability function describing the behavior of the vector $(Y_1, Y_2, \ldots, Y_k)$ is called the *multinomial distribution,* and

$$P(Y_1 = y_1, \ldots, Y_k = y_k) = \frac{n!}{y_1!\, y_2! \cdots y_k!} \cdot p_1^{y_1} p_2^{y_2} \cdots p_k^{y_k};$$

$$y_i = 0, 1, \ldots, n, \; i = 1, 2, \ldots, k; \quad \sum_{i=1}^{k} y_i = n$$

Comment

When $k = 2$ and x_1 and x_2 refer to success and failure, respectively, then $p_1 = p$, $p_2 = 1 - p = q$, and the multinomial reduces to the binomial.

Example 3.6.5

In 1927 when he hit 60 home runs, Babe Ruth compiled the following statistics (71):

G	AB	H	2B	3B	HR	R	RBI	BB	SO	SB	BA
151	540	192	29	8	60	158	164	138	89	7	.356

Suppose he had five official at bats in a certain game. Estimate his probability of getting two singles, a home run, and two outs (not necessarily in that order).

To put this question in the framework of Theorem 3.6.2, we first need to find the probabilities associated with the designated outcomes. Since his batting average was .356, his probability of making an out was 0.644 ($1 - 0.356$). By inspection, his chance of getting a home run was $\frac{60}{540}$ or 0.111. Finally, if his doubles, triples, and home runs are subtracted from his total hits, we find that he had 95 singles ($192 - 29 - 8 - 60$), so the probability he got a base hit was $\frac{95}{540} = 0.176$. Let

$$Y_1 = \text{number of singles}$$

$$Y_2 = \text{number of home runs}$$

$$Y_3 = \text{number of outs}$$

Given that he had five at bats,

$$P(Y_1 = 2, Y_2 = 1, Y_2 = 2) = \frac{5!}{2!\,1!\,2!}\,(0.176)^2(0.111)^1(0.644)^2$$

$$= 0.043$$

(In terms simply of hits and outs, what was Ruth's most-likely set of outcomes in five official at bats?)

Example 3.6.6

Recall the Alaskan pipeline problem discussed in Example 2.5.1. A probability function, $f(x)$, was given that described the hypothetical points of impact of air-to-surface missiles fired at the pipeline (x is the perpendicular distance from the pipeline to where the missile hits). Figure 3.6.2 is a sketch of that $f(x)$. Suppose six missiles are fired. What is the probability that two land within 20 ft to the left of the pipeline and three land within 30 ft to the right?

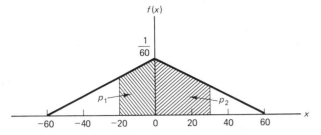

Figure 3.6.2

Notice here that the statement of the question forces the outcome of each firing to be in one of three categories (or x-values):

Category 1 (x_1): Missile lands within 20 ft to the left of the pipeline.
Category 2 (x_2): Missile lands within 30 ft to the right of the pipeline.
Category 3 (x_3): Missile lands farther to the left of -20 or farther to the right of $+30$.

If Y_i denotes the number of missiles falling into category i, $i = 1, 2, 3$, we are being asked to find $P(Y_1 = 2, Y_2 = 3, Y_3 = 1)$. The p_i values in this case are integrals of $f(x)$:

$$p_1 = P(\text{missile is in category 1}) = \int_{-20}^{0} \frac{60 + x}{3600}\, dx$$

$$= \frac{5}{18}$$

$$p_2 = P(\text{missile is in category 2}) = \int_{0}^{30} \frac{60 - x}{3600}\, dx$$

$$= \frac{3}{8}$$

$$p_3 = P(\text{missile is in category 3}) = 1 - p_1 - p_2$$

$$= 1 - \frac{5}{18} - \frac{3}{8} = \frac{25}{72}$$

Substituting into the multinomial formula gives

$$P(Y_1 = 2,\ Y_2 = 3,\ Y_3 = 1) = \frac{6!}{2!\ 3!\ 1!} \left(\frac{5}{18}\right)^2 \left(\frac{3}{8}\right)^3 \left(\frac{25}{72}\right)^1$$

$$= 0.085$$

CASE STUDY 3.2

Webster's dictionary defines *contagion* as "the transmission of a disease by direct or indirect contact." For diseases where the probability of an infected person transmitting the condition to an uninfected person is relatively high, contagion can be a (painfully) obvious phenomenon. Until recently, measles was a familiar case in point. But if the transmission probability is small, demonstrating the existence of contagion can prove to be a very difficult statistical problem.

Leukemia is a rare disease that may or may not be contagious—evidence has been presented on both sides, and while most medical researchers now believe it is *not*, the controversy has not been fully resolved to everyone's satisfaction. One of the earliest analytical attempts to address the question is pertinent to our discussion in this section because it made direct use of the multinomial distribution.

Ederer, Myers, and Mantel (21) studied the onset times of 333 cases of childhood leukemia reported among the 169 towns (counties) in Connecticut during the 15-year period from 1945 to 1959. Their objective was to devise an index that would reflect any tendency in these cases to cluster.

To understand the rationale behind their model, it will help if we begin with a numerical example. Suppose a total of $r = 6$ cases of leukemia occur in a given city over a period of 5 years. Let Y_i denote the number of cases having onsets in year i, $i = 1, 2, 3, 4, 5$. If the cases are assumed to have occurred at random with respect to time, it follows from Theorem 3.6.2 that

$$P(Y_1 = y_1,\ \ldots\ ,\ Y_5 = y_5) = \frac{6!}{y_1!\ y_2!\ \cdots\ y_5!} \left(\frac{1}{5}\right)^6$$

As a measure of contagion, Ederer, Myers, and Mantel define the quantity M_1, where

$$M_1 = \text{maximum number of cases occurring in a single}$$
$$\text{year of a 5-year period}$$

Thus if the year-by-year distribution of the six cases was, for example, (2, 1, 0, 3, 0), the value of M_1 would be 3.

The probability distribution for M_1 is tedious to work out, but it can be gotten by making a *second* application of Theorem 3.6.2. Consider, for example, the problem of finding the probability that $M_1 = 4$. A distribution of $r = 6$ cases will have $M_1 = 4$ if its annual frequencies are either 4, 2, 0, 0, 0 *in some order* or 4, 1, 1, 0, 0 *in some order*. But the number of ways to arrange the frequencies 4, 2, 0, 0, 0 is $5!/1!\ 1!\ 3!$. Similarly, there are $5!/1!\ 2!\ 2!$ ways to permute the numbers 4, 1, 1, 0,

0. It follows that the probability of six cases occurring in 5 years showing a maximum of four cases in any one year is given by

$$P(M_1 = 4) = P(\{4, 2, 0, 0, 0\} \text{ distribution}) + P(\{4, 1, 1, 0, 0\} \text{ distribution})$$

$$= \frac{5!}{1! \, 1! \, 3!} \cdot \frac{6!}{4! \, 2! \, 0! \, 0! \, 0!} \left(\frac{1}{5}\right)^6 + \frac{5!}{1! \, 2! \, 2!} \cdot \frac{6!}{4! \, 1! \, 1! \, 0! \, 0!} \left(\frac{1}{5}\right)^6$$

$$= 0.0768$$

Table 3.6.2 shows the entire distribution of M_1 (for $r = 6$).

TABLE 3.6.2

m_1	$P(M_1 = m_1)$
2	$P(\{2, 2, 2, 0, 0\}) + P(\{2, 2, 1, 1, 0\}) + P(\{2, 1, 1, 1, 1\}) = 0.51840$
3	$P(\{3, 3, 0, 0, 0\}) + P(\{3, 2, 1, 0, 0\}) + P(\{3, 1, 1, 1, 0\}) = 0.39680$
4	$P(\{4, 2, 0, 0, 0\}) + P(\{4, 1, 1, 0, 0\}) = 0.07680$
5	$P(\{5, 1, 0, 0, 0\}) = 0.00768$
6	$P(\{6, 0, 0, 0, 0\}) = 0.00032$

Initially, the Connecticut data consisted of a listing, by year, of all leukemia cases occurring in each of that state's 169 towns over a period of 15 years. For epidemiological reasons, it was decided to subdivide the 15-year interval into three 5-year periods, thus forming a total of 3×169, or 507, 5-year-town units. Of course, not all 507 of the 5-year-town units contributed to the search for evidence of contagion—attention needed to be restricted to those for which $r \geq 2$. Table 3.6.3 shows the breakdown by r and M_1 for the 73 multiply-occupied 5-year-town units.

TABLE 3.6.3

r	1	2	3	M_1 4	5	6	Total
2	33	6					39
3	6	8	1				15
4		3					3
5	1	3	1				5
6		3	2				5
7		1	1				2
8				1			1
9				1			1
10				1			1
14						1	$\underline{1}$
							73

At this point, the argument advanced by Ederer, Myers, and Mantel becomes decidedly statistical. If contagion has influenced the positioning of cases within a 5-year-town-unit, we would expect large M_1 values to occur disproportionately often. Thus, it becomes necessary to examine what *did* happen (Table 3.6.3) within the context of the probability function for M_1. For example, there were five instances of six cases of leukemia occurring within a 5-year-town-unit. For three of those five, M_1 equaled 2; for the other two, M_1 was 3. Is that particular set of M_1 values compatible with the theoretical distribution given in Table 3.6.2? Coming to grips with that question requires a statistical procedure known as a chi-square test. When applied here, its conclusion was unmistakable: The distribution of clusters in Table 3.6.3 is totally compatible with the hypothesis that no contagion was present and the cases had occurred randomly.

Question 3.6.20 Prove Theorem 3.6.2.

Question 3.6.21 Repair calls for a central air conditioner fall into four general categories: coolant leakage, compressor failure, fuse blowout, and miscellaneous. From past records, the probabilities associated with these four are 0.4, 0.3, 0.1, and 0.2, respectively. In a repairperson's next 10 service calls, what are the chances that half will involve the coolant and half will involve the compressor?

Question 3.6.22 Suppose a loaded die is tossed 12 times, where

$$p_i = P(i \text{ spots appear}) = ki, \qquad i = 1, 2, \ldots, 6$$

Let Y_i denote the number of times the face with i spots appears, $i = 1, 2, \ldots, 6$. Find the probability that all the Y_i's equal 2.

Question 3.6.23 After being scaled, scores on a standard personality inventory test are known from past experience to be distributed according to the probability function,

$$f(x) = 6x(1 - x), \qquad 0 \le x \le 1$$

Let Y_i denote the number of subjects getting scores in the interval $[(i - 1)/4, i/4)$, $i = 1, 2, 3, 4$. If nine people take the test, find the probability that two get scores in the $0 \le x < 0.25$ range, three each in the $0.25 \le x < 0.50$ and $0.50 \le x < 0.75$ ranges, and one in the $0.75 \le x < 1.00$ range.

Question 3.6.24 Recall Example 2.2.6 and the $1 : 2 : 1$ set of ratios proposed for the inheritance of extreme : mild : normal phenotypes in frizzle chickens. Suppose 93 crossings of hybrid roosters (F, f) with hybrid hens (F, f) are made. If the alleles recombine randomly, write down a formula giving the probability that 23 of the progeny will be extreme frizzles, 50 will be mild frizzles, and the remaining 20 will be normal.

Many probability problems can be modeled by the simple experiment of drawing objects out of an urn. Such a model has two important variations, though, depending on whether the sampling is done *with replacement* or *without replacement*. Imagine an urn containing N chips, of which r are red and w are white. Suppose we draw a chip out at random, note its color, return it to the urn, draw a second chip out, note *its* color, return it to the urn, and so on. Following such a procedure n times would be called taking a sample of size n *with replacement*. If we let Y denote the total number of red chips in the n trials, its distribution will obviously be binomial:

$$P(Y = k) = \binom{n}{k}\left(\frac{r}{N}\right)^k\left(\frac{w}{N}\right)^{n-k}, \qquad k = 0, 1, \ldots, n$$

The alternative is *not* to return a chip after it has been drawn, or, equivalently, to reach in and draw the entire sample of size n all at once—that is, *without replacement*. What we want to investigate in the remainder of this section is the effect of such a sampling scheme on the probability distribution of Y.

Theorem 3.6.3. Suppose an urn contains r red chips and w white chips $(r + w = N)$. If n chips are drawn at random, without replacement, and Y denotes the total number of red chips selected, then Y is said to have a *hypergeometric distribution* and

$$P(Y = k) = \frac{\binom{r}{k}\binom{w}{n-k}}{\binom{N}{n}}, \qquad k = 0, 1, \ldots, \min(r, n)$$

Proof. Consider the chips to be distinguishable. Then the total number of ways to select a sample of size n is $\binom{N}{n}$. By the same reasoning, there are $\binom{r}{k}$ ways to select k red chips, and, for each of those, $\binom{w}{n-k}$ ways to select enough white chips $(n - k)$ to fill out the sample. Since each selection is presumed equally likely, it follows that the probability of drawing k red chips is the product $\binom{r}{k}\binom{w}{n-k}$ divided by the total number of samples, $\binom{N}{n}$. (A second proof is outlined in Question 3.6.27.)

Comment

The name *hypergeometric* derives from a series introduced by the Swiss mathematician and physicist, Leonhard Euler, in 1769:

$$1 + \frac{ab}{c}x + \frac{a(a + 1)b(b + 1)}{2! \, c(c + 1)}x^2 +$$

$$\frac{a(a + 1)(a + 2)b(b + 1)(b + 2)}{3! \, c(c + 1)(c + 2)}x^3 + \cdots$$

This is an expansion of considerable flexibility: Given appropriate values for a, b, and c, it reduces to many of the standard infinite series used in analysis. In particular, if a is set equal to 1 and b and c are set equal to each other, it reduces to the familiar *geometric* series,

$$1 + x + x^2 + x^3 + \cdots$$

hence the name *hypergeometric*. The relationship of the probability function of Theorem 3.6.3 to Euler's series becomes apparent if we set $a = -n$, $b = -r$, $c = w - n + 1$, and multiply the series by $\binom{w}{n}/\binom{N}{n}$. Then the coefficient of x^k will be

$$\frac{\binom{r}{k}\binom{w}{n-k}}{\binom{N}{n}}$$

the value the theorem gives for $P(Y = k)$.

Example 3.6.7

Nevada Keno is among the most popular games in Las Vegas, even though it is one of the least fair, in the sense that the odds are overwhelmingly in favor of the house. (Betting on Keno is only a little less foolish than playing a slot machine!) A Keno card has 80 numbers, 1 through 80, from which the player selects a sample of size k, where k can be anything from 1 to 15. The caller then announces 20 winning numbers, chosen at random from the 80. If—and how much—the player wins depends on how many of his or her numbers match the 20 identified by the caller. Suppose a player bets on a 10-spot ticket. What is the probability of "catching" five numbers?

Imagine an urn containing 80 numbers, 20 of which are winners and 60 are losers (see Figure 3.6.3). By betting on a 10-spot ticket, the player, in effect, is drawing a sample of size 10 from that urn. Let Y denote the number of winning numbers included among the player's 10 selections. What we are trying to find is $P(Y = 5)$. But by Theorem 3.6.3 (with $r = 20$, $w = 60$, $n = 10$, $N = 80$, and $k = 5$),

$$P(Y = 5) = \frac{\binom{20}{5}\binom{60}{5}}{\binom{80}{10}}$$

or, approximately, 0.05.

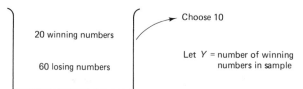

Figure 3.6.3

Example 3.6.8

Muffy is studying for a history exam covering the French Revolution that will consist of 5 essay questions selected at random from a list of 10 the professor has handed out to the class in advance. Not exactly a Napoleon buff, Muffy would like to avoid researching all 10 questions but still be reasonably assured of getting a fairly good grade. Specifically, she wants to have at least an 85% chance of getting at least 4 of the 5 questions right. Will it be sufficient if she studies 8 of the 10 questions?

No. Think of the 10 questions as being two kinds of chips in an urn—there are the 8 whose answers Muffy will know and the two for which she will be unprepared (see Figure 3.6.4). In making out the test, the professor is drawing a random sample of size 5. Let Y denote the number in the sample coming from the 8 questions Muffy will have prepared. Unfortunately,

$$P(Y \geq 4) = P(Y = 4) + P(Y = 5)$$

$$= \frac{\binom{8}{4}\binom{2}{1}}{\binom{10}{5}} + \frac{\binom{8}{5}\binom{2}{0}}{\binom{10}{5}} = \frac{196}{252}$$

$$= 0.78$$

so it's back to the books! By studying 8 questions she has only a 78% chance (rather than 85%) of getting at least four correct. (Would Muffy satisfy her 85% requirement if she prepared for 9 of the questions?)

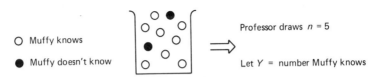

O Muffy knows

● Muffy doesn't know

Professor draws $n = 5$

Let Y = number Muffy knows

Figure 3.6.4

Example 3.6.9

Urn I contains five red chips and four white chips; urn II contains four red chips and five white chips. Two chips are to be transferred from urn I to urn II. Then a single chip is to be drawn from urn II (see Figure 3.6.5). What is the probability the chip drawn from the second urn will be white?

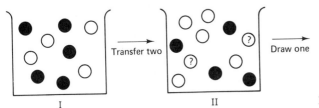

Transfer two

Draw one

I

II

Figure 3.6.5

Let W be the event "white chip is drawn from urn II." Let A_i, $i = 0, 1, 2$, denote the event "i white chips are transferred from urn I to urn II." By Theorem 2.6.1,

$$P(W) = P(W|A_0)P(A_0) + P(W|A_1)P(A_1) + P(W|A_2)P(A_2)$$

Note that $P(W|A_i) = (5 + i)/11$ and that $P(A_i)$ is gotten directly from Theorem 3.6.3. Therefore,

$$P(W) = \left(\frac{5}{11}\right)\frac{\binom{4}{0}\binom{5}{2}}{\binom{9}{2}} + \left(\frac{6}{11}\right)\frac{\binom{4}{1}\binom{5}{1}}{\binom{9}{2}} + \left(\frac{7}{11}\right)\frac{\binom{4}{2}\binom{5}{0}}{\binom{9}{2}}$$

$$= \left(\frac{5}{11}\right)\left(\frac{10}{36}\right) + \left(\frac{6}{11}\right)\left(\frac{20}{36}\right) + \left(\frac{7}{11}\right)\left(\frac{6}{36}\right)$$

$$= \frac{53}{99}$$

Example 3.6.10

The hypergeometric distribution is a key result in the field of *acceptance sampling*. Consider the plight of a manufacturer who orders a supply of parts from an outside contractor. When the shipment arrives, she would like to have some assurance that the parts meet her specifications. She could, of course, inspect each and every item; but that would be costly, time-consuming, and, in some cases, impossible (what if the part were a flashbulb?). A more-reasonable approach is to take a random sample from the shipment, inspect each member of the sample, and then accept the shipment (as being of sufficiently high quality), only if the number of defectives found is fewer than some specified number.

As a simple example of this technique, suppose the shipment contains 100 items, out of which she selects a sample of size $n = 2$. Let Y be the number of defectives found in the sample. She decides—more or less arbitrarily—that if $Y \geq 1$, she will return the shipment. What is the probability she will accept a shipment that is 10% defective?

Since the manufacturer intends to reject the shipment when $Y \geq 1$, she will accept it when $Y = 0$. Therefore,

$$P(\text{she accepts shipment}) = P(Y = 0) = \frac{\binom{90}{2}\binom{10}{0}}{\binom{100}{2}}$$

$$= \frac{(4005)(1)}{4950} = 0.81$$

If the shipment were 20% defective,

$$P(\text{she accepts shipment}) = \frac{\binom{80}{2}\binom{20}{0}}{\binom{100}{2}}$$

$$= 0.64$$

Figure 3.6.6 is a graph plotting the manufacturer's acceptance probability

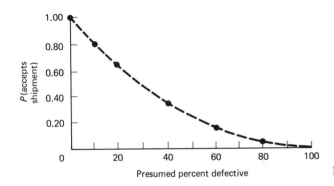

Figure 3.6.6

P(accepts shipment)

Presumed percent defective

as a function of the shipment percent defective. Graphs of this sort are referred to as *operating characteristic curves;* they show the sensitivity of the sampling plan in detecting lapses in shipment quality.

Included in the information that Figure 3.6.6 summarizes is the clear indication that the suggested sampling plan is really not very good. If the shipment were 60% defective, for example—certainly an intolerable state of affairs for the customer—the plan would still recommend acceptance 15% of the time. (How can we get a "steeper" operating characteristic curve? See Question 3.6.32.)

Example 3.6.11

Estimating the size of a wildlife population is a problem often confronting ecologists. Among the procedures for getting such information is the *capture-recapture* method, which is based on the hypergeometric distribution. Chattergee (10) describes a study where the objective was to estimate the number of largemouth bass in Dryden Lake, a small body of water in central New York. First, a total of $T = 213$ largemouth bass were caught, tagged, and released. Then, after sufficient time had been allowed for the tagged fish to mix with their untagged companions, a second sample of $k = 104$ bass were caught. Among those 104 were $r = 13$ that had been previously tagged. What is a reasonable estimate for the number of largemouth bass in Dryden Lake?

Let N denote the unknown number we are looking for and $\hat{N}$ denote our estimate of that number. If $\hat{N}$ were set equal to, for instance, 500, the probability of recapturing only 13 out of 213 tagged fish (with a sample of 104) would be

$$\frac{\binom{213}{13}\binom{287}{91}}{\binom{500}{104}}$$

But this is a very small probability, less than 10^{-10}, and does not lend much credence to the speculation that N equals 500. At the other extreme, had we estimated N to be very large—5000, for example—the analogous probability would be

$$\frac{\binom{213}{13}\binom{4787}{91}}{\binom{5000}{104}}$$

also a miniscule figure.

The two hypergeometric probabilities just computed should suggest that as an estimation procedure it would not be unreasonable to select as $\hat{N}$ the value that maximizes the likelihood of what has occurred. That is, given T, r, and k, we should maximize

$$\frac{\binom{T}{r}\binom{N-T}{k-r}}{\binom{N}{k}}$$

as a function of N. We can do that by considering the *ratio* of the probabilities for two successive values of N:

$$\frac{\binom{T}{r}\binom{N-T}{k-r}}{\binom{N}{k}} \Big/ \frac{\binom{T}{r}\binom{N-1-T}{k-r}}{\binom{N-1}{k}} = \frac{(N-T)(N-k)}{N(N-T-k-r)}$$

Note that this ratio is larger than 1 (that is, the probabilities are increasing with N) if and only if

$$(N-T)(N-k) > N(N-T-k-r)$$

or, equivalently, if and only if

$$\frac{kT}{r} > N$$

As a first guess, then, it makes sense to set

$$\hat{N} = \frac{kT}{r}$$

Here, $\hat{N} = (213)(104)/13$, or 1704. (Does the formula $\hat{N} = kT/r$ seem intuitively reasonable? Explain.)

The last example in this section describes an application of the hypergeometric distribution to the problem of determining whether or not a bullet was fired from a certain gun. Central to the solution is the notion of *statistical significance*. Put simply, the statistical significance of an event is the probability (relative to some hypothesis) of observing a result as extreme or more extreme than what actually happened. If that probability turns out to be very small—less than 0.05, for example, we will conclude that the initial hypothesis was incorrect.

Example 3.6.12

When a bullet is fired it becomes scored with minute striations produced by imperfections in the gun barrel. Appearing as a series of parallel lines, these striations have long been recognized as a basis for matching a bullet with a gun, since repeated firings of the same weapon will produce bullets having substantially the same configuration of markings. Until recently, deciding how close two patterns had to be before it could be concluded the bullets came from the same weapon was largely subjective. A ballistics expert would simply look at the two bullets under a microscope and make an informed judgment based on past experience. Today, criminologists are beginning to address the problem more quantitatively, partly with the help of the hypergeometric distribution (30).

Suppose a bullet is recovered from the scene of a crime, along with the suspect's gun. Under a microscope, a grid of m cells, numbered 1 to m, is superimposed over the bullet. If m is chosen large enough so the width of the cells is sufficiently small, each of that evidence bullet's n_e striations will fall into a different cell (see Figure 3.6.7(a)). Then the suspect's gun is fired, yielding a test bullet, which will have a total of n_t striations located in a possibly different set of cells (see Figure 3.6.7(b)). Our objective is to make some sort of probability statement about the similarity in cell locations for the two striation patterns.

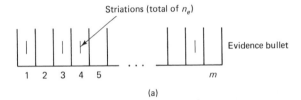

(a)

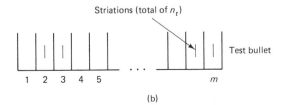

(b)

Figure 3.6.7

As a model for the striation pattern on the evidence bullet, imagine an urn containing m numbered chips, with the n_e corresponding to the striation locations shaded red. Now think of the striation pattern on the test bullet as representing a sample of size n_t drawn from the evidence urn. Let Y denote the number of cell locations shared by the two striation patterns. From Theorem 3.6.3,

$$P(Y = k) = \frac{\binom{n_e}{k}\binom{m - n_e}{n_t - k}}{\binom{m}{n_t}}$$

A numerical example should clarify how the formula for $P(Y = k)$ can be applied in this particular context. Suppose the grid to be used has $m = 25$ cells, the evidence bullet has $n_e = 4$ striations, the test bullet has $n_t = 3$ striations, and the location for one of the test bullet's striations matches up with the location of one of the evidence bullet's. What statistical significance should we attach to the number 1, and what should we conclude about the origin of the two bullets?

By definition, the statistical significance of the event "1 match" is the probability of getting *1 or more* matches. But

$$P(1 \text{ or more matches}) = P(Y \geq 1)$$

$$= \frac{\binom{4}{1}\binom{21}{2}}{\binom{25}{3}} + \frac{\binom{4}{2}\binom{21}{1}}{\binom{25}{3}} + \frac{\binom{4}{3}\binom{21}{0}}{\binom{25}{3}}$$

$$= 0.42$$

If $P(Y \geq 1)$ had equaled something small—a number less than 0.05, for example—we would have concluded the two bullets came from the same gun. But with $P(Y \geq 1)$ being so large, there is no way to rule out the possibility that the bullets were fired from different guns and the observed match was simply due to chance.

Question 3.6.25 A Scrabble set consists of 54 consonants and 44 vowels. What is the probability that your initial draw (of seven letters) will be all consonants? Six consonants and one vowel? Five consonants and two vowels?

Question 3.6.26 The Admissions Committee of a medical school has 10 applicants, 8 white and 2 members of ethnic minorities, from which the remaining n positions in the first-year class are to be filled. If the Committee chooses at random from the 10, what is the smallest value of n that will ensure a probability greater than $\frac{1}{2}$ of admitting at least one minority applicant?

Question 3.6.27 Show directly that the set of probabilities associated with the hypergeometric model sum to 1. (*Hint:* Expand the identity

$$(1 + \mu)^N = (1 + \mu)^r(1 + \mu)^{N-r}$$

and equate coefficients.)

Question 3.6.28 Urn I contains four red chips, three white chips, and two blue chips. Urn II has three red, four white, and five blue. Two chips are drawn at random and without replacement from each urn. What is the probability that all four chips are the same color?

Question 3.6.29 In a draft lottery, all 366! possible birthday sequences are meant to be equally likely. Suppose that you and two of your friends all have different birthdays. What is the probability that exactly two of you will have your birthdays drawn in the first 10 rounds? (Good intentions notwithstanding, efforts to achieve an equally likely model *in practice* have not always proved successful. See (48) for a discussion of the 1969 lottery.)

Question 3.6.30 Six terminals, numbered 1 through 6, are on-line to a DEC-10 computer; all are ready to execute their programs. You and a friend are working on terminals 2 and 5. At random, the computer selects three terminals and advances them in the access priority queue. What is the probability that both your terminal and your friend's terminal were among the three selected to be advanced?

Question 3.6.31 An urn contains eight chips numbered 1 through 8. Four are selected, without replacement. Let X represent the number of the second smallest chip drawn. Find the probability that $X = 3$.

Question 3.6.32 Recall the acceptance sampling discussion of Example 3.6.10. Suppose that *five*, rather than two, items are to be sampled from the shipment (of 100), with acceptance requiring that Y, the number of defectives, be less than or equal to 1. Construct the corresponding operating characteristic curve and compare it to Figure 3.6.6.

Question 3.6.33 A bleary-eyed freshman awakens one morning, late for an 8:00 class, and pulls two socks out of a drawer that contains two black, six brown, and two blue socks, all randomly arranged. What is the probability that the two he draws are a matched pair?

Question 3.6.34 A camera manufacturer receives a shipment of 100 semiautomatic lens housings. For his sampling plan, he decides to select 10 of the housings at random and accept the shipment if no more than one is defective. Construct the corresponding operating characteristic curve. For approximately what incoming quality will he accept the shipment 50% of the time?

Question 3.6.35 Show that as $N \rightarrow \infty$ such that $r/N \rightarrow p$, sampling without replacement is equivalent to sampling with replacement. That is, show that each hypergeometric probability converges to a corresponding binomial probability.

3.7 PRINCIPLE OF INCLUSION AND EXCLUSION

One of the first rules of probability introduced in Chapter 2 was the addition law: If A and B are any two events defined on the same sample space, $P(A \cup B) = P(A) + P(B) - P(A \cap B)$. The rationale behind that proof was simple. We were looking for the probability associated with all the outcomes in the union, $A \cup B$. But if we just added $P(A)$ and $P(B)$, any outcomes satisfying both events would have their probabilities counted twice. To avoid that, we needed to subtract $P(A \cap B)$ once.

There is a more-general result similar to this in combinatorics, known as the *principle of inclusion and exclusion*. As the name implies, its proof follows along the same lines as the derivation of the addition law.

Theorem 3.7.1. Let $a_1, a_2, \ldots, a_r$ be a set of r properties. Let $N(a_i)$ be the number of objects possessing property a_i, $i = 1, 2, \ldots, r$. Let $N(a_i a_j)$ be the number of objects possessing properties a_i and a_j, $i \neq j$. Define $N(a_i a_j a_k), \ldots,$ $N(a_1 a_2 \ldots a_r)$ similarly. Let N denote the number of objects possessing *at least one* of the properties. Then

$$N = \sum_{i=1}^{r} N(a_i) - \sum_{i<j} N(a_i a_j) + \sum_{i<j<k} N(a_i a_j a_k) - \cdots$$

$$+ (-1)^{r+1} N(a_1 a_2 \cdots a_r)$$

Proof. Consider the set of all objects possessing k arbitrary specified properties—say, $a_{i_1}, a_{i_2}, \ldots, a_{i_k}$—and *no others*. We need to show that those objects are counted once and only once by the right-hand side of the statement of the theorem. If we can do that, the result follows, since k was arbitrary and the properties were arbitrary.

Note that the objects in question get counted $\binom{k}{1}$ times in $\Sigma_{i=1}^{r} N(a_i)$, $\binom{k}{2}$ times in $\Sigma_{i<j} N(a_i a_j)$, $\binom{k}{3}$ times in $\Sigma_{i<j<k} N(a_i a_j a_k)$, and so on (why?). Therefore, according to the theorem, they will be counted a total of

$$\binom{k}{1} - \binom{k}{2} + \binom{k}{3} - \cdots + (-1)^{k+1}\binom{k}{k}$$

times. But

$$(1-1)^k = 0^k = \sum_{j=0}^{k} \binom{k}{j}(-1)^j (1)^{k-j}$$

$$= \binom{k}{0} - \binom{k}{1} + \binom{k}{2} - \cdots + (-1)^k\binom{k}{k}$$

which implies that

$$\binom{k}{1} - \binom{k}{2} + \cdots + (-1)^{k+1}\binom{k}{k} = \binom{k}{0} = 1$$

and the theorem is proved.

Example 3.7.1

A contemporary of Archimedes, Eratosthenes was a Greek astronomer who today is probably best remembered as the first person to devise a method for estimating the circumference of the earth. But Eratosthenes also did some mathematics and is credited with discovering an easy-to-use algorithm for identifying all the primes less than or equal to some given number N.

List the integers from 1 to N. Start with the number 2 and remove all the multiples of 2 from the list, except for 2 itself. Do the same for 3. Continue in this fashion, removing multiples of primes, until the multiples of r have been removed, where r is the largest prime less than or equal to $\sqrt{n}$. It can be proved (see (33)) that the only numbers remaining at that point will be primes. Figure 3.7.1 shows this *sieve of Eratosthenes* applied to the integers from 1 to 100.

Related to the question of identifying primes is the problem of *counting* them. Specifically, can we determine the number of primes between 1 and n without going to the trouble of listing them? The answer is yes, if we make the proper use of Theorem 3.7.1.

Let

a_2 be the property: integer is divisible by 2

a_3 be the property: integer is divisible by 3

1 (2) (3) 4̸ (5) 6̸ (7) 8̸ 9̸ 1̸0̸ (11) 1̸2̸ (13) 1̸4̸ 1̸5̸
1̸6̸ (17) 1̸8̸ (19) 2̸0̸ 2̸1̸ 2̸2̸ (23) 2̸4̸ 2̸5̸ 2̸6̸ 2̸7̸ 2̸8̸ (29)
3̸0̸ (31) 3̸2̸ 3̸3̸ 3̸4̸ 3̸5̸ 3̸6̸ (37) 3̸8̸ 3̸9̸ 4̸0̸ (41) 4̸2̸ (43)
4̸4̸ 4̸5̸ 4̸6̸ (47) 4̸8̸ 4̸9̸ 5̸0̸ 5̸1̸ 5̸2̸ (53) 5̸4̸ 5̸5̸ 5̸6̸ 5̸7̸
5̸8̸ (59) 6̸0̸ (61) 6̸2̸ 6̸3̸ 6̸4̸ 6̸5̸ 6̸6̸ (67) 6̸8̸ 6̸9̸ 7̸0̸ (71)
7̸2̸ (73) 7̸4̸ 7̸5̸ 7̸6̸ 7̸7̸ 7̸8̸ (79) 8̸0̸ 8̸1̸ 8̸2̸ (83) 8̸4̸ 8̸5̸
8̸6̸ 8̸7̸ 8̸8̸ (89) 9̸0̸ 9̸1̸ 9̸2̸ 9̸3̸ 9̸4̸ 9̸5̸ 9̸6̸ (97) 9̸8̸ 9̸9̸
1̸0̸0̸

Figure 3.7.1

a_5 be the property: integer is divisible by 5

$\vdots$

a_r be the property: integer is divisible by r, where r is the largest prime $\leq \sqrt{n}$

Note that

$$N(a_i) = \text{number of integers from 1 to } n \text{ that are divisible by } i$$

$$= \left[\frac{n}{i}\right]$$

the greatest integer in n/i. Also,

$$N(a_i a_j) = \text{number of integers from 1 to } n \text{ that are divisible}$$
$$\text{by both } i \text{ and } j$$

$$= \text{number of integers from 1 to } n \text{ that are divisible by } ij$$

$$= N(a_{ij})$$

$$= \left[\frac{n}{ij}\right]$$

By the same argument, $N(a_i a_j a_k) = [n/ijk]$, and so on. Substituting into Theorem 3.7.1, we get an expression for N:

$$N = \sum_i \left[\frac{n}{i}\right] - \sum_{i<j} \left[\frac{n}{ij}\right] + \sum_{i<j<k} \left[\frac{n}{ijk}\right] - \cdots + (-1)^{r+1} \left[\frac{n}{2 \cdot 3 \cdots r}\right]$$

It follows that

$$\text{number of primes from 1 to } n = n - N - 1 - t \qquad (3.7.1)$$

where t is the number of primes less than or equal to $\sqrt{n}$ (1 is subtracted because we are not counting 1 as a prime).

We can test Equation 3.7.1 on the grid shown in Figure 3.7.1. For $n = 100$, the t primes that need to be considered are 2, 3, 5, and 7. Therefore,

$$N = \sum_i N(a_i) - \sum_{i<j} N(a_i a_j) + \sum_{i<j<k} N(a_i a_j a_k) - N(a_2 a_3 a_5 a_7)$$

$$= N(a_2) + N(a_3) + N(a_5) + N(a_7) - N(a_6) - N(a_{10}) - N(a_{14})$$

$$- N(a_{15}) - N(a_{21}) - N(a_{35}) + N(a_{30}) + N(a_{42}) + N(a_{70})$$

$$+ N(a_{105}) - N(a_{210})$$

$$= 50 + 33 + 20 + 14 - 16 - 10 - 7 - 6 - 4 - 2 + 3 + 2 + 1$$

$$+ 0 - 0$$

$$= 78$$

According to Equation 3.7.1, then, the number of primes from 1 to 100 should be

$$n - N - 1 - t = 100 - 78 - 1 + 4$$

$$= 25$$

and that agrees with a direct count made in Figure 3.7.1.

Example 3.7.2

A *derangement* of n distinct ordered objects is a rearrangement where no object appears in its original position. Thus (B, C, A) is a derangement of (A, B, C), but (C, B, A) is not. For arbitrary n, how many derangements are possible?

Let $(s_1, s_2, \ldots, s_n)$ denote any specified arrangement of n distinct objects. Consider the set of $n!$ permutations of $(s_1, s_2, \ldots, s_n)$. We will say that a permutation possesses property a_i if its ith component is the same as the ith component in $(s_1, s_2, \ldots, s_n)$. If N denotes the number of permutations where *at least one* of the n components has remained in its original position, what we are looking for is the difference $n! - N$. That difference will be the total number of possible derangements.

In the notation of Theorem 3.7.1,

$$N = \sum_{i=1}^{n} N(a_i) - \sum_{i<j} N(a_i a_j) + \cdots + (-1)^{n+1} N(a_1 a_2 \cdots a_n)$$

Note that $N(a_i)$ will equal the number of ways to permute all the components in $(s_1, s_2, \ldots, s_n)$ *except* s_i. That is, $N(a_i) = (n - 1)!$ for all i. Similarly, $N(a_i a_j) = (n - 2)!$, $N(a_i a_j a_k) = (n - 3)!$, and so on. Therefore,

$$N = \binom{n}{1}(n - 1)! - \binom{n}{2}(n - 2)! + \binom{n}{3}(n - 3)!$$

$$- \cdots + (-1)^{n+1}\binom{n}{n}(n - n)!$$

so the total number of derangements, D, reduces to

$$D = n! - N = n!\left[1 - \frac{1}{1!} + \frac{1}{2!} - \frac{1}{3!} + \cdots + (-1)^n\frac{1}{n!}\right]$$

The first two columns of Table 3.7.1 give values of D for n ranging from 2 to 9.

TABLE 3.7.1

n	D	$n!\, e^{-1}$	Percent error
2	1	0.736	26.4
3	2	2.207	10.4
4	9	8.829	1.9
5	44	44.145	0.33
6	265	264.873	0.035
7	1854	1854.110	0.0059
8	14833	14832.881	0.00080
9	133496	133495.931	0.000052

Although D gets to be somewhat cumbersome to compute for anything other than the smallest values for n, it can be easily approximated. Notice that what is inside the brackets in the equation defining D are the first n terms in the Taylor expansion for e^{-1}. Because that series converges to its limit very rapidly, a good approximation for D is

$$D \doteq n!\, e^{-1}$$

The last two columns of Table 3.7.1 show values of $n!\, e^{-1}$ and $100 \times |D - n!\, e^{-1}|/D$. As predicted, the agreement between D and $n!\, e^{-1}$ is excellent, even for values of n as small as 5 or 6.

Comment

Theorem 3.7.1 deals with the *number* of objects having at least one of r specified properties. By making some minor wording changes in the proof, we can get an equivalent statement about the *probability* of the union of a set of r events. That is, if $A_1, A_2, \ldots, A_r$ are any r events defined on a sample space S, it can be shown that

$$P(A_1 \cup A_2 \cup \cdots \cup A_r) = \sum_i P(A_i)$$
$$- \sum_{i<j} P(A_i \cap A_j) + \sum_{i<j<k} P(A_i \cap A_j \cap A_k)$$
$$- \cdots + (-1)^{r+1} P(A_1 \cap A_2 \cap \cdots \cap A_r)$$

(Recall that this was the statement of Theorem 2.3.7.) Some applications of the principle of inclusion and exclusion are formulated in terms of numbers of outcomes, as was true for the derangement example. Others, like the one that follows, are more easily expressed in terms of probabilities.

Example 3.7.3

A bubble gum company is printing a special series of baseball cards featuring r of the greatest base-stealers of the past decade. Each player appears on the same number of cards, and the cards are randomly distributed to retail outlets. If a collector buys n ($\geq r$) packs of gum (each containing one card), what is the probability she gets a complete set of the special series?

Let A_i be the event the collector has *no* card for player i, $i = 1, 2, \ldots,$ r, and define A to be the union, $A = \bigcup_{i=1}^{r} A_i$. Then

$$P(\text{collector has at least one card of each player}) = 1 - P(A)$$

To begin the derivation of $P(A)$, notice that for any value of k, $k = 1, 2, \ldots, r$,

$$P(A_1 \cap A_2 \cap \cdots \cap A_k) = \left(1 - \frac{k}{r}\right)^n \qquad \text{(Why?)}$$

Therefore, from the general addition law of Theorem 2.3.3,

$$P\left(\bigcup_{i=1}^{r} A_i\right) = P(A) = \sum_{i=1}^{r}\left(1 - \frac{1}{r}\right)^n - \sum_{i<j}\left(1 - \frac{2}{r}\right)^n$$

$$+ \sum_{i<j<k}\left(1 - \frac{3}{r}\right)^n - \cdots + (-1)^{r+1} \cdot 0$$

$$= \binom{r}{1}\left(1 - \frac{1}{r}\right)^n - \binom{r}{2}\left(1 - \frac{2}{r}\right)^n$$

$$+ \binom{r}{3}\left(1 - \frac{3}{r}\right)^n - \cdots + (-1)^{r+1} \cdot 0$$

More concisely,

$$P(A) = \sum_{k=1}^{r} (-1)^{k+1}\binom{r}{k}\left(1 - \frac{k}{r}\right)^n$$

Table 3.7.2 shows the evaluation of $1 - P(A)$ for several sets of values for n and r.

TABLE 3.7.2

r	n	$1 - P(A)$
2	2	0.500
	4	0.875
	6	0.969
4	4	0.094
	6	0.381
	8	0.623
	10	0.781
	12	0.875
	14	0.929
6	6	0.016
	10	0.272
	14	0.583
	18	0.785
	22	0.893

Thus if there are $n = 4$ different cards available and the collector buys eight packs of gum, she has a 62.3% chance of getting a complete set of base-stealers.

Comment

There is a more-general formulation of the principle of inclusion and exclusion than Theorem 3.7.1. What we have derived is an expression for the number of objects, N, having at least one of r specified properties (or the probability that at least one of r events occurs). It is also possible to get an equation for the number of objects possessing exactly m of the properties. See, for example, (26) or (51).

Question 3.7.1 The problem of counting derangements as discussed in Example 3.7.2 is a special case of the famous "matching problem" (or *probleme des rencontres*) first posed by DeMoivre in 1718. The language used in DeMoivre's version is interesting (66):

> Any number of letters a, b, c, d, e, f, etc. all of them different, being taken promiscuously as it happens; to find the Probability that some of them shall be found in their places according to the rank they obtain in the alphabet and that others of them should at the same time be displaced.

Find an approximate answer to DeMoivre's question. That is, find the probability that a rearrangement of a permutation of n distinct objects will leave exactly r of the objects in their original positions. (*Hint:* Consider the simpler problem where exactly r objects remain unchanged in the rearrangement and those r are located in positions 1 through r.)

Question 3.7.2 A disgruntled secretary stuffs n letters at random into n envelopes. What is the probability that no letter gets put into its proper envelope?

Question 3.7.3 Suppose eight married couples sign up for a dance class. If the instructor pairs the eight men and eight women at random, what is the probability that at least one husband will be dancing with his wife?

Question 3.7.4 In how many ways can the numbers $1, 2, 3, \ldots, 2n$ be permuted so that no odd digit is in its natural position?

Question 3.7.5 How many ways can eight rooks be put on a chessboard so that none is on the main diagonal (the diagonal going from the upper left-hand corner to the lower right-hand corner) and no two can take each other?

Question 3.7.6 At a political fundraiser attended by 200 elected officials from all branches of government, there are 75 Republicans, none of whom ever lies or cheats. Among the others present are 65 Democrats, 85 liars, and 70 cheats. Also, there are 45 liars who are Democrats, 40 Democrats who are cheats, 30 cheats who are liars, 30 liars who are neither Democrats nor cheats, 20 cheats who are neither Democrats nor liars, and 20 Democrats who are liars and cheats. How many Democrats are neither liars nor cheats?

Question 3.7.7 Suppose a code for the numbers 0 through 9 is to be devised where each digit is to be replaced by a different one. How many such codes could be constructed?

Question 3.7.8 How many prime numbers are between 1 and 150?

Question 3.7.9 How many ways can the digits $1, 2, 3, 4, 5$, and 6 be permuted so that none of the patterns 12, 34, or 56 appears?

Question 3.7.10 Prove the principle of inclusion and exclusion by induction.

Question 3.7.11 A stamp collector is trying to get a set of five U.S. commemoratives honoring the Southern agrarian movement. He asks his friends who correspond with him to mail their letters with stamps from this series. Assuming all five commemoratives are equally available to his friends, how many letters will he have to receive to have an a priori probability of at least 0.90 of getting the complete set?

Question 3.7.12 If an ordered arrangement of n distinct digits is permuted, show that the probability of exactly r of the objects remaining in place is approximately $e^{-1}/r!$.

Question 3.7.13 Show that the number of ways to permute the n numbers $1, 2, \ldots, n$ so that none of the $n - 1$ pairs $(1, 2), (2, 3), \ldots, (n - 1, n)$ occurs is

$$(n - 1)! \left[n - \frac{n - 1}{1!} + \frac{n - 2}{2!} - \frac{n - 3}{3!} + \cdots + \frac{(-1)^{n-1}}{(n - 1)!} \right]$$

Question 3.7.14 The general addition law is also true if unions and intersections are interchanged. Verify that statement for three events, A, B, and C. That is, prove the identity,

$$P(A \cap B \cap C) = P(A) + P(B) + P(C) - P(A \cup B)$$

$$- P(A \cup C) - P(B \cup C) + P(A \cup B \cup C)$$

4

Random Variables

> There were never in the world two opinions alike, any more than two hairs or two grains; the most universal quality is diversity.
>
> *M. de Montaigne*

4.1 INTRODUCTION

Throughout most of Chapter 2, probability functions were defined in terms of the elementary outcomes making up an experiment's sample space. Thus, if two fair dice were tossed, a P-value was assigned to each of the 36 possible pairs of upturned faces: $P((3, 2)) = \frac{1}{36}$, $P((2, 3)) = \frac{1}{36}$, $P((4, 6)) = \frac{1}{36}$, and so on. We have already seen, though, that in certain situations some attribute of an outcome may hold more interest for the experimenter than the outcome itself. A craps player, for example, may be concerned only about throwing a 7 and not whether the 7 was the result of a 5 and a 2, a 4 and a 3, or a 6 and a 1. Similarly, a virologist conducting a clinical trial in which n subjects are inoculated with a new influenza vaccine would probably focus on the *number* of those subjects developing flulike symptoms: Knowing that Mr. R and Ms. W became ill while Mr. T remained healthy, for example, would probably be peripheral to the evaluation of the vaccine's effectiveness.

In cases such as these, it could be argued that the original sample space is needlessly complicated and not appropriately attuned to the experiment's objectives. For craps, why not replace the 36-member sample space of (x, y)-pairs with the less complicated 11-member set of all possible two-dice *sums*, $\{2, 3, \ldots, 12\}$? Likewise, the analysis of the vaccine trial would be facilitated if the original sample space of all possible 2^n success-failure n-tuples were replaced by the set of integers $0, 1, \ldots, n$, representing the $n + 1$ possible numbers of subjects showing signs of influenza. In general, rules for making numerical assignments of this sort are known as *random variables*. As we will see, they play a very special role in the theory of probability.

Definition 4.1.1. A real-valued function whose domain is the sample space S is known as a *random variable*. We denote random variables by uppercase letters, often X, Y, or Z.

Comment

In theory, there is a stipulation that must be appended to Definition 4.1.1. A real-valued function Y qualifies as a random variable only if $Y^{-1}(B)$, where B is any Borel set, is in the σ-algebra on which the probability function is defined (see (2) or (90)). For the kinds of problems we will encounter, though, any random variable that we might want to define will have that property.

Example 4.1.1

For the game of craps (recall Example 2.9.2), the "sum" random variable is of obvious interest since the shooter wins or loses according to the total of the two faces showing. The sample space corresponding to a roll of two dice has 36 outcomes: $S = \{(1, 1), (1, 2), \ldots , (6, 6)\}$. The appropriate random variable Y would be defined (on S) according to the equation $Y(a, b) = a + b$. This would make $Y(1, 4) = 5$, $Y(6, 3) = 9$, and so on. (If the dice were fair, what proportion of the time would Y equal 7?)

Example 4.1.2

Suppose a penny is flipped independently four times. We win $5 for each head and lose $2 for each tail. Define a random variable whose value describes our net gain after four such tosses.

The sample space here can be represented as a set of 16 (2^4) four-tuples, where the ith component is either a 0 or a 1—a 1 if a head occurred on the ith flip; a 0, otherwise. In set notation, $S = \{(0000), (0001), \ldots , (1111)\}$. Let (a, b, c, d) be an arbitrary member of S. It follows that the amount of money we stand to gain if (a, b, c, d) occurs is given by the random variable $Y = Y(a, b, c, d)$, where

$$Y(a, b, c, d) = 5(a + b + c + d) - 2[4 - (a + b + c + d)]$$
$$= 7(a + b + c + d) - 8$$

(Write out the possible values of Y if the penny were tossed *twice*. What probabilities would you associate with the different values of Y if the penny is fair?)

Example 4.1.3

Let S be the subset of the Cartesian plane defined by $S = \{(x, y) \mid x^2 + y^2 \leq 1\}$.

How can a random variable D be defined to give the distance between the origin and a randomly chosen point?

Let (a, b) denote the coordinates of an arbitrary point in S (see Figure 4.1.1). The Euclidean distance between points $(0, 0)$ and (a, b) can be represented by

$$D = D(a, b) = \sqrt{a^2 + b^2}$$

(If the points in S are uniformly distributed, what probability would be associated with the event "$D \leq t$," for $0 < t < 1$?)

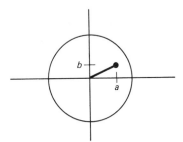

Figure 4.1.1

Example 4.1.4

Random variables are being developed formally for the first time in this chapter, but they were actually introduced—without any fanfare—somewhat earlier. In defining the binomial probability model (recall Theorem 3.6.1), we let Y represent the number of successes in n independent Bernoulli trials, and we wrote

$$P(Y = k) = \binom{n}{k} p^k (1 - p)^{n-k}, \qquad k = 0, 1, \ldots, n$$

If the sample space for that situation is thought of as the set of all 2^n possible n-tuples of 0s and 1s,

$$S = \{(0, 0, \ldots, 0), (1, 0, \ldots, 0), \ldots, (1, 1, \ldots, 1)\}$$

or

$$S = \{(x_1, x_2, \ldots, x_n): x_i = 0 \text{ or } 1\}$$

then for any $s \in S$, $Y = \sum_{i=1}^{n} x_i$. (Have we used random variable notation in connection with any model other than the binomial?)

As a conceptual framework, random variables are of fundamental importance in the theory of statistics: They provide a single rubric under which *all* probability problems may be brought. Even in cases where the original sample space needs no redefinition—that is, when the measurement recorded is the measurement of interest (see, for instance, Example 2.5.4)—the concept still applies: We simply take the random variable to be the identity mapping.

On the whole, this is a nuts-and-bolts chapter. Its primary purpose is to introduce the bits and pieces of random variable terminology and technique that make up the mathematics of probability and statistics; the really useful applications of these notions will come later.

4.2 DENSITIES AND DISTRIBUTIONS

If a random variable does, as claimed, measure some important aspect of a sample observation, it seems only reasonable that we would want to know the probability of that variable taking on certain values. What are the chances, for example, that the honor count in a bridge hand [$4 \cdot$ (number of aces) $+ 3 \cdot$ (number of kings) $+ 2 \cdot$ (number of queens) $+ 1 \cdot$ (number of jacks)] equals 16 or the probability that a craps player rolls a 7? In point of fact, such questions arise sufficiently often that to deal with them conveniently requires some special notation. The effect of Definitions 4.2.1–4.2.3 is to transfer our attention, once and for all, from the experiment's original sample space to the sample space of the associated random variable.

Definition 4.2.1. Let Y be a random variable defined on a sample space S with probability function P. For any real number t, the *cumulative distribution function of Y* (hereafter written $F_Y(t)$ and referred to as the *cdf*) is the probability of the set of all those sample points in S whose Y-values are less than or equal to t. Formally,

$$F_Y(t) = P(\{s \in S \mid Y(s) \leq t\})$$

Comment

The probabilities associated with all the sample points in S whose Y-value is (1) *greater than t* or (2) *greater than a but less than or equal to b* can also be expressed in terms of the cdf:

$$P(\{s \in S \mid Y(s) > t\}) = 1 - F_Y(t) \tag{4.2.1}$$

$$P(\{s \in S \mid a < Y(s) \leq b\}) = F_Y(b) - F_Y(a) \tag{4.2.2}$$

Equations 4.2.1 and 4.2.2 follow immediately from the properties of P.

Comment

To simplify cdf notation, we find it convenient to suppress the implicit relationship between Y and S. Thus $P(\{s \in S \mid Y(s) \leq t\})$ is written $P(Y \leq t)$, $P(\{s \in S \mid Y(s) > t\})$ becomes $P(Y > t)$, and $P(\{s \in S \mid a < Y(s) \leq b\})$ abbreviates to $P(a < Y \leq b)$.

The functional properties of a cdf, beyond those self-evident from Definition 4.2.1, depend on the nature of the random variable Y. Before elaborating on that statement mathematically, we look at several examples that will prove useful as stereotypes of cdf behavior.

Example 4.2.1

Consider the random variable defined for the game of craps in Example 4.1.1: $Y((a, b)) = a + b$. How would the cdf for Y be computed?

To calculate $F_Y(t)$ we simply add up the probabilities associated with all dice sums less than or equal to t. For example, $F_Y(4) = \frac{1}{6}$:

$$F_Y(4) = P(\{s \in S \mid Y(s) \le 4\}) = P(\{(a, b) \in S \mid a + b \le 4\})$$

$$= P(\{(1, 1), (1, 2), (2, 1), (1, 3), (3, 1), (2, 2)\})$$

$$= \tfrac{1}{36} + \tfrac{1}{36} + \tfrac{1}{36} + \tfrac{1}{36} + \tfrac{1}{36} + \tfrac{1}{36}$$

$$= \tfrac{1}{6}$$

Note that $F_Y(4.7)$ is also $\frac{1}{6}$ since $\{s \in S \mid Y(s) \le 4\} = \{s \in S \mid Y(s) \le 4.7\}$, there being no s with $4 < Y(s) \le 4.7$. Indeed, $F_Y(y)$ is constant in the interval $[4, 5)$.

Example 4.2.2

Imagine a fair die being rolled four times. For all t, find and graph the cdf of Y, the number of sixes that appear.

First, from Theorem 3.6.1, we see that Y has a binomial distribution with the probability of success at each of the four Bernoulli trials being $\frac{1}{6}$:

$$P(Y = k) = \binom{4}{k}\left(\frac{1}{6}\right)^k\left(\frac{5}{6}\right)^{4-k} \qquad \text{for } k = 0, 1, 2, 3, 4$$

For any t in the semiopen interval $[0, 1)$, $F_Y(t) = P(Y \le t) = P(Y = 0) = \left(\frac{5}{6}\right)^4$. Similarly, for any $1 \le t < 2$, $F_Y(t) = P(Y = 0) + P(Y = 1) = \left(\frac{5}{6}\right)^4 + 4\left(\frac{1}{6}\right)\left(\frac{5}{6}\right)^3$. Continuing this argument establishes $F_Y(t)$ to be a step function with jumps at the points $t = 0, 1, 2, 3,$ and 4:

$$F_Y(t) = \begin{cases} 0 & t < 0 \\ \left(\frac{5}{6}\right)^4 & 0 \le t < 1 \\ \left(\frac{5}{6}\right)^4 + 4\left(\frac{1}{6}\right)\left(\frac{5}{6}\right)^3 & 1 \le t < 2 \\ \left(\frac{5}{6}\right)^4 + 4\left(\frac{1}{6}\right)\left(\frac{5}{6}\right)^3 + 6\left(\frac{1}{6}\right)^2\left(\frac{5}{6}\right)^2 & 2 \le t < 3 \\ \left(\frac{5}{6}\right)^4 + 4\left(\frac{1}{6}\right)\left(\frac{5}{6}\right)^3 + 6\left(\frac{1}{6}\right)^2\left(\frac{5}{6}\right)^2 + 4\left(\frac{1}{6}\right)^3\left(\frac{5}{6}\right) & 3 \le t < 4 \\ 1 & 4 \le t \end{cases}$$

Figure 4.2.1 is a graph of $F_Y(t)$. (Sketch the graph of $F_Y(t)$ for the random variable in Example 4.2.1.)

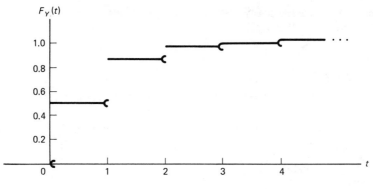

Figure 4.2.1

Example 4.2.3

Recall the distance random variable of Example 4.1.3. Find $F_D(t)$, assuming the points in S are uniformly distributed.

For any t, $0 \le t \le 1$,

$$F_D(t) = P(D \le t) = P(\{(x, y) \,|\, \sqrt{x^2 + y^2} \le t\})$$

$$= \frac{1}{\pi} \cdot (\text{area of circle of radius } t) \qquad (\text{Why?})$$

$$= \frac{1}{\pi} \cdot \pi t^2 = t^2$$

For points outside the unit interval, the cdf is either 0 or 1:

$$F_D(t) = \begin{cases} 0 & t < 0 \\ t^2 & 0 \le t \le 1 \\ 1 & 1 < t \end{cases}$$

(see Figure 4.2.2). Here, note that $F_D(t)$ is a continuous function; the cdf's in Examples 4.2.1 and 4.2.2 were not. (Use the cdf in Figure 4.2.2 to show *graphically* the probability that D lies between $\frac{1}{4}$ and $\frac{1}{2}$.)

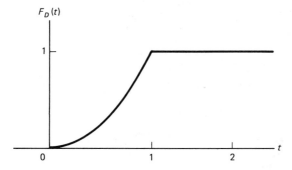

Figure 4.2.2

Example 4.2.4

The distance X that a molecule of gas travels before colliding with another molecule was discussed in Example 2.5.2. We assumed at the time that the distribution of X could be described by an exponential probability function,

$$f(x) = \left(\frac{1}{\lambda}\right)e^{-x/\lambda}, \qquad x > 0$$

where λ is a function of the particular gas being considered. Given that such a model is valid, what is the cdf for X?

If $t < 0$, $F_X(t)$ is clearly 0. A short integration gives $F_X(t)$ for $t \geq 0$:

$$F_X(t) = P(X \leq t) = \int_0^t \left(\frac{1}{\lambda}\right)e^{-x/\lambda}\, dx = 1 - e^{-t/\lambda}$$

(Note that the appropriate random variable here is the identity mapping. The original sample space (of intercollision distances) is the set S of nonnegative real numbers. To couch the problem in random variable terminology, we simply define $X(s) = s$, where $s \in S$.)

Example 4.2.5

Consider again the identity mapping $X(s) = s$ described in Example 4.2.4. Suppose the probability function defined on S is not the exponential, but the $f(x)$ given below:

$$f(x) = \begin{cases} 0 & x < 0 \\ 2x & 0 \leq x \leq \frac{1}{2} \\ 6 - 6x & \frac{1}{2} < x \leq 1 \\ 0 & x > 1 \end{cases}$$

(see Figure 4.2.3). Find the cdf for the random variable X.

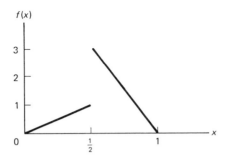

Figure 4.2.3

Looking at the form of $f(x)$, we see that $F_X(t)$ will have four different functional expressions, depending on whether $t < 0$, $0 \leq t \leq \frac{1}{2}$, $\frac{1}{2} < t \leq 1$, or $t > 1$. For $t < 0$, $F_X(t) = P(X \leq t) = 0$. For $0 \leq t \leq \frac{1}{2}$,

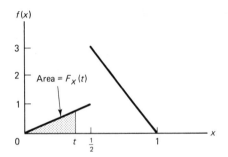

Figure 4.2.4   **Figure 4.2.5**

$$F_X(t) = P(X \le t) = \int_{-\infty}^{0} 0 \, dx + \int_{0}^{t} 2x \, dx$$

$$= t^2$$

(see Figure 4.2.4). For $\frac{1}{2} < t \le 1$,

$$F_X(t) = P(X \le t) = \int_{-\infty}^{0} 0 \, dx + \int_{0}^{\frac{1}{2}} 2x \, dx + \int_{\frac{1}{2}}^{t} (6 - 6x) \, dx$$

$$= \tfrac{1}{4} + (6x - 3x^2) \Big|_{\frac{1}{2}}^{t}$$

$$= 6t - 3t^2 - 2$$

(see Figure 4.2.5). Of course, for $t > 1$, $F_X(t) = 1$. Putting all these cases together gives

$$F_X(t) = \begin{cases} 0 & t < 0 \\ t^2 & 0 \le t \le \frac{1}{2} \\ 6t - 3t^2 - 2 & \frac{1}{2} < t \le 1 \\ 1 & t > 1 \end{cases}$$

Figure 4.2.6 is a graph of $F_X(t)$. (Let $A = \{s \in S \mid \frac{3}{4} < X(s) \le 1\}$. Find $P(A)$ two different ways—first by using $f(x)$ and then by using $F_X(t)$.)

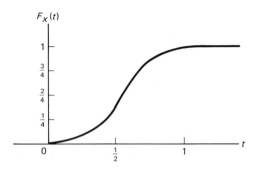

Figure 4.2.6

CASE STUDY 4.1

Consider again the evidence presented in *People v. Collins,* as outlined in Example 2.8.10. The prosecution's case rested on the unlikelihood that a given couple would match up with the six characteristics reported by the several eyewitnesses to the crime. It was estimated that the joint occurrence of a white female with blond hair combed in a ponytail riding in a yellow car with a black male having a beard and a mustache was on the order of 1 in 12 million—a number so small, the prosecution contended, that Ms. Collins and her male friend were clearly guilty, a classic open-and-shut case. Not so, argued the counsel for the defense. By approaching the same data from a slightly different perspective, they were able to show that "reasonable doubt" had not really been eliminated, despite the apparent odds of 12 million to 1. What finally convinced the judge that Ms. Collins and her friend might not be guilty was a probability argument based on a ratio of cdf's.

Suppose N is the total number of couples who could conceivably have been in the area and perpetrated the crime, and p is the probability that any such couple would share the six characteristics introduced by the prosecution as evidence. Define the random variable Y to be the number of couples matching up with the eyewitness accounts. It is not unreasonable to assume that Y is binomial, in which case

$$P(Y = k) = \binom{N}{k} p^k (1 - p)^{N-k}, \qquad k = 0, 1, \ldots, N$$

Therefore,

$$P(Y = 1) = Npq^{N-1}$$

and

$$P(Y \geq 1) = 1 - F_Y(0) = 1 - P(Y = 0) = 1 - (1 - p)^N$$

Also,

$$P(Y > 1) = 1 - F_Y(1) = 1 - (1 - p)^N - Npq^{N-1}$$

Now, consider the ratio

$$\frac{P(Y > 1)}{P(Y \geq 1)} = \frac{1 - (1 - p)^N - Npq^{N-1}}{1 - (1 - p)^N} = \frac{1 - F_Y(1)}{1 - F_Y(0)} \qquad (4.2.3)$$

$$= P(\text{more than one of the } N \text{ couples}$$
$$\text{fit the description given that}$$
$$\text{at least one does)}$$

$$= P(\text{there is at least one other couple who}$$
$$\text{could have committed the crime})$$

If $P(Y > 1)/P(Y \geq 1)$ is anything other than a very small number, we would have to accept the possibility that Ms. Collins and her friend have a pair of lookalikes and that perhaps the lookalikes were the culprits.

Table 4.2.1 shows the value of the cdf ratio (Equation 4.2.3) for various values of N befitting a large metropolitan area and for the prosecutor's estimate of

TABLE 4.2.1

p	N	$\dfrac{P(Y > 1)}{P(Y \geq 1)}$
$\dfrac{1}{12,000,000}$	1,000,000	0.0402
$\dfrac{1}{12,000,000}$	2,000,000	0.0786
$\dfrac{1}{12,000,000}$	5,000,000	0.1875
$\dfrac{1}{12,000,000}$	10,000,000	0.3479

$p(1/12,000,000)$. What the last column makes clear is that $P(Y > 1)/P(Y \geq 1)$ is *not* a particularly small number.

From this viewpoint, the data are certainly not as incriminating as the "1-in-12,000,000" argument would have us believe. At least that was the opinion of the California Supreme Court: Based on the numbers shown in Table 4.2.1, it over-turned the initial guilty verdict that had been handed down by the Superior Court of Los Angeles County.

Question 4.2.1 In a simulated baseball game, the spinner for Babe Ruth is based on his 1927 statistics (summarized in Example 3.6.5). The circle traced by the spinner is marked off so that each spin approximates Ruth's probabilities of reaching first, second, third, or home on his very next at bat. Define an appropriate random variable and sketch its cdf.

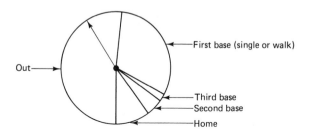

Question 4.2.2 Refer to Table 2.5.2. Suppose a state is selected at random and we let X denote its traffic fatality rate. Compute $F_X(4.7)$.

Question 4.2.3 The pth percentile of a distribution (denoted by x_p) is that value of x such that $F_X(x_p) = p/100$. Find x_{10} for the data of Table 2.5.2 on traffic fatality rates.

Question 4.2.4 Let the random variable X denote the length of a given word in the first stanza of Edgar Allan Poe's *Ulalume:*

> The skies they were ashen and sober;
>> The leaves they were crisped and sere,
>> The leaves they were withering and sere;

It was night in the lonesome October
Of my most immemorial year;
It was hard by the dim lake of Auber,
In the misty mid region of Weir:
It was down by the dank tarn of Auber,
In the ghoul-haunted woodland of Weir.

Find and graph the cdf, $F_X(x)$. What proportion of words in the next stanza would we estimate to have more than seven letters?

Question 4.2.5 Recall Example 4.2.3. What is the probability a point chosen at random in the circle is closer to the center than it is to the circumference?

Question 4.2.6 In Example 2.5.4, an exponential probability model was fit to the distribution of lifetimes of a sample of 903 V805 radar tubes. Observed relative frequencies and calculated probabilities were given in Table 2.5.1 for intervals 0–40, 40–80, and so on. If $F_X(x)$ and $G_X(x)$ denote the cdf's for the observed relative frequencies and the computed probabilities, respectively, find $F_X(200)$ and $G_X(200)$. Also, interpolate to find the values x and x' for which $F_X(x) = 0.50$ and $G_X(x') = 0.50$.

Question 4.2.7 Graph the cdf for X, the number of aces received in a five-card poker hand.

Question 4.2.8 An urn contains nine chips, four red and five black. Three are drawn without replacement. Let Y denote the number of black chips drawn. Find and graph $F_Y(y)$.

Question 4.2.9 An insurance company runs a trainee program for actuaries that has a 20% attrition rate. Suppose five persons enter the program. Graph the cdf of the number who finish.

Question 4.2.10 An urn contains n chips numbered 1 through n. Suppose the probability of any chip, k, being drawn is proportional to k. If a single chip is selected, what is the probability its number is less than or equal to m?

Question 4.2.11 Find the cdf for a random variable distributed uniformly over the interval (a, b).

Question 4.2.12 Find and graph the cdf for the missile-to-pipeline distance described in Example 2.5.1. Use the cdf to determine the probability that a missile will land within 20 ft of the pipeline.

Question 4.2.13 The time interval X between the purchase of a certain video cassette recorder (VCR) and its first trip to the repair shop is approximated by the exponential probability function

$$f(x) = e^{-x}, \qquad x > 0$$

where x is in years. Find the cdf for X and use it to estimate the probability that the VCR will work for at least 2 years without requiring any service.

Question 4.2.14 The normal distribution was introduced in Example 2.5.5. The cdf for a *standard* normal distribution (one where $\mu = 0$ and $\sigma = 1$) is tabulated in the appendix. If Z denotes the standard normal, then $F_Z(-1.17) = 0.1210$, $F_Z(0.95) = 0.8289$, and so on. Use the cdf to find (a) $P(Z \le 1.61)$, (b) $P(0.49 < Z \le 1.16)$, and (c) $P(Z > -1.23)$. What value of z has the property that $P(|Z| > z) = 0.05$?

Question 4.2.15 A random variable X has a cdf given by

$$F_X(x) = \begin{cases} 0 & -\infty < x < 0 \\ x^2 & 0 \le x < 1 \\ 1 & 1 \le x < \infty. \end{cases}$$

Find $P(\frac{1}{2} < X \le \frac{3}{4})$.

Question 4.2.16 Find the cdf for a random variable X whose probability function is given by

$$f(x) = \begin{cases} 0 & |x| > 1 \\ 1 - |x| & |x| \le 1 \end{cases}$$

Question 4.2.17 The probability function for a random variable is shown in the figure. Find $F_X(x)$ and sketch its graph.

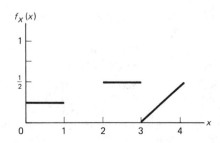

Question 4.2.18 Suppose the probability function describing the proportion of right answers a student gets on a chemistry midterm is

$$f(x) = 6x(1 - x), \qquad 0 \le x \le 1$$

Anyone scoring less than 60% fails.
a. Find $F_X(0.60)$.
b. If 10 students take the test, what is the probability that at least 8 pass?

Question 4.2.19 In the past, persons afflicted with a certain neurological disease have had a 30% chance of complete recovery. A radically different therapy has been tested on 10 patients, and 8 have recovered. Let the random variable X denote the number who recover when given the new therapy. Find $1 - F_X(7)$ under the assumption that the new therapy is no better than the old. In very general terms, how would the magnitude of $1 - F_X(7)$ be used to draw an inference about the efficacy of the new treatment?

On an intuitive level, Examples 4.2.1—4.2.5 suggest some of the salient features of distribution functions. Before extending these ideas any further, though, we examine the cdf a bit more rigorously from a mathematical standpoint.

Theorem 4.2.1. Let Y be a random variable with cdf $F_Y(y)$. Then:

1. $F_Y(y)$ is monotonically nondecreasing: $F_Y(s) \le F_Y(t)$ if $s < t$.
2. $F_Y(y)$ is right-continuous: $\lim_{y \to t^+} F_Y(y) = F_Y(t)$.
3. $\lim_{y \to -\infty} F_Y(y) = 0$, $\lim_{y \to +\infty} F_Y(y) = 1$.
4. $P(Y < t) = \lim_{y \to t^-} F_Y(t)$.

Proof

1. Since $(Y \le s) \subset (Y \le t)$, $P(Y \le s) \le P(Y \le t)$ by Theorem 2.3.3. Thus $F_Y(s) = P(Y \le s) \le P(Y \le t) = F_Y(t)$.

2. To establish right continuity, it suffices to show that $\lim_{n\to\infty} F_Y(t_n) = F_Y(t)$, where $\{t_n\}$ is any sequence with $t < \cdots < t_2 < t_1$ and $t_n \to t$. By definition,

$$F_Y(t_n) = P(Y \le t_n) = P(Y \le t) + P(t < Y \le t_n)$$
$$= F_Y(t) + P(t < Y \le t_n)$$

But $\lim_{n\to\infty} P(t < Y \le t_n) = 0$ by Axiom 4 (in Section 2.3).

3. Let t be $-\infty$ in the proof of (2) to establish the limit at $-\infty$. For the limit at ∞, observe that

$$\lim_{t\to\infty} F_Y(t) = 1 - \lim_{t\to-\infty} F_Y(t)$$

4. The proof is left as an exercise.

Corollary. The probability that a random variable Y lies in any given interval can always be determined from its cdf, $F_Y(y)$. The exact formulation depends on the nature of the interval:

$$P((a, b]) = F_Y(b) - F_Y(a)$$
$$P((a, b)) = \lim_{y\to b^-} F_Y(y) - F_Y(a)$$
$$P([a, b]) = F_Y(b) - \lim_{y\to a^-} F_Y(y)$$
$$P([a, b)) = \lim_{y\to b^-} F_Y(y) - \lim_{y\to a^-} F_Y(y)$$

Even though any function with Properties (1)–(3) of Theorem 4.2.1 is a cdf, most cdf's are one of two types. Recall from Sections 2.4 and 2.5 that, depending on the nature of an experiment's sample space, probability functions fall into one of two categories—they are either *discrete* or *continuous*. The very same dichotomy applies to random variables. (For the sake of comparison, the random variables in Examples 4.1.1 and 4.1.2 are discrete; the one in Example 4.1.3 is continuous.)

The remainder of this section examines the mathematical structure characterizing these two kinds of random variables. It also looks at the relationship between a random variable's *cdf* and what we define to be its *pdf*.

Definition 4.2.2. A *discrete random variable* Y is a real-valued function that can assume at most a countably infinite set of values. The *probability density function* for Y, written $f_Y(y)$ and abbreviated *pdf*, is the probability that Y takes on the value y:

$$f_Y(y) = P(Y = y)$$

If we know a random variable's pdf, we can easily determine its cdf, and vice versa. Theorem 4.2.2 details the nature of the association for the case of a discrete random variable.

> **Theorem 4.2.2.** Let Y be a discrete random variable with cdf $F_Y(y)$. Then:
>
> 1. $F_Y(t) = \sum_{y \le t} f_Y(y) = P(Y \le t)$.
> 2. $f_Y(t) = F_Y(t) - \lim_{y \to t^-} F_Y(y)$.
> 3. $P(Y = t) = f_Y(t) = 0$ if and only if $F_Y(y)$ is left-continuous—and, hence, continuous—at t.
> 4. If Y does not take on any value between points s and t, $s < t$, then $f_Y(t) = F_Y(t) - F_Y(s)$.

Proof

1. For a fixed point t, let $\{y_1, y_2, \ldots\}$ be the set of values of Y such that $y_i < t$ and $f_Y(y_i) \ne 0$. Then

$$F_Y(t) = P(Y \le t) = P(\{t, y_1, y_2, \ldots\})$$

$$= \sum_{i=1}^{\infty} P(Y = y_i) + P(Y = t)$$

$$= \sum_{i=1}^{\infty} f_Y(y_i) + f_Y(t) = \sum_{y \le t} f_Y(y)$$

2. $f_Y(t) = P(Y = t) = P(Y \le t) - P(Y < t) = F_Y(t) - \lim_{y \to t^-} F_Y(y)$, the last equality being a consequence of Property 4 of Theorem 4.2.1.
3. This follows immediately from (2) and the definition of left-continuity.
4. If Y takes on no value between s and t, then $\lim_{y \to t^-} F_Y(y) = F_Y(s)$. An application of (2) completes the proof.

We have already seen an application of Property 1 of Theorem 4.2.2: The cdf in Example 4.2.2 was found by summing the corresponding pdf. Going the other direction, from the cdf to the pdf, is just as easy. Look again at Example 4.2.2. For any y-value that is not a jump discontinuity of $F_Y(y)$—such as $y = 1.5$—the probability density function (from Property 2) will be 0:

$$f_Y(1.5) = F_Y(1.5) - \lim_{y \to 1.5^-} F_Y(1.5)$$
$$= F_Y(1.5) - F_Y(1.5) = 0$$

On the other hand, for $y = 2$, for example,

$$f_Y(2) = F_Y(2) - \lim_{y \to 2^-} F_Y(2)$$
$$= F_Y(2) - F_Y(1)$$
$$= (\tfrac{5}{6})^4 + 4(\tfrac{1}{6})(\tfrac{5}{6})^3 + 6(\tfrac{1}{6})^2(\tfrac{5}{6})^2 - (\tfrac{5}{6})^4 - 4(\tfrac{1}{6})(\tfrac{5}{6})^3$$
$$= 6(\tfrac{1}{6})^2(\tfrac{5}{6})^2 = (\tfrac{4}{2})(\tfrac{1}{6})^2(\tfrac{5}{6})^2$$

We recognize the latter to be the correct expression for the probability that a binomial random variable with $n = 4$ takes on the value $Y = 2$.

The next example shows a less-trivial application of this important pdf-cdf relationship.

Example 4.2.6

An urn contains n chips, numbered 1 through n. Suppose we draw a sample of k chips without replacement. Let Y denote the highest-numbered chip among those drawn. Find $f_Y(t)$.

Note that for any integer t, $1 \le t \le n$,

$$F_Y(t) = P(Y \le t) = \frac{\binom{t}{k}}{\binom{n}{k}}$$

Therefore,

$$f_Y(t) = \frac{\binom{t}{k} - \lim_{y \to t^-} F_Y(y)}{\binom{n}{k}} = \frac{\binom{t}{k} - \binom{t-1}{k}}{\binom{n}{k}}$$

Of course, $f_Y(t) = 0$ for any other t. (Find $f_Y(t)$ *directly*, without first determining $F_Y(t)$.)

Example 4.2.7

One of the most useful of all discrete pdf's is the *Poisson,* which was introduced in Example 2.4.4: For $\lambda > 0$,

$$P(X = x) = f_X(x) = \frac{e^{-\lambda}\lambda^x}{x!}, \qquad x = 0, 1, 2, \ldots$$

For theoretical reasons (see Chapter 6), the Poisson pdf is often used as a model for the occurrence of rare events. Historically, one of its earliest applications was in connection with radioactive decay. Suppose we know that a source emits alpha particles at the rate of $\lambda = 3$ per second. What is the probability that fewer than 5 will be emitted during the next second? More than 2?

Let X denote the number of particles emitted during the next second. Then

$$P(X < 5) = P(X \le 4) = F_X(4) = \sum_{x=0}^{4} f_X(x)$$

$$= \sum_{x=0}^{4} \frac{e^{-3}3^x}{x!}$$

$$= 0.82$$

Similarly, the probability that more than 2 emissions occur is 0.58:

$$P(X > 2) = P(X \ge 3) = 1 - F_X(2) = 1 - \sum_{x=0}^{2} \frac{e^{-3}3^x}{x!}$$

$$= 1 - 0.42$$

$$= 0.58$$

(Use $F_X(x)$ to evaluate $P(3 \le X \le 4)$.)

Definition 4.2.3. A random variable Y is said to be *continuous* if $F_Y(y)$ is continuous and $F'_Y(y)$ exists at all but a finite number of points. The probability density function for Y, written $f_Y(y)$, is the derivative $F'_Y(y)$.

Example 4.2.8

Suppose a random variable X has the cdf shown in Figure 4.2.7. In functional form,

$$F_X(t) = \begin{cases} 0 & t < 0 \\[6pt] \dfrac{3t}{4} & 0 \le t \le 1 \\[10pt] \dfrac{3}{4} & 1 < t \le 2 \\[10pt] \dfrac{t}{4} + \dfrac{1}{4} & 2 < t \le 3 \\[10pt] 1 & t > 3 \end{cases}$$

Find the corresponding pdf.

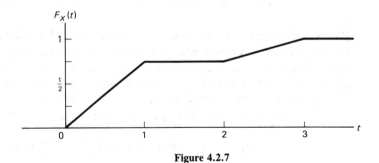

Figure 4.2.7

Since $F_X(t)$ is everywhere continuous and $F'_X(t)$ exists at all but four points, we can use Definition 4.2.3 to find $f_X(t)$. Differentiating $F_X(t)$ gives

$$f_X(t) = F'_X(t) = \begin{cases} 0 & t < 0 \\ \frac{3}{4} & 0 \le t \le 1 \\ 0 & 1 < t \le 2 \\ \frac{1}{4} & 2 < t \le 3 \\ 0 & t > 3 \end{cases}$$

which shows that $f_X(t)$ is a simple step-type function (see Figure 4.2.8). (Find $P(\frac{1}{2} < X < \frac{3}{4})$ in terms of the cdf; then check your answer by using the pdf.)

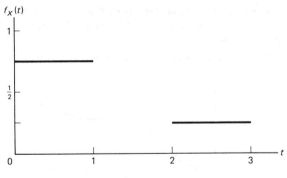

Figure 4.2.8

Example 4.2.9

For persons infected with a certain form of malaria, the length of time, X, they spend in remission is described by the pdf, $f_X(x) = (\frac{1}{9})x^2$, $0 < x \leq 3$. Find the cdf for X and use it to compute the probability that a malaria patient's remission lasts less than 1 year.

For $t \leq 0$, the cdf is 0. For $0 < t \leq 3$,

$$F_X(t) = P(X \leq t) = \int_0^t \frac{1}{9}x^2 \, dx = \frac{t^3}{27}$$

For $t > 3$, of course, the cdf is 1. It follows that the probability a remission lasts less than a year is $\frac{1}{27}$:

$$P(X < 1) = P(X \leq 1) = F_X(1) = \frac{1}{27}$$

(What is the probability that exactly two out of four malaria patients are in remission for less than a year?)

Comment

It may be helpful to think of the pdf for a continuous random variable, $f_Y(y)$, as being that function which, when integrated between any two values a and b, gives the probability that Y assumes a value somewhere between those two points. That is, $f_Y(y)$ is the function describing the behavior of Y in the sense that

$$P(a \leq Y \leq b) = \int_a^b f_Y(y) \, dy \qquad (4.2.4)$$

for any a and b. Unlike what is true for discrete random variables, $f_Y(y)$ is *not* the probability that $Y = y$ if Y is continuous.

Question 4.2.20 Suppose X is a continuous random variable defined over the interval A. Describe an event B having the property that the probability of B is 1, but B is not certain to occur. Describe an event C for which $P(C) = 0$, yet C is not impossible.

Question 4.2.21 A random variable is said to be *mixed* if its cdf can be written as a linear combination of a discrete cdf and a continuous cdf. Sketch a graph showing what the cdf of a mixed random variable might look like.

Question 4.2.22 A cdf has the given graph (for the interval $0 < t \le \frac{1}{2}$, $F_X(t) = t^2$).

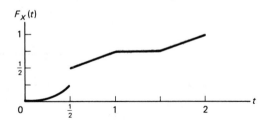

$F_X(t)$

Find each of the following values.
a. $P(X = \frac{1}{2})$
b. $P(\frac{1}{2} < X < 1\frac{1}{2})$
c. $P(1\frac{1}{2} < X < 2\frac{1}{2})$
d. $f_X(t)$

Question 4.2.23 A circular dartboard having a 1-ft radius is divided into three regions by two concentric circles, as shown. A dart landing in the outer ring earns 10 points; in the middle ring, 5 points; in the center, 15 points. If Y denotes the number of points earned on a given throw, find $f_Y(y)$. Assume the player is not very good and the throws tend to be uniformly distributed over the board. If the player throws three times, what is the probability exactly two land in the middle ring?

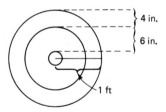

4 in.
6 in.
1 ft

Question 4.2.24 Suppose $f_X(x)$ is a symmetric continuous pdf: $f_X(x) = f_X(-x)$ for $x > 0$. Show that

$$P(-a < X < a) = 2 \cdot F_X(a) - 1$$

for $a > 0$.

Question 4.2.25 A random variable X has cdf

$$F_X(x) = \begin{cases} 0 & x < 1 \\ \ln x & 1 < x \le e \\ 1 & e < x \end{cases}$$

Find each of the following values.
a. $P(X < 2)$
b. $P(2 < X \le 2\frac{1}{2})$
c. $P(2 < X < 2\frac{1}{2})$
d. $f_X(x)$

Question 4.2.26 Among the most-famous meteor showers are the Perseids, which occur each year in early August. In some areas the frequency of visible Perseids can be as high as 10 per quarter hour. Assume the number of meteors a person can sight per quarter hour is a random variable described by a Poisson pdf with λ equal to 10 (recall Example 4.2.7). If a person watches the skies for 15 min, what is the probability he or she will see at least half as many meteors as expected?

Question 4.2.27 Hemophilia is a recessive sex-linked disease. If a woman who is a carrier—that is, who has the recessive hemophilia gene but not the symptoms of the disease—has a child with a man who is normal, the probability is $\frac{1}{4}$ that their next child will be hemophilic. Suppose the couple intend to have three children. Let the random variable X denote the number who will be hemophilic. Find $f_X(x)$.

Question 4.2.28 Suppose the length of time, X, that a customer has to stand in line in front of a bank teller's window before being served can be described quite well by the pdf

$$f_X(x) = (\tfrac{1}{5}) \cdot e^{-x/5}, \qquad x > 0$$

where X is in minutes. However, if he or she is not waited on within 10 min of arrival, the customer will leave. What is the probability of that event? Suppose the customer intends to make five trips to the bank in the next month. Let Y denote the number of times the customer will leave before getting waited on. Find $f_Y(2)$. What is the probability the customer will leave the line at least once?

Question 4.2.29 The cdf for a random variable X is defined by $F_X(t) = 0$ for $t < 0$; $F_X(t) = 4t^3 - 3t^4$ for $0 \leq t \leq 1$; and $F_X(t) = 1$ for $t > 1$. Find $P(\tfrac{1}{4} \leq X \leq \tfrac{3}{4})$.

Question 4.2.30 An insurance company has records showing that the length of time taken by relatives to file for death benefits, X, is described by the pdf $f_X(x) = 4x^2 e^{-2x}$, $x > 0$, where x is measured in days.
a. Find $F_X(t)$.
b. Verify that $\lim_{t \to \infty} F_X(t) = 1$ and $\lim_{t \to 0} F_X(t) = 0$.
c. What proportion of the company's policyholders file claims within 3 days of a relative's death?

Question 4.2.31 Suppose n battery-operated toy robots are turned on and left running for 5 h. If the battery life, X, is described by the pdf $f_X(x) = (1/\lambda)e^{-x/\lambda}$, $\lambda > 0$, find an expression for the probability that exactly r robots are still working after the 5 h have elapsed.

Question 4.2.32 If X denotes a family's annual income, $P(X > x)$ is the proportion of families earning more than x dollars per year. Vilfredo Pareto, an Italian economist (1848–1923), noticed that in certain situations, $P(X > x)$, where $x \geq x_0 > 0$ can be approximated by the function $cx^{-\alpha}(\alpha > 0)$. Find an expression for c and state the corresponding pdf for X. (The $f_X(x)$ being asked for here is known as *Pareto's distribution*—it is one of the more useful continuous pdf's.)

CASE STUDY 4.2

The time it takes for a piece of machinery or electronic equipment to fail is frequently modeled as a continuous random variable. The exponential pdf is a case in point. If a random variable X denotes failure time, we often assume that

$$f_X(x) = \left(\frac{1}{\lambda}\right)e^{-x/\lambda}, \qquad x > 0 \qquad (4.2.5)$$

where λ is set equal to the equipment's *average* life span (recall Example 2.5.4). We can gain some insight into *why* a particular pdf, such as Equation 4.2.5, is so effective at describing distributions of failure times by defining a related function, the hazard rate.

Let X be a continuous random variable representing the time to failure (or death) of some organism or piece of equipment. Characterizing the distribution of X is a conditional probability function called the *hazard rate* (actuaries know it by a more-poetic name, the *force of mortality*). A hazard rate is an instantaneous rate of failure (or death) at time t *given that the item has already survived until t*. Mathematically, we define the hazard rate, $h(t)$, as a limit:

$$h(t) = \lim_{\Delta t \to 0} \frac{P(t < T \leq t + \Delta t \,|\, T \geq t)}{\Delta t}$$

$$= \lim_{\Delta t \to 0} \frac{P(t < T \leq t + \Delta t)}{\Delta t \cdot P(T \geq t)} \qquad \text{(Why?)}$$

$$= \lim_{\Delta t \to 0} \frac{F_X(t + \Delta t) - F_X(t)}{\Delta t \cdot [1 - F_X(t)]}$$

$$= \frac{f_X(t)}{1 - F_X(t)}$$

(Note that the probability of failure in the interval $(t, t + \Delta t)$ *given survival until t* is approximately equal to $h(t) \cdot \Delta t$.)

For the exponential model of Equation 4.2.5, the hazard function is independent of t:

$$h(t) = \frac{(1/\lambda)e^{-t/\lambda}}{1 - (1 - e^{-t/\lambda})} = 1/\lambda$$

This means the exponential pdf is a "no-aging" model—it predicts that an item's probability of survival to time $t + s$ (given survival to t) is the same for all t. At first glance, a constant $h(t)$ seems like a contradiction in terms. Not so. As a first approximation over certain ranges of t, no wearout is sometimes a very reasonable assumption.

More typically, though, $h(t)$ *will* increase as a function of t. For those situations, polynomial hazard functions of the form

$$h(t) = \left(\frac{\beta}{\lambda}\right) \cdot t^{\beta-1} \tag{4.2.6}$$

often provide a more realistic picture of an item's survival behavior. (Equation 4.2.6 is associated with the *Weibull distribution,* a two-parameter pdf that has proved to be especially useful for engineering and physical science applications.)

Rather than pursue an application of Equation 4.2.6, we examine more closely a familiar phenomenon giving rise to a very different sort of hazard function. Businesses, like organisms, live and die—but unlike their biological counterparts, the ability of businesses to survive *increases* as they age. Their hazard functions, then, must *decrease*.

The first two columns of Table 4.2.2 show the length of life of retail businesses established in Poughkeepsie, N.Y., between 1844 and 1926 (53). To model these data, we choose a general form for $h(t)$ and work backwards.

TABLE 4.2.2 LIFETIMES OF BUSINESSES ESTABLISHED IN POUGHKEEPSIE, N.Y., BETWEEN 1844 AND 1926

Life (rounded to the nearest year)	Observed number	Number predicted by model
2 or less	2189	2204.8
3	470	432.5
4	310	321.2
5	245	247.4
6	205	196.2
7	155	159.1
8	130	131.6
9	105	110.5
10	100	94.0
11 or more	1089	1100.7
Total	4998	4998

Since we are speculating that $h(t)$ decreases with t, a simple choice for the hazard function is

$$h(t) = \frac{b}{a + t} \qquad (4.2.7)$$

where a and b are positive constants. Whether or not Equation 4.2.7 is appropriate, of course, depends on how well its associated pdf fits the data. Recovering $f_X(t)$ (or $F_X(t)$) from $h(t)$ is straightforward. First, note that

$$h(t) = \frac{f_X(t)}{1 - F_X(t)} = -\frac{d}{dt} \ln[1 - F_X(t)]$$

Then

$$\int_0^x h(t)\, dt = -\ln[1 - F_X(t)] \Big|_{t=0}^{t=x}$$

$$= -\ln[1 - F_X(t)]$$

since $F_X(0) = 0$. Also,

$$e^{-\int_0^x h(t)\, dt} = 1 - F_X(x)$$

or

$$F_X(x) = 1 - e^{-\int_0^x h(t)\, dt} \qquad (4.2.8)$$

When $b/(a + t)$ is substituted for $h(t)$, Equation 4.2.8 simplifies to

$$F_X(x) = 1 - \left(\frac{a}{a + x}\right)^b, \qquad x > 0 \tag{4.2.9}$$

The third column in Table 4.2.2 lists the set of expected frequencies based on Equation 4.2.9. (Estimates for a and b, computed using standard statistical techniques, were 3.67 and 1.12, respectively.) The agreement between Columns 2 and 3 is remarkably good, lending credence to our choice of $h(t)$.

Question 4.2.33 The *Weibull pdf* referred to in Case Study 4.2 is given by

$$f_X(x) = \left(\frac{\beta}{\lambda}\right) \cdot x^{\beta-1} e^{-x^\beta/\lambda}, \qquad x > 0$$

Show that the corresponding hazard function has the form given in Equation 4.2.6.

Question 4.2.34 One of the most useful characteristics of the Weibull distribution (see Question 4.2.33) is the wide range of shapes its hazard function can assume, depending on the value of β. Plot $h(t) = (\beta/\lambda) \cdot t^{\beta-1}$ for $\beta = 0.50$, 1, 1.5, and 2.

Question 4.2.35 An external heat monitor on a space capsule is provided with a backup unit. If the primary monitor malfunctions, the backup is immediately brought on line. If the lifetime, X, of the primary monitor is described by an exponential pdf with parameter $\lambda_1, f_X(x) = \lambda_1 e^{-\lambda_1 x}$, and the lifetime, Y, of the backup is also exponential but with a different parameter, $g_Y(y) = \lambda_2 e^{-\lambda_2 y}$, the lifetime of the *system*, $V = X + Y$, will be modeled by a *hypoexponential pdf*, $k_V(v)$, where

$$k_V(v) = \frac{\lambda_1 \lambda_2}{\lambda_2 - \lambda_1} (e^{-\lambda_1 v} - e^{-\lambda_2 v}), \qquad v > 0$$

Show that the hazard function for the system is

$$h(t) = \frac{\lambda_1 \lambda_2 (e^{-\lambda_1 t} - e^{-\lambda_2 t})}{\lambda_2 e^{-\lambda_1 t} - \lambda_1 e^{-\lambda_2 t}}$$

Question 4.2.36 Let the random variable X denote an item's lifetime. The *reliability* of that item, $R(t)$, is defined to be the probability that X exceeds t. Suppose the hazard function for the item is $h(t) = \lambda \sqrt{t}$, where $\lambda > 0$. Find its reliability.

Question 4.2.37 Recall that any increasing $F(x)$ with $F(0) = 0$ and $\lim_{x\to\infty} F(x) = 1$ can be considered a cdf. Show that any $F(x)$ defined by

$$F(x) = 1 - e^{-\int_0^x h(t)\, dt}$$

will have these properties if and only if $h(t) \geq 0$, for all t, and $\int_0^\infty h(t)\, dt = \infty$.

4.3 JOINT DENSITIES

Section 4.2 introduced the basic terminology for describing the probabilistic behavior of a *single* random variable, whether that variable is discrete or continuous. Such information, while adequate for many problems, is insufficient in situations where more than one random variable affects the outcome of an experiment. For example, consider an electronic system containing two components, one for backup, but both under load. Suppose the only way the system will fail is if both components cease to function. The distribution of Z, the system's life, depends *jointly* on the distributions of X and Y, the component lives. Knowing only $F_X(x)$ and $F_Y(y)$, though, will not necessarily provide us with enough information to determine $F_Z(z)$. What we need is a probability function giving the "simultaneous" behavior of X and Y.

More generally, we may need to deal with the joint behavior of n random variables. The management of a fast-food chain, for example, may want to predict their total sales volume coming from five local franchises, each competing to some extent for the same pool of customers. Or consider the problem faced by medical researchers in trying to sort out which of several factors—serum cholesterol, triglyceride level, blood pressure, genetic predisposition, and so on—contribute to a patient's risk of a coronary, and to what extent. As might be expected, when the number of random variables increases, the mathematical task of describing their joint behavior becomes much more difficult. To keep technical details from obscuring underlying concepts, we concentrate first on the two-variable case. At the end of the section, analogous results for the n-variable case are outlined.

Definition 4.3.1. Suppose X and Y are two random variables defined on the same sample space S. The *joint cdf of X and Y* is the function $F_{X,Y}(t, u)$, where

$$F_{X,Y}(t, u) = P(X \le t, Y \le u)$$
$$= P(\{s \in S \mid X(s) \le t \text{ and } Y(s) \le u\})$$

The domain of $F_{X,Y}(t, u)$ is the set of all pairs of real numbers.

Example 4.3.1

Consider the simple experiment of simultaneously tossing a fair coin and rolling a fair die. The resulting sample space S consists of 12 equally likely outcomes, each having two components:

$S = \{(H, 1), (H, 2), (H, 3), (H, 4), (H, 5), (H, 6),$

$$(T, 1), (T, 2), (T, 3), (T, 4), (T, 5), (T, 6)\}$$

Define the following two random variables on S:

$$X = \text{number of heads showing on coin}$$

$$Y = \text{number of spots showing on die}$$

Find $F_{X,Y}(1, 2)$. (For the outcome $s = (H, 3)$, we would write $X(s) = 1$ and $Y(s) = 3$.)

By definition, $F_{X,Y}(1, 2)$ is the sum of the probabilities associated with all the points in S for which $X(s) \leq 1$ and $Y(s) \leq 2$. A simple inspection gives $F_{X,Y}(1, 2) = \frac{1}{3}$:

$$F_{X,Y}(1, 2) = P(X \leq 1, Y \leq 2) = P\{(T, 1), (T, 2), (H, 1), (H, 2)\}$$
$$= \tfrac{1}{12} + \tfrac{1}{12} + \tfrac{1}{12} + \tfrac{1}{12}$$
$$= \tfrac{1}{3}$$

(What would the graph of $F_{X,Y}(t, u)$ look like?)

The mathematical properties of single-variable cdf's all have their counterparts in the bivariate case. Note the similarities between Theorem 4.3.1 and 4.2.1.

Theorem 4.3.1. The following statements are true for any joint cdf $F_{X,Y}$:

1. If $t_1 \leq t_2$ and $u_1 \leq u_2$, then $F_{X,Y}(t_1, u_1) \leq F_{X,Y}(t_2, u_2)$
2. $\lim_{\substack{x \to t^+ \\ y \to u^+}} F_{X,Y}(x, y) = F_{X,Y}(t, u)$
3. $\lim_{t \to -\infty} F_{X,Y}(t, u) = \lim_{u \to -\infty} F_{X,Y}(t, u) = 0$, $\lim_{\substack{t \to \infty \\ u \to \infty}} F_{X,Y}(t, u) = 1$

Proof

1. Since $(X \leq t_1, Y \leq u_1) \subset (X \leq t_2, Y \leq u_2)$,

$$F_{X,Y}(t_1, u_1) = P(X \leq t_1, Y \leq u_1) \leq P(X \leq t_2, Y \leq u_2)$$
$$= F_{X,Y}(t_2, u_2)$$

2. Let $\{(t_n, u_n)\}$ be any sequence of pairs of numbers with $t_n \to t$, $t < \cdots < t_2 < t_1$, and $u_n \to u$, $u < \cdots < u_2 < u_1$. It suffices to show that $\lim_{n \to \infty} F_{X,Y}(t_n, u_n) = F_{X,Y}(t, u)$. But

$$F_{X,Y}(t_n, u_n) = P(X \leq t_n, Y \leq u_n)$$
$$= P(X \leq t, Y \leq u) + P(t < X \leq t_n, u < Y \leq u_n)$$
$$= F_{X,Y}(t, u) + P(t < X \leq t_n, u < Y \leq u_n)$$

and $\lim_{n \to \infty} P(t < X \leq t_n, u < Y \leq u_n) = 0$ by Axiom 4 in Section 2.3.

3. The proof is left as an exercise.

Question 4.3.1 An urn contains four chips numbered 1 through 4. Two are drawn out at random without replacement. Let X denote the number on the first one drawn and Y denote the number on the second.
a. Find $F_{X,Y}(3, 3)$.
b. Compute the probability of the complement of the event: $X \leq 3$ and $Y \leq 2$.

Question 4.3.2 An urn contains 12 chips—4 red, 3 black, and 5 white. A sample of size 4 is to be drawn without replacement. Let X denote the number of white chips in the sample and Y denote the number of red chips. Find $F_{X,Y}(1, 2)$.

Question 4.3.3 Let the random variables X and Y denote the number of letters and the number of vowels, respectively, in each word of the following quotation:

Saepe creat molles aspera spina rosas

Tabulate $F_{X,Y}(t, u)$.

Question 4.3.4 Does the function

$$F(x, y) = 1 - e^{-x-y}, \qquad x > 0, y > 0$$

qualify as a joint cdf?

Question 4.3.5 Prove that $F_X(x) = \lim_{y \to \infty} F_{X,Y}(x, y)$ and, similarly, that $F_Y(y) = \lim_{x \to \infty} F_{X,Y}(x, y)$.

Question 4.3.6 Prove that

$$P(a < X \le b, c < Y \le d) = F_{X,Y}(b, d) - F_{X,Y}(a, d) - F_{X,Y}(b, c) + F_{X,Y}(a, c)$$

Next we consider the appropriate definitions for the bivariate pdf and investigate the relationship between joint pdf's and joint cdf's. Note, again, the similarity between these results and what was established earlier for the one-variable situation.

Definition 4.3.2. Let X and Y be discrete random variables. The *joint probability density function of X and Y*, denoted by $f_{X,Y}(x, y)$, is given by

$$f_{X,Y}(x, y) = P(X = x, Y = y)$$

The *joint cumulative distribution function of X and Y*, written $F_{X,Y}(t, u)$, is the sum

$$F_{X,Y}(t, u) = \sum_{x \le t} \sum_{y \le u} f_{X,Y}(x, y)$$

Example 4.3.2

A supermarket has two express lines. Let X and Y denote the numbers of customers in the first and in the second, respectively, at any given time. During nonrush hours, the joint pdf of X and Y is estimated to be the following:

		\(X\)			
		0	1	2	3
Y	0	0.1	0.2	0	0
	1	0.2	0.25	0.05	0
	2	0	0.05	0.05	0.025
	3	0	0	0.025	0.05

Find $P(|X - Y| = 1)$, the probability that X and Y differ by exactly 1. By definition,

$$P(|X - Y| = 1) = \sum_{|x-y|=1} \sum f_{X,Y}(x, y)$$

$$= f_{X,Y}(0, 1) + f_{X,Y}(1, 0) + f_{X,Y}(1, 2)$$

$$+ f_{X,Y}(2, 1) + f_{X,Y}(2, 3) + f_{X,Y}(3, 2)$$

$$= 0.2 + 0.2 + 0.05 + 0.05 + 0.025 + 0.025$$

$$= 0.45$$

(Would you expect $f_{X,Y}(x, y)$ to be symmetric? Would you expect the event $|X - Y| \geq 2$ to have zero probability?)

For X and Y both continuous, it seems reasonable (by analogy with the single-variable case) to try to define $f_{X,Y}(x, y)$ in terms of differentiability conditions on $F_{X,Y}(x, y)$. With that in mind, the formulation given in Definition 4.3.3 should come as no surprise.

Definition 4.3.3. A joint cdf, $F_{X,Y}(x, y)$, is said to be *continuous* if it is everywhere a continuous function and if

$$f_{X,Y}(x, y) = \frac{\partial^2}{\partial x \, \partial y} F_{X,Y}(x, y)$$

exists and is continuous, except possibly on a one-dimensional set. (The exceptional set is usually a finite union of horizontal and vertical straight lines.) We call $f_{X,Y}(x, y)$ the *joint pdf* of X and Y. Equivalently, $f_{X,Y}(x, y)$ is said to be the (*continuous*) *joint pdf* of X and Y if for all t and u,

$$F_{X,Y}(t, u) = \int_{-\infty}^{t} \int_{-\infty}^{u} f_{X,Y}(x, y) \, dy \, dx$$

Comment

We will have to accept without proof the equivalence of the two definitions just given for a continuous joint pdf: A formal verification would be mathematically demanding and somewhat peripheral for our purposes.

Example 4.3.3

A study shows that the daily number of hours X a teenager watches television and the daily number of hours Y he or she works on homework are approximated by the joint pdf

$$f_{X,Y}(x, y) = xye^{-(x+y)}, \qquad x > 0, y > 0$$

What is the probability a teenager chosen at random spends at least twice as much time watching television as working on homework?

The region R in the xy-plane corresponding to the event "$X \geq 2Y$" is

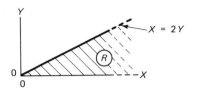

Figure 4.3.1

shown in Figure 4.3.1. It follows that $P(X \geq 2Y)$ is the volume under $f_{X,Y}(x, y)$ above the region R (recall the discussion of two-variable geometric probability in Section 2.5):

$$P(X \geq 2Y) = \int_0^\infty \int_0^{x/2} xye^{-(x+y)} \, dy \, dx$$

Separating variables, we can write

$$P(X \geq 2Y) = \int_0^\infty xe^{-x} \left[\int_0^{x/2} ye^{-y} \, dy \right] dx$$

The double integral eventually reduces to 7/27:

$$P(X \geq 2Y) = \int_0^\infty xe^{-x} \left[1 - \left(\frac{x}{2} + 1 \right) e^{-x/2} \right] dx$$

$$= \int_0^\infty xe^{-x} \, dx - \int_0^\infty \frac{x^2}{2} e^{-3x/2} \, dx - \int_0^\infty xe^{-3x/2} \, dx$$

$$= 1 - \frac{16}{54} - \frac{4}{9}$$

$$= \frac{7}{27}$$

(Derive an expression for the joint cdf, $F_{X,Y}(t, u)$.)

Question 4.3.7 An urn contains four red chips, three white chips, and two blue chips. A random sample of size 3 is to be drawn without replacement. Let X denote the number of white chips in the sample and Y denote the number of blue.
a. Write a formula giving the joint pdf of X and Y.
b. Tabulate the joint cdf of X and Y.

Question 4.3.8 An advisor looks over the schedules of her 50 students to see how many math and science courses each has registered for in the coming semester. The tally is shown below.

		Number of math courses (X)		
		0	1	2
Number	0	11	6	4
of science	1	9	10	3
courses (Y)	2	5	0	2

What is the probability a student selected at random will have signed up for more math courses than science courses?

Question 4.3.9 A point is chosen at random from the interior of a circle whose equation is $x^2 + y^2 \leq 4$. Let the random variables X and Y denote the x- and y-coordinates of the sampled point. Find $f_{X,Y}(x, y)$.

Joint pdf's and single-variable pdf's are, themselves, intimately related: Given an $f_{X,Y}(x, y)$, we can "recover" the individual pdf's for X and Y by integrating out (or summing over) the unwanted variable. The necessity to do this arises quite often in some of the proofs we will see in later chapters.

> **Theorem 4.3.2.** Let X and Y be discrete random variables with joint pdf $f_{X,Y}(x, y)$. The *individual* pdf's for X and Y—$f_X(x)$ and $g_Y(y)$, respectively—can be determined from the joint pdf by an appropriate summation:
>
> $$f_X(x) = \sum_{\text{all } y} f_{X,Y}(x, y) \qquad f_Y(y) = \sum_{\text{all } x} f_{X,Y}(x, y)$$
>
> Similarly, if X and Y are both continuous, $f_X(x)$ and $f_Y(y)$ can be determined from $f_{X,Y}(x, y)$ by an appropriate integration:
>
> $$f_X(x) = \int_{-\infty}^{\infty} f_{X,Y}(x, y)\, dy \qquad f_Y(y) = \int_{-\infty}^{\infty} f_{X,Y}(x, y)\, dx$$

Proof. The proof will be given for the discrete case only. The analogous result for continuous X and Y is derived in much the same way.

By definition,

$$f_X(x) = P(X = x) = P\left(\bigcup_{\text{all } y} (X = x, Y = y) \right)$$

But the events $(X = x, Y = y_i)$ and $(X = x, Y = y_j)$ are mutually exclusive for $i \neq j$, implying that

$$P\left(\bigcup_{\text{all } y} (X = x, Y = y) \right) = \sum_{\text{all } y} P(X = x, Y = y)$$

Therefore,

$$f_X(x) = \sum_{\text{all } y} P(X = x, Y = y) = \sum_{\text{all } y} f_{X,Y}(x, y)$$

Example 4.3.4

Consider an experiment consisting of three flips of a fair coin. Let X denote the number of heads on the last flip and Y denote the total number of heads for the three tosses. Table 4.3.1 shows the (x, y)-value associated with each of the eight possible outcomes. Find $f_X(x)$ and $f_Y(y)$.

TABLE 4.3.1

Outcome	(x, y)	$P(X = x, Y = y)$
(H, H, H)	$(1, 3)$	$\frac{1}{8}$
(T, H, H)	$(1, 2)$	$\frac{1}{8}$
(H, T, H)	$(1, 2)$	$\frac{1}{8}$
(H, H, T)	$(0, 2)$	$\frac{1}{8}$
(T, T, H)	$(1, 1)$	$\frac{1}{8}$
(T, H, T)	$(0, 1)$	$\frac{1}{8}$
(H, T, T)	$(0, 1)$	$\frac{1}{8}$
(T, T, T)	$(0, 0)$	$\frac{1}{8}$

By appropriately combining the second and third columns of Table 4.3.1, we can write the joint pdf of X and Y as the 2×4 matrix in Table 4.3.2.

TABLE 4.3.2

				Y		
		0	1	2	3	$f_X(x)$
	0	$\frac{1}{8}$	$\frac{1}{4}$	$\frac{1}{8}$	0	$\frac{1}{2}$
X	1	0	$\frac{1}{8}$	$\frac{1}{4}$	$\frac{1}{8}$	$\frac{1}{2}$
	$f_Y(y)$	$\frac{1}{8}$	$\frac{3}{8}$	$\frac{3}{8}$	$\frac{1}{8}$	

Note that summing across the rows gives $f_X(x)$, while summing down the columns yields $f_Y(y)$.

Comment

When $f_{X,Y}(x, y)$ is written as a matrix, the individual densities will appear, as they do in Table 4.3.2, as "margins." For this reason, $f_X(x)$ and $f_Y(y)$, in the context of a joint pdf, are often referred to as *marginal densities*. The use of the word *marginal*, though, is solely for emphasis and clarity—there is absolutely no difference between a pdf and a marginal pdf.

Comment

There are cdf's not explicitly covered by Definitions 4.3.2 and 4.3.3. For example, X may be continuous and Y discrete. In such situations, we make the obvious modifications. In particular, the joint cdf of a continuous X and a discrete Y is given by

$$F_{X,Y}(t, u) = \int_{-\infty}^{t} \left(\sum_{y \le u} f_{X,Y}(x, y) \right) dx$$

where $f_{X,Y}(x, y)$ is the joint pdf of X and Y. Case Study 4.3 in Section 4.7 examines a pair of variables of this sort in more detail.

Question 4.3.10 Suppose two fair dice are tossed one time. Let X denote the number of 2s that appear and Y be the number of 3s. Write out the matrix giving the joint probability density function for X and Y. Suppose a third random variable Z is defined, where $Z = X + Y$. Use $f_{X,Y}(x, y)$ to find $f_Z(z)$. (We will examine this sort of problem—finding the pdf for a sum—in considerable detail in Section 4.5.)

Question 4.3.11 From Table 4.3.2, find (a) $P(X = 0$ and $Y \leq 1)$ and (b) $P(XY \leq 1)$.

Question 4.3.12 Suppose $f_{X,Y}(x, y) = cxy$ at the points $(1, 1)$, $(2, 1)$, $(2, 2)$, and $(3, 1)$ and is 0 elsewhere. Tabulate $f_X(x)$.

Question 4.3.13 Find $f_X(x)$ and $f_Y(y)$ if the joint pdf for X and Y is

$$f_{X,Y}(x, y) = \frac{1}{x}, \qquad 0 < y < x, 0 < x < 1$$

Question 4.3.14 Suppose X and Y have a joint pdf given by

$$f_{X,Y}(x, y) = 6x, \qquad 0 < x < 1, 0 < y < 1 - x$$

Find $f_X(x)$ and $f_Y(y)$. Sketch (a) the joint pdf and (b) the two marginals.

Question 4.3.15 For each of the following joint pdf's, $f_{X,Y}(x, y)$, find $f_X(x)$ and $f_Y(y)$.
 a. $f_{X,Y}(x, y) = \frac{1}{2}$, $0 < x < 2, 0 < y < 1$
 b. $f_{X,Y}(x, y) = xye^{-(x+y)}$, $x > 0, y > 0$
 c. $f_{X,Y}(x, y) = 2e^{-(x+y)}$, $0 < x < y, 0 < y$
 d. $f_{X,Y}(x, y) = \frac{3}{2}y^2$, $0 \leq x \leq 2, 0 \leq y \leq 1$
 e. $f_{X,Y}(x, y) = c(x + y)$, $0 < x < 1, 0 < y < 1$
 f. $f_{X,Y}(x, y) = 2$, $0 < x < y < 1$
 g. $f_{X,Y}(x, y) = ye^{-xy-y}$, $x > 0, y > 0$
 h. $f_{X,Y}(x, y) = 4xy$, $0 < x < 1, 0 < y < 1$
 i. $f_{X,Y}(x, y) = \frac{2}{3}(x + 2y)$, $0 < x < 1, 0 < y < 1$

Question 4.3.16 Find $f_X(x)$ if

$$f_{X,Y}(x, y) = \frac{1}{\sqrt{4xy}}$$

for $0 < x < 1$ and $0 < y < x$.

Question 4.3.17 Complete the square in the exponent to find the marginal pdf's of X and Y if their joint pdf has the form

$$f_{X,Y}(x, y) = k \cdot e^{-(x^2-xy+y^2)/2}, \qquad -\infty < x < \infty, -\infty < y < \infty$$

Question 4.3.18 The campus recruiter for an international conglomerate classifies the large number of students she interviews into three categories—the lower quarter, the middle half, and the upper quarter. If she meets six students on a given morning, what is the probability they will be evenly divided among the three categories? What is the marginal probability that exactly two will belong to the middle half? What implicit assumption are you making here?

Question 4.3.19 Suppose (X, Y) and (U, V) are two sets of jointly distributed random variables. If $f_X(x) = f_U(u)$ and $f_Y(y) = f_V(v)$, will it necessarily be true that $f_{X,Y}(x, y) = f_{U,V}(u, v)$?

The next several examples point out some of the typical integration problems encountered when dealing with continuous bivariate pdf's. The first two are framed in the context of a two-component electronic system. Notice how the assumptions made about the relationship between the components affect the limits of the double integral of $f_{X,Y}(x, y)$.

Example 4.3.5

Consider the electronic system mentioned in the beginning of this section. Suppose the two components operate independently and have identical performance characteristics. Let X and Y be random variables denoting their life spans. Experience with wear-out times suggests that in certain cases a good choice for $f_{X,Y}(x, y)$ would be

$$f_{X,Y}(x, y) = \begin{cases} \lambda^2 e^{-\lambda(x+y)} & x \geq 0, y \geq 0 \\ 0 & \text{otherwise} \end{cases}$$

where λ is some constant greater than 0.

Suppose the manufacturer advertises a money-back guarantee if the system fails to last for more than 1000 h. What are the chances that a given system will be returned for a refund?

Since the system fails only if both components fail,

$$P(\text{refund}) = P(X \leq 1000, Y \leq 1000)$$

$$= \int_0^{1000} \int_0^{1000} \lambda^2 e^{-\lambda(x+y)} \, dy \, dx$$

$$= \int_0^{1000} \left(\int_0^{1000} \lambda^2 e^{-\lambda x} e^{-\lambda y} \, dy \right) dx$$

The integration in parentheses is done with respect to y, with x being treated as a constant. The expression $\lambda e^{-\lambda x}$ can therefore be factored out of the y-integration, so

$$\int_0^{1000} \left(\int_0^{1000} \lambda^2 e^{-\lambda x} e^{-\lambda y} \, dy \right) dx = \int_0^{1000} \lambda e^{-\lambda x} \left(\int_0^{1000} \lambda e^{-\lambda x} \, dy \right) dx$$

$$= \int_0^{1000} \lambda e^{-\lambda x} (1 - e^{-\lambda \cdot 1000}) \, dx$$

$$= (1 - e^{-\lambda \cdot 1000}) \int_0^{1000} \lambda e^{-\lambda x} \, dx$$

Evaluating the final integral and multiplying by the factor that precedes it, we find

$$P(\text{refund}) = (1 - e^{-1000\lambda})^2$$

Question 4.3.20 Suppose X and Y have a bivariate uniform density over the unit square:

$$f_{X,Y}(x, y) = c$$

for $0 < x < 1$ and $0 < y < 1$.
a. Find c.
b. Find $P(0 < X < \frac{1}{2}, 0 < Y < \frac{1}{4})$.

Question 4.3.21 Suppose X and Y are jointly distributed according to the pdf

$$f_{X,Y}(x, y) = \frac{x^2}{52} + \frac{y}{78}, \qquad 0 \le x \le 3, 0 \le y \le 4$$

a. Find $P(|X - 1| \le \frac{1}{2})$.
b. Find $P(0 \le X \le 1, 0 \le Y \le 2)$.

Question 4.3.22 Let X and Y be continuous with joint pdf

$$f_{X,Y}(x, y) = 2e^{-x-y}, \qquad 0 < x < y$$

Find $P(1 < x < 2, 1 < y < 2)$.

Question 4.3.23 Suppose $f_{X,Y}(x, y) = xye^{-(x+y)}$, $x > 0$ and $y > 0$. Prove for any real numbers a, b, c, and d that

$$P(a < X < b, c < Y < d) = P(a < X < b) \cdot P(c < Y < d)$$

Example 4.3.5 was not particularly unpleasant, as double integrals go, because $f_{X,Y}(x, y)$ factored into a product with one factor involving only x and the other depending solely on y. Another simplifying feature was the rectangularity of the region of integration. A more-complicated region of integration emerges if one component is kept in reserve, being activated only when the first one fails.

Example 4.3.6

Let X, Y, and $f_{X,Y}(x, y)$ be defined as they were in Example 4.3.5. However, suppose the system itself is modified so that one component is kept on reserve and activated only when the other needs replacing. As before, the system fails only when both components burn out, but now the component lives are cumulative. Again we seek the probability that the system fails in 1000 h or less. Translated into a statement about X and Y, the probability of a refund reduces to

$$P(\text{refund}) = P(X + Y \le 1000) = \iint_R \lambda^2 e^{-\lambda(x+y)} \, dy \, dx$$

where R is the shaded region shown in Figure 4.3.2.

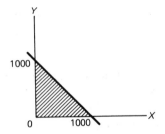

Figure 4.3.2

As in the previous example, the double integral is evaluated by first holding x constant. Here, though, the upper limit of the integration for y depends on x: for x fixed, y varies between 0 and $1000 - x$ (see Figure 4.3.3). Thus,

$$\iint_R f_{X,Y}(x, y) \, dy \, dx = \int_0^{1000} \left(\int_0^{1000-x} \lambda^2 e^{-\lambda x} e^{-\lambda y} \, dy \right) dx$$

$$= \int_0^{1000} \lambda e^{-\lambda x} \left(\int_0^{1000-x} \lambda e^{-\lambda y} \, dy \right) dx$$

$$= \int_0^{1000} \lambda e^{-\lambda x} (1 - e^{-\lambda(1000-x)}) \, dx$$

$$= \int_0^{1000} (\lambda e^{-\lambda x} - \lambda e^{-\lambda \cdot 1000}) \, dx$$

$$= 1 - e^{-\lambda \cdot 1000} - 1000 \lambda e^{-\lambda \cdot 1000}$$

Collecting terms gives

$$P(\text{system fails in 1000 h or less}) = 1 - (1 + 1000\lambda)e^{-1000\lambda}$$

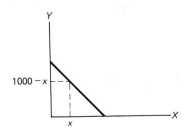

Figure 4.3.3

Question 4.3.24 The random variables X and Y are uniformly distributed over a rectangle whose corners have coordinates $(2, 2)$, $(4, 2)$, $(2, 6)$, and $(4, 6)$. Find the probability that X is at least twice as large as Y.

Question 4.3.25 Find $P(X + Y \geq \frac{1}{2})$ if X and Y have the joint pdf

$$f_{X,Y}(x, y) = 4xy$$

for $0 \leq x \leq 1$ and $0 \leq y \leq 1$.

Question 4.3.26 Find $P(X + Y \leq 1)$ if X and Y have the joint cdf

$$F_{X,Y}(x, y) = (1 - e^{-x})(1 - e^{-y})$$

for $x > 0$ and $y > 0$.

Question 4.3.27 Let X and Y have the bivariate uniform density over the unit square. Let $Z = X + Y$. Find $f_Z(z)$. (*Hint:* Consider two cases, $0 < z \leq 1$ and $1 < z \leq 2$. For each case, draw a diagram showing the region of integration.)

Qeustion 4.3.28 Suppose X and Y have the joint pdf

$$f_{X,Y}(x, y) = xe^{-x} \cdot \tfrac{1}{2}e^{-y/2}, \qquad x > 0, y > 0$$

Find $P(X + Y \leq 1)$.

Question 4.3.29 A number is chosen at random from the interval $(0, 3)$. A second number is chosen at random from the interval $(0, 4)$. What is the 80th percentile of the sum of the two numbers?

The last example in this section deals with a joint pdf that does *not* factor into a function of x and a function of y. (Whether or not a joint pdf can be factored has important implications. If it can be, the two random variables are said to be *independent*, meaning that a knowledge of one is of no help in predicting the value of the other. The notion of independence is introduced formally in Section 4.4; its consequences will be evident throughout the remainder of the text.)

Example 4.3.7

Suppose X and Y are two random variables jointly distributed over the first quadrant of the xy-plane according to the pdf

$$f_{X,Y}(x, y) = \begin{cases} y^2 e^{-y(x+1)} & x \geq 0, y \geq 0 \\ 0 & \text{elsewhere} \end{cases}$$

Show that $f_{X,Y}(x, y) \neq f_X(x) \cdot f_Y(y)$.
First, consider $f_X(x)$:

$$f_X(x) = \int_{-\infty}^{\infty} f_{X,Y}(x, y)\, dy = \int_0^{\infty} y^2 e^{-y(x+1)}\, dy$$

In the integrand, substitute,

$$u = y(x + 1)$$

making $du = (x + 1)\, dy$. This gives

$$f_X(x) = \frac{1}{x + 1} \int_0^{\infty} \frac{u^2}{(x + 1)^2} e^{-u}\, du = \frac{1}{(x + 1)^3} \int_0^{\infty} u^2 e^{-u}\, du$$

After applying integration by parts (twice) to $\int_0^{\infty} u^2 e^{-u}\, du$, we obtain

$$f_X(x) = \frac{1}{(x + 1)^3}[-u^2 e^{-u} - 2ue^{-u} - 2e^{-u}]_0^{\infty}$$

$$= \frac{1}{(x + 1)^3} \left[2 - \lim_{u \to \infty} \left(\frac{u^2}{e^u} + \frac{2u}{e^u} + \frac{2}{e^u} \right) \right]$$

$$= \frac{2}{(x + 1)^3}$$

Finding $f_Y(y)$ is a bit easier:

$$f_Y(y) = \int_{-\infty}^{\infty} f_{X,Y}(x, y) \, dx = \int_0^{\infty} y^2 e^{-y(x+1)} \, dx$$

$$= y^2 e^{-y} \int_0^{\infty} e^{-yx} \, dx = y^2 e^{-y} \left(\frac{1}{y} \right) [-e^{-yx}]_0^{\infty}$$

$$= y e^{-y}$$

By inspection, then, $f_{X,Y}(x, y) \neq f_X(x) \cdot f_Y(y)$.

Question 4.3.30 Let X and Y be random variables with the joint pdf

$$f_{X,Y}(x, y) = \begin{cases} \dfrac{1}{x} & 0 < y < x, 0 < x < 1 \\ 0 & \text{elsewhere} \end{cases}$$

Verify that the double integral over $f_{X,Y}(x, y)$ is 1 and show that X and Y are not independent (see Question 4.3.13).

Question 4.3.31 Given the joint pdf $f_{X,Y}(x, y) = 2x + y - 2xy$ for $0 < x < 1$ and $0 < y < 1$, find numbers a, b, c, and d such that

$$P(a < X < b, c < Y < d) \neq P(a < X < b) \cdot P(c < Y < d)$$

The generalization of the two-variable case to the n-variable case is conceptually straightforward, although the mathematics becomes much more formidable. In what follows, we will be assuming that each of n random variables, $Y_1, Y_2, \ldots, Y_n$, is defined on the same sample space S.

Definition 4.3.4. The *joint cdf* of $Y_1, Y_2, \ldots, Y_n$ is the probability that each Y_i is less than or equal to a specified y_i:

$$F_{Y_1, Y_2, \ldots, Y_n}(y_1, y_2, \ldots, y_n) = P(Y_1 \leq y_1, Y_2 \leq y_2, \ldots, Y_n \leq y_n)$$

$$= \prod_{i=1}^{n} P(Y_i \leq y_i)$$

If the Y_i's are discrete, their *joint pdf* is given by

$$f_{Y_1, Y_2, \ldots, Y_n}(y_1, \ldots, y_n) = P(Y_1 = y_1, Y_2 = y_2, \ldots, Y_n = y_n)$$

We call the Y_i's *jointly continuous* if $F_{Y_1, Y_2, \ldots, Y_n}$ is continuous everywhere and the partial derivative $f_{Y_1, Y_2, \ldots, Y_n} = \dfrac{\partial^n}{\partial y_1 \partial y_2 \cdots \partial y_n} F_{Y_1, \ldots, Y_n}$ exists and is continuous, except possibly in a set of dimension 1. When the derivative does exist,

$$P((Y_1, Y_2, \ldots, Y_n) \in B) = \iint \cdots \int_B f_{Y_1, Y_2, \ldots, Y_n}(y_1, y_2, \ldots, y_n) \, dy_1 \cdots dy_n$$

for any set B for which the multiple integral is defined. The function $f_{Y_1, Y_2, \ldots, Y_n}$ is the *joint pdf* of the Y_i's.

Theorem 4.3.3. Let $F_{Y_1, Y_2, \ldots, Y_n} (y_1, y_2, \ldots, y_n)$ be a joint cdf.

1. If $t_i \le u_i$, $i = 1, \ldots, n$, then
$$F_{Y_1, Y_2, \ldots, Y_n} (t_1, t_2, \ldots, t_n) \le F_{Y_1, Y_2, \ldots, Y_n} (u_1, u_2, \ldots, u_n)$$

2. $\lim\limits_{\substack{y_i \to t_i^+ \\ i=1, \ldots, n}} F_{Y_1, Y_2, \ldots, Y_n} (y_1, \ldots, y_n) = F_{Y_1, Y_2, \ldots, Y_n} (t_1, t_2, \ldots, t_n).$

3. For any i, $1 \le i \le n$, $\lim\limits_{y_i \to -\infty} F_{Y_1, Y_2, \ldots, Y_n} (y_1, \ldots, y_i, \ldots, y_n) = 0.$

4. $\lim\limits_{\substack{y_i \to \infty \\ i=1, \ldots, n}} F_{Y_1, Y_2, \ldots, Y_n} (y_1, y_2, \ldots, y_n) = 1.$

In the case of two random variables, we found the marginal pdf for Y, by summing or integrating over X. Given n random variables, the analogous result would be a procedure for recovering the marginal pdf of some subset of r of those n. Theorem 4.3.4 confirms the generalization our intuition would suggest. Its proof is left as an exercise.

Theorem 4.3.4. Partition the set $\{1, 2, \ldots, n\}$ into $\{i_1, i_2, \ldots, i_r\}$ and $\{j_1, j_2, \ldots, j_{n-r}\}$. In the discrete case, the marginal pdf of $Y_{i_1}, Y_{i_2}, \ldots, Y_{i_r}$ is the $(n - r)$-fold summation

$$f_{Y_{i_1}, Y_{i_2}, \ldots, Y_{i_r}} = \sum_{y_{j_1}} \sum_{y_{j_2}} \cdots \sum_{y_{j_{n-r}}} f_{Y_1, Y_2, \ldots, Y_n}$$

In the continuous case,

$$f_{Y_{i_1}, Y_{i_2}, \ldots, Y_{i_r}} = \int_{-\infty}^{\infty} \int_{-\infty}^{\infty} \cdots \int_{-\infty}^{\infty} f_{Y_1, Y_2, \ldots, Y_n} \, dj_1 dj_2 \cdots dj_{n-r}$$

Question 4.3.32 A certain brand of fluorescent bulbs will last, on the average, 1000 h. Suppose four of these bulbs are installed in an office. What is the probability that all four are still functioning after 1050 h? If X_i denotes the ith bulb's life, assume that

$$f_{X_1, X_2, X_3, X_4}(x_1, x_2, x_3, x_4) = \prod_{i=1}^{4} \left(\frac{1}{1000} \right) e^{-x_i/1000}$$

for $x_i > 0$, $i = 1, 2, 3, 4$.

Question 4.3.33 A hand of six cards is dealt from a standard poker deck. Let X denote the number of aces, Y the number of kings, and Z the number of queens.

a. Write a formula for $f_{X,Y,Z}(x, y, z)$.

b. Find $f_{X,Y}(x, y)$ and $f_{X,Z}(x, z)$.

Question 4.3.34 Suppose the random variables X, Y, and Z have the multivariate pdf

$$f_{X,Y,Z}(x, y, z) = (x + y)e^{-z}$$

for $0 < x < 1$, $0 < y < 1$, and $z > 0$. Find (a) $f_{X,Y}(x, y)$, (b) $f_{Y,Z}(y, z)$, and (c) $f_Z(z)$.

Question 4.3.35 Fifty observations are drawn from the pdf

$$f_Y(y) = 6y(1 - y), \qquad 0 < y < 1$$

Let X_i be the number of observations lying in the interval $((i - 1)/4, i/4)$, $i = 1, 2, 3, 4$. Write a formula for $f_{X_1,X_2,X_3,X_4}(10, 15, 15, 10)$.

Question 4.3.36 The four random variables W, X, Y, and Z have the multivariate pdf

$$f_{W,X,Y,Z}(w, x, y, z) = 16wxyz$$

for $0 < w < 1$, $0 < x < 1$, $0 < y < 1$, and $0 < z < 1$. Find the marginal pdf, $f_{W,X}(w, x)$, and use it to compute $P(0 < W < \frac{1}{2}, \frac{1}{2} < X < 1)$.

Question 4.3.37 The joint pdf of X, Y, and Z is

$$f_{X,Y,Z}(x, y, z) = \frac{1}{24} \cdot e^{-(x/2)-(y/3)-(z/4)}$$

for $x > 0$, $y > 0$, and $z > 0$. Find $P(X + Y + Z \le 1)$.

Question 4.3.38 Give a formula for

$$P(a_1 < Y_1 \le b_1, a_2 < Y_2 \le b_2, \ldots, a_n < Y_n \le b_n)$$

in terms of $F_{Y_1, Y_2, \ldots, Y_n}$. (*Hint:* Use Theorem 3.7.1.)

4.4 INDEPENDENT RANDOM VARIABLES

The concept of independent events that was introduced in Section 2.8 leads quite naturally to a similar definition for independent random variables.

Definition 4.4.1. Random variables X and Y are said to be *independent* if for any two intervals A and B,

$$P(X \in A, Y \in B) = P((X, Y) \in A \otimes B) = P(X \in A) \cdot P(Y \in B)$$

Although it *defines* independence, Definition 4.4.1 is of little value in *establishing* independence. Given a continuous X and a continuous Y, it would be impossible to check all intervals A and B to verify the relationship between $P(X \in A, Y \in B)$ and $P(X \in A) \cdot P(Y \in B)$. A more-workable characterization of independence is provided in Theorem 4.4.1.

> **Theorem 4.4.1.** Two random variables X and Y are independent if and only if
>
> $$f_{X,Y}(x, y) = f_X(x) \cdot f_Y(y)$$
>
> for all x and y.

Proof. We will show that if X and Y are independent, their joint pdf factors into a product of their marginal pdf's. The converse is left as an exercise.

Suppose X and Y are discrete and independent; let a and b be any two numbers. Then

$$f_{X,Y}(a, b) = P(X = a, Y = b) = P(X = a) \cdot P(Y = b) = f_X(a) \cdot f_Y(b)$$

Verifying the result in the continuous case is a bit more difficult. Let X and Y be continuous and independent, and take A and B to be any two intervals. We can write

$$\int_A \left(\int_B f_{X,Y}(x, y) \, dy \right) dx = \iint_{A \otimes B} f_{X,Y}(x, y) \, dy \, dx = P(X \in A, Y \in B)$$

$$= P(X \in A) \cdot P(Y \in B)$$

$$= \int_A f_X(x) \, dx \cdot \int_B f_Y(y) \, dy$$

$$= \int_A \left(\int_B f_X(x) \cdot f_Y(y) \, dy \right) dx$$

Note that the functions

$$\int_B f_{X,Y}(x, y) \, dy \quad \text{and} \quad \int_B f_X(x) f_Y(y) \, dy$$

have equal integrals over every interval A; therefore, by a theorem in calculus, the two are equal. But then it follows that $f_{X,Y}(x, y)$ and $f_X(x) \cdot f_Y(y)$ have equal integrals over every interval B, so by the same theorem, *they* are equal.

In Chapter 2, extending the notion of independence from *two* events to n events proved to be something of a problem: The independence of each subset of the n events had to be checked separately (recall Definition 2.8.2). This is not necessary in the case of random variables; to generalize Theorem 4.4.1, we need only establish that the joint pdf factors into a product of the n marginals.

> **Definition 4.4.2.** The n random variables $X_1, X_2, \ldots, X_n$ are said to be *independent* if, for all $x_1, x_2, \ldots, x_n$,
>
> $$f_{X_1,X_2,\ldots,X_n}(x_1, x_2, \ldots, x_n) = f_{X_1}(x_1) \cdot f_{X_2}(x_2) \cdots f_{X_n}(x_n)$$

An important special case of Definition 4.4.2 arises when all n independent random variables have the same pdf, $f_X(x)$. The set $X_1, X_2, \ldots, X_n$ is then called a *random sample of size n.* If S is the sample space for each of the X_i's, the sample space T for a random sample of size n is the Cartesian product introduced in Section 2.9: $T = S \otimes S \otimes \cdots \otimes S$ (n times). For each i, X_i is defined on T according to the relationship

$$X_i((s_1, s_2, \ldots, s_i, \ldots, s_n)) = X(s_i) \qquad (4.4.1)$$

Example 4.4.1 illustrates Equation 4.4.1 and Definition 4.4.2 for a Bernoulli random sample of size 2.

Example 4.4.1

Let $S = (s, f)$ be the usual Bernoulli sample space with $P(\{s\}) = p$. Define $X(s) = 1$ and $X(f) = 0$. Suppose $n = 2$ such trials are observed, with the number of successes on the ith trial being denoted by the random variable X_i, $i = 1, 2$. Set $T = S \otimes S$ and let the product probability function be defined as it was in Section 2.7—that is, $P((s, f)) = pq$, $P((s, s)) = p^2$, and so on.

We define X_1 on T by the equations $X_1((s, f)) = X_1((s, s)) = X(s) = 1$ and $X_1((f, s)) = X_1((f, f)) = X(f) = 0$. A similar set of equations define X_2: $X_2((f, s)) = X_2((s, s)) = X(s) = 1$ and $X_2((s, f)) = X_2((f, f)) = X(f) = 0$. It is a simple matter to show that

$$f_{X_1}(x) = f_{X_2}(x) = f_X(x)$$

and

$$f_{X_1, X_2}(x_1, x_2) = f_{X_1}(x_1) \cdot f_{X_2}(x_2) = f_X(x_1) \cdot f_X(x_2)$$

(Verify this last result numerically for all x_1 and x_2 when $p = \frac{1}{4}$.)

Example 4.4.2

A chip is to be drawn at random from each of k urns, each holding n chips numbered 1 through n. What is the probability all k chips will bear the same number?

If $X_1, X_2, \ldots, X_k$ denote the numbers on the 1st, 2nd, $\ldots$, kth chips, respectively, we are looking for the probability that $X_1 = X_2 = \cdots X_k$. In terms of the joint pdf,

$$P(X_1 = X_2 = \cdots = X_k) = \sum_{x_1 = x_2 = \cdots = x_k} f_{X_1, X_2, \ldots, X_k}(x_1, x_2, \ldots, x_k)$$

Each of the selections is obviously independent of all the others, so the joint pdf factors according to Definition 4.4.2, and we can write

$$P(X_1 = X_2 = \cdots = X_k) = \sum_{i=1}^{n} f_{X_1}(x_i) \cdot f_{X_2}(x_i) \cdots f_{X_k}(x_i)$$

$$= n \cdot \left(\frac{1}{n} \cdot \frac{1}{n} \cdots \frac{1}{n} \right) = \frac{1}{n^{k-1}}$$

An interesting application of Example 4.4.2 occurs in the field of cryptanalysis, the science of decoding secret writing. Suppose we imagine each of the chips in an urn to be one of the 26 letters from A to Z. If one such chip is drawn from each of *two* urns, the probability of the letters matching is $26(\frac{1}{26})(\frac{1}{26}) = (\frac{1}{26})$, or 0.0385. If, on the other hand, the chips are *not* equally likely but have probabilities of being drawn equal to their relative frequencies in English text (recall Table 2.4.1), the probability of matching increases to 0.0664:

$$P(\text{two letters match}) = P(\text{two } A\text{'s}) + P(\text{two } B\text{'s}) + \cdots + P(\text{two } Z\text{'s})$$

$$= P(A \text{ on 1st}) \cdot P(A \text{ on 2nd}) + \cdots$$

$$+ P(Z \text{ on 1st}) \cdot P(Z \text{ on 2nd})$$

$$= (0.0788)(0.0788) + \cdots + (0.0006)(0.0006)$$

$$= 0.0664$$

Now, recall the method of polyalphabetic substitution described in Case Study 3.1. The encoder defines a cycle (of length n) of permutations of the alphabet. Call the permutations $\alpha_1, \alpha_2, \ldots, \alpha_n$. The first letter of the message is encoded using permutation α_1; the second letter, α_2, and so on. (The permutation α_1, for example, might be a shift two letters to the right so an A in the original text becomes a C in the ciphertext, a B becomes a D, and so on.) The $(n + 1)$st letter would again use permutation α_1, and the cycle would repeat. Decoding a polyalphabetic substitution is obviously a formidable task, but the match probability of 0.0664 just derived can be helpful in getting the cryptanalyst started.

Suppose two messages are received that are believed to have been encoded using the same permutation cycle, $\alpha_1, \ldots \alpha_n$. However, the starting point, α_i, for each message is not known. As a way of bringing the two messages in phase, cryptanalysts will line the two messages up letter for letter and count the matches (see Figure 4.4.1). They will then slide one message relative to the other until the frequency of matches is on the order of 7 (rounding up from 6.64) per 100 letters. There are no guarantees, of course, but if the plaintext is in English, that particular alignment may very well be the point at which the permutation cycles are in phase. The two messages can then be appropriately joined to form one long message. (See (44) for a detailed history of cryptanalysis and its impact on diplomacy and warfare.)

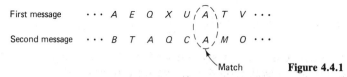

First message $\cdots$ A E Q X U A T V $\cdots$

Second message $\cdots$ B T A Q C A M O $\cdots$

Match   **Figure 4.4.1**

Question 4.4.1 Write the joint probability density function for a random sample of size n drawn from the exponential pdf $f_Y(y) = (1/\lambda)e^{-y/\lambda}$, $y > 0$.

Question 4.4.2 Two fair dice are tossed. Let X denote the number appearing on the first die and Y the number on the second. Show that X and Y are independent.

Question 4.4.3 A joint pdf is given by

$$f_{X_1, X_2}(x_1, x_2) = 12 x_1 x_2 (1 - x_2)$$

for $0 < x_1 < 1$ and $0 < x_2 < 1$. Are X_1 and X_2 independent?

Question 4.4.4 Let X_1 and X_2 be independent random variables each having the pdf

$$f_X(x) = \frac{x}{2}, \qquad 0 < x < 2$$

Find the joint cdf for X_1 and X_2.

Question 4.4.5 Let X and Y be random variables with joint pdf

$$f_{X,Y}(x, y) = k, \qquad 0 \le x \le 1; 0 \le y \le 1; 0 \le x + y \le 1$$

Give a geometric argument to show that X and Y are not independent.

Question 4.4.6 Prove that if X and Y are two independent random variables over a bounded region, then $U = g(X)$ and $V = h(Y)$ are also independent.

Question 4.4.7 If two random variables X and Y are defined over a region in the XY-plane that is *not* a rectangle (or union of rectangles) with sides parallel to the coordinate axes, can X and Y be independent?

4.5 COMBINING AND TRANSFORMING RANDOM VARIABLES

Frequently the random variables being measured in an experiment are not, themselves, the researcher's ultimate objective. What *is* of primary interest is some function of those variables. Example 4.3.6 is a case in point: there the life, Z, of an electronic system depended on the *sum*, $X + Y$, of two component lives. At issue, of course, was $f_Z(z)$. Other situations may require a change in the *scale* of a random variable—for example, suppose X is originally calibrated in degrees Fahrenheit but the forces of metric conversion demand its reexpression in degrees Celsius. How does the pdf of Y, the Celsius random variable, compare to that of X?

In cases such as these it is inefficient to compute the pdf's of the "new" random variables by returning to elementary principles. Easier methods are available. If Y is a given function of X—say, $Y = u(X)$—and $f_X(x)$ is known, we can find $f_Y(y)$ by examining the relationship between the pdf's or cdf's of X and Y.

The most-commonly occurring conversions from X to Y are the linear and quadratic transformations $Y = aX + b$ and $Y = X^2$. Formulas for the corresponding $f_Y(y)$'s are given in the next two theorems.

Theorem 4.5.1. Let X be a random variable with probability density function $f_X(x)$. Let $a \ne 0$ and b be constants, and define $Y = aX + b$.

1. If X is discrete,

$$f_Y(y) = f_X\left(\frac{y - b}{a}\right)$$

---

2. If X is continuous,

$$f_Y(y) = \frac{1}{|a|}f_X\left(\frac{y-b}{a}\right)$$

Proof. When X is discrete, we can get the pdf for Y almost immediately by substituting $aX + b$ for Y and simplifying:

$$f_Y(y) = P(Y = y) = P(aX + b = y)$$

$$= P\left(X = \frac{y-b}{a}\right) = f_X\left(\frac{y-b}{a}\right)$$

The continuous case is somewhat different and requires that the pdf be approached via the cdf. Clearly,

$$F_Y(y) = P(Y \le y) = P(aX + b \le y)$$

$$= P\left(X \le \frac{y-b}{a}\right) = F_X\left(\frac{y-b}{a}\right), \qquad \text{if } a > 0$$

Differentiating both sides of the equation

$$F_Y(y) = F_X\left(\frac{y-b}{a}\right)$$

gives the desired result,

$$f_Y(y) = \frac{1}{a}f_X\left(\frac{y-b}{a}\right)$$

The proof for $a < 0$ is left as an exercise.

Example 4.5.1

Suppose a random variable X has pdf

$$f_X(x) = \begin{cases} 6x(1-x) & 0 < x < 1 \\ 0 & \text{elsewhere} \end{cases}$$

Let Y be a second random variable defined by

$$Y = 2X + 1$$

What is the pdf for Y?

From Theorem 4.5.1,

$$f_Y(y) = \frac{1}{2} \cdot 6\left(\frac{y-1}{2}\right)\left(1 - \frac{y-1}{2}\right) = \frac{3}{4}(-y^2 + 4y - 3) \qquad (4.5.1)$$

Note that Equation 4.5.1 does not hold for *all* y: The two conditions $0 < x < 1$ and $Y = 2X + 1$ imply that $f_Y(y)$ will be nonzero only for y between 1 and 3.

The proper expression for $f_Y(y)$ is thus

$$f_Y(y) = \begin{cases} \frac{3}{4}(-y^2 + 4y - 3) & 1 < y < 3 \\ 0 & \text{elsewhere} \end{cases}$$

Question 4.5.1 An urn contains four red chips and three white chips. Two are drawn out at random without replacement. Let X denote the number of red chips in the sample. Find the pdf for Y, where $Y = 2X - 1$.

Question 4.5.2 The number of flaws, X, in a 1-ft length of metal cooling pipes has pdf $f_X(x) = e^{-\lambda} \cdot \lambda^x/x!$, $x = 0, 1, \ldots$. If each flaw requires a weld costing 10¢, then $Y = 0.10X$ represents the cost in dollars to repair a foot of piping. Find $f_Y(y)$.

Question 4.5.3 Suppose the random variable X takes on the values -1, 1, and 4, each with probability $\frac{1}{3}$, and is 0 elsewhere. Find the pdf of $Y = 3X + 2$.

Question 4.5.4 Let X have the uniform pdf over the unit interval. Find $f_Y(y)$, where $Y = -3X - 4$.

Question 4.5.5 If X has the exponential pdf,

$$f_X(x) = 2e^{-2x}, \qquad x > 0$$

find the cdf for $Y = 2X + 4$.

Question 4.5.6 If $f_X(x) = 6x(1 - x)$, $0 < x < 1$, find $f_Y(y)$, where $Y = 2X - 3$.

Question 4.5.7 Suppose X has the uniform pdf over the interval (a, b). What linear transformation of X represents a random variable having the uniform pdf over $(0, 1)$?

Theorem 4.5.2. Let X be a continuous random variable with pdf $f_X(x)$. Let $Y = X^2$. For $y > 0$,

$$f_Y(y) = \frac{1}{2\sqrt{y}} \left(f_X(\sqrt{y}) + f_X(-\sqrt{y}) \right)$$

Proof. Reprising our strategy in deriving Theorem 4.5.1, we will find a relationship between $F_X(x)$ and $F_Y(y)$ and then differentiate the latter to obtain $f_Y(y)$. Starting with the cdf for Y, we can write

$$F_Y(y) = P(Y \leq y) = P(X^2 \leq y) = P(-\sqrt{y} \leq X \leq \sqrt{y})$$
$$= P(-\sqrt{y} < X \leq \sqrt{y}) \qquad \text{(Why?)}$$
$$= F_X(\sqrt{y}) - F_X(-\sqrt{y})$$

Differentiating both sides of the last equation gives the result

$$f_Y(y) = f_X(\sqrt{y}) \cdot \frac{1}{2\sqrt{y}} - f_X(-\sqrt{y}) \cdot \frac{-1}{2\sqrt{y}}$$

$$= \frac{1}{2\sqrt{y}} \left(f_X(\sqrt{y}) + f_X(-\sqrt{y}) \right)$$

Example 4.5.2

Let X have the uniform density over the interval $(-1, 2)$:

$$f_X(x) = \begin{cases} \frac{1}{3} & -1 < x < 2 \\ 0 & \text{elsewhere} \end{cases}$$

Find the probability density function for Y, where $Y = X^2$.

This is not as straightforward a problem as it might first seem to be. From Theorem 4.5.2 we can easily write an expression for $f_Y(y)$:

$$f_Y(y) = \frac{1}{2\sqrt{y}} \left(f_X(\sqrt{y}) + f_X(-\sqrt{y}) \right) \tag{4.5.2}$$

but we need to be very careful in evaluating the two terms involving f_X. From what was given, $f_X(\sqrt{y})$ will be nonzero only if $-1 < \sqrt{y} < 2$ or, equivalently, when y is between 0 and 4. Similarly, $f_X(-\sqrt{y})$ will be nonzero when $-1 < -\sqrt{y} < 2$, but the latter translates into y-values between 0 and 1. To apply Equation 4.5.2, we need to consider separately the cases $0 \le y < 1$ and $1 \le y < 4$.

For $0 \le y < 1$,

$$f_Y(y) = \frac{1}{2\sqrt{y}} \left(\frac{1}{3} + \frac{1}{3} \right) = \frac{1}{3\sqrt{y}}$$

For $1 \le y < 4$,

$$f_Y(y) = \frac{1}{2\sqrt{y}} \left(\frac{1}{3} + 0 \right) = \frac{1}{6\sqrt{y}}$$

The final form for $f_Y(y)$ has three different expressions, depending on the value of y:

$$f_Y(y) = \begin{cases} \dfrac{1}{3\sqrt{y}} & 0 \le y < 1 \\ \dfrac{1}{6\sqrt{y}} & 1 \le y < 4 \\ 0 & \text{elsewhere} \end{cases}$$

Question 4.5.8 A random variable X has density function $f_X(x) = (\frac{3}{8})x^2$ over the interval $0 < x < 2$ (and 0 elsewhere). Suppose a circle is generated by a radius whose length is the value of X. Find the density function for Y, the *area* of the circle.

Question 4.5.9 If X has pdf

$$f_X(x) = \frac{1 + x}{2}, \quad -1 < x < 1$$

find $f_Y(y)$, where $Y = X^2$.

Question 4.5.10 If $f_X(x) = 2xe^{-x^2}$, for $x > 0$, find $f_Y(y)$, where $Y = X^2$.

Question 4.5.11 Let X have pdf $f_X(x) = x^2/9$, $0 < x < 3$. Let $Y = X^2$. Find $f_Y(y)$.

For single-variable transformations other than $Y = aX + b$ or $Y = X^2$, the best way to derive $f_Y(y)$ is to follow an argument similar to what was used in the proof of Theorem 4.5.1: Find an expression for $F_Y(y)$ by appropriately integrating f_X and then differentiate. The next two examples are cases in point.

Example 4.5.3

Assume the velocity of a gas molecule of mass m is a random variable X with pdf $f_X(x) = ax^2e^{-bx^2}$, $x > 0$, where a and b are constants depending on the gas. Find the pdf for the kinetic energy Y of such a molecule, where $Y = (m/2)X^2$.
Let $F_Y(y)$ be the cdf for Y. Then

$$F_Y(y) = P(Y \le y) = P\left(\frac{m}{2}X^2 \le y\right)$$

$$= P\left(X^2 \le \frac{2y}{m}\right)$$

Figure 4.5.1 is a diagram of the X-axis and the Y-axis. Note the interval along the X-axis that gets mapped into the event "$Y \le y$." It follows that

$$F_Y(y) = P\left(X^2 \le \frac{2y}{m}\right) = P\left(X \le \sqrt{\frac{2y}{m}}\right)$$

$$= \int_0^{\sqrt{2y/m}} ax^2e^{-bx^2}\, dx$$

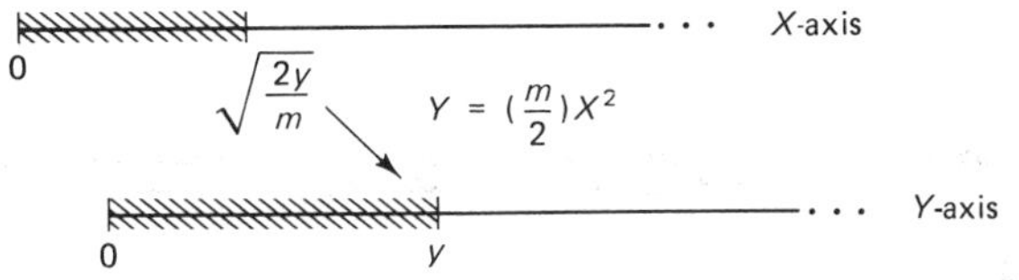

$$Y = \left(\frac{m}{2}\right)X^2$$

Figure 4.5.1

Finally, from a theorem in calculus,

$$f_Y(y) = F_Y'(y) = a\left(\sqrt{\frac{2y}{m}}\right)^2 e^{-b(\sqrt{2y/m})^2} \cdot \frac{d}{dy}\left(\sqrt{\frac{2y}{m}}\right)$$

$$= a \cdot \frac{2y}{m} \cdot e^{-b(2y/m)} \cdot \sqrt{\frac{2}{m}} \cdot \frac{1}{2} \cdot y^{-1/2}$$

$$= \frac{a\sqrt{2}}{m^{3/2}} \cdot \sqrt{y}\, e^{-2by/m}, \qquad y > 0$$

Example 4.5.4

In doing Monte Carlo studies, it is sometimes necessary to generate a series of exponential random variables, Y_1, Y_2, . . . , where $f_{Y_i}(y) = (1/\lambda)e^{-y/\lambda}$, $y > 0$. Many computers do not have built-in programs for generating such variables, but they do have subroutines for generating random *uniform* variables. By using an appropriate transformation, we can use the latter to construct the former.

Let X denote a random variable uniformly distributed over the unit interval. Define the transformation $Y = -\lambda \ln X$. We will show that the pdf for Y is exponential with parameter λ.

By definition,

$$F_Y(y) = P(Y \le y) = P(-\lambda \ln X \le y)$$

$$= P\left(\ln X > -\frac{y}{\lambda} \right) = P(X > e^{-y/\lambda})$$

Thus, the probability that Y is less than or equal to y is equal to the probability associated with that portion of the X-axis lying to the *right* of $e^{-y/\lambda}$ (see Figure 4.5.2). But since X is uniform over $(0, 1)$,

$$F_Y(y) = P(X > e^{-y/\lambda}) = \int_{e^{-y/\lambda}}^{1} 1 \cdot dx = 1 - e^{-y/\lambda}$$

Therefore,

$$f_Y(y) = F_Y'(y) = \left(\frac{1}{\lambda} \right) e^{-y/\lambda}$$

the result we wished to prove.

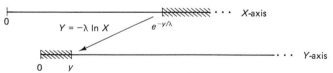

Figure 4.5.2

Question 4.5.12 Suppose the random variable X has pdf

$$f_X(x) = 6x(1 - x), \qquad 0 < x < 1$$

Define $Y = X^3$. Find the pdf for Y.

Question 4.5.13 Suppose X has pdf $f_X(x) = \lambda e^{-\lambda x}$, $x > 0$. Let $Y = X^{-t}$ for $t > 0$. Find the pdf for Y.

Question 4.5.14 Let $f_X(x) = e^{-x}$, $x > 0$. Find $f_Y(y)$ for $Y = \ln X$.

Question 4.5.15 Let X be a continuous random variable with cdf $F_X(x)$. Define $Y = F_X(x)$. Show that Y has a uniform distribution over the unit interval.

The linear and quadratic transformations covered in Theorems 4.5.1 and 4.5.2

are the most important single-variable transformations in probability. In problems involving n random variables, though, the most frequently encountered transformation is the *sum*, $Z = X_1 + X_2 + \cdots + X_n$. Theorem 4.5.3 gives a formula for the simplest of these problems, finding the pdf for the sum of *two* (independent) random variables.

Theorem 4.5.3. Let X and Y be independent random variables with pdf's $f_X(x)$ and $f_Y(y)$, respectively. Let $Z = X + Y$.

1. If X and Y are discrete,

$$f_Z(z) = \sum_{\text{all } x} f_X(x) \cdot f_Y(z - x)$$

2. If X and Y are continuous,

$$f_Z(z) = \int_{-\infty}^{\infty} f_X(x) f_Y(z - x) \, dx$$

Proof

1. When X and Y are discrete, the probability that $Z = z$ can be written first as a union and then as a sum:

$$P(Z = z) = P(X + Y = z) = P\left(\bigcup_{\text{all } x} (X = x, Y = z - x) \right)$$

$$= \sum_{\text{all } x} P(X = x, Y = z - x) = \sum_{\text{all } x} P(X = x) \cdot P(Y = z - x)$$

$$= \sum_{\text{all } x} f_X(x) f_Y(z - x)$$

2. When X and Y are continuous, the pdf for Z can be gotten by differentiating the cdf for Z. By definition,

$$F_Z(z) = P(Z \le z) = P(X + Y \le z)$$

$$= \iint_A f_{X,Y}(x, y) \, dy \, dx$$

where A is the region shown in Figure 4.5.3.

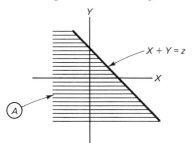

Figure 4.5.3

Since X and Y are independent, the double integral over A can be expressed as a product of single integrals:

$$F_Z(z) = \int_{-\infty}^{\infty} \int_{-\infty}^{z-x} f_{X,Y}(x, y) \, dy \, dx$$

$$= \int_{-\infty}^{\infty} \int_{-\infty}^{z-x} f_X(x) \cdot f_Y(y) \, dy \, dx$$

$$= \int_{-\infty}^{\infty} f_X(x) \cdot \left(\int_{-\infty}^{z-x} f_Y(y) \, dy \right) dx$$

It follows that

$$f_Z(z) = F_Z'(z) = \int_{-\infty}^{\infty} f_X(x) \left(\frac{d}{dz} \int_{-\infty}^{z-x} f_Y(y) \, dy \right) dx$$

$$= \int_{-\infty}^{\infty} f_X(x) f_Y(z - x) \, dx$$

Comment

An integral of the form

$$\int_{-\infty}^{\infty} f_X(x) f_Y(z - x) \, dx$$

is referred to as the *convolution* of the functions f_X and f_Y. Besides their frequent appearances in random variable problems, convolutions have applications in many areas of applied and theoretical mathematics.

Example 4.5.5

Suppose X and Y are independent binomial random variables, each with the same success probability p but defined on two possibly different numbers of Bernoulli trials. Specifically,

$$f_X(k) = \binom{m}{k} p^k q^{m-k}, \qquad 0 \le k \le m$$

and

$$f_Y(k) = \binom{n}{k} p^k q^{n-k}, \qquad 0 \le k \le n$$

Let $Z = X + Y$. Find $f_Z(z)$.

Where we have to be careful here is in the evaluation of the expression *all* x that appears in the statement of Theorem 4.5.3. We can write

$$f_Z(z) = \sum_{\text{all } x} f_X(x) \cdot f_Y(z - x)$$

but $f_X(x)$ vanishes for $x < 0$, as does $f_Y(z - x)$ for $z - x < 0$ (or $x > z$). Therefore,

$$f_Z(z) = \sum_{x=0}^{z} f_X(x) \cdot f_Y(z - x)$$

$$= \sum_{x=0}^{z} \binom{m}{x} p^x q^{m-x} \cdot \binom{n}{z - x} p^{z-x} q^{n-z+x}$$

$$= \sum_{x=0}^{z} \binom{m}{x} \binom{n}{z - x} p^z q^{m+n-z}$$

To simplify this last expression, we can first factor the terms in p and q outside the summation. Also, the sum of the products of the two combinatorial terms that remain gives rise to a well-known identity:

$$\sum_{x=0}^{z} \binom{m}{x} \binom{n}{z - x} = \binom{m+n}{z}$$

(see (73)). It follows, then, that

$$P(Z = z) = f_Z(z) = \binom{m + n}{z} p^z q^{m+n-z}, \qquad 0 \le z \le m + n$$

Notice that the binomial distribution "reproduces" itself—that is, if X and Y are binomial, so is their sum, $Z = X + Y$. Not all random variables share this property, as we shall see in Example 4.5.6.

Example 4.5.6

Suppose X and Y are two independent exponential random variables for which

$$f_X(t) = f_Y(t) = e^{-t}, \qquad t > 0$$

What is $f_Z(z)$, where $Z = X + Y$?

To use Theorem 4.5.3, we must find the values of X for which $f_X(x) \cdot f_Y(z - x) \ne 0$. But the first factor is nonzero if and only if $x > 0$, and the second is nonzero if and only if $z - x > 0$ or $x < z$. By Theorem 4.5.3 $f_Z(z) = ze^{-z}$ for $z > 0$:

$$f_Z(z) = \int_{-\infty}^{\infty} f_X(x) \cdot f_Y(z - x) \, dx$$

$$= \int_{0}^{z} e^{-x} \cdot e^{-(z-x)} \, dx$$

$$= \int_{0}^{z} e^{-z} \, dx$$

$$= ze^{-z}$$

From the form of $f_Z(z)$, we see that the sum of two independent exponentials is *not* exponential. (Is the sum of two independent uniforms uniform?)

Question 4.5.16 Let X and Y be two independent random variables with probability density functions

$$f_X(x) = \frac{e^{-r}r^x}{x!}, \qquad x = 0, 1, 2, \ldots$$

and

$$f_Y(y) = \frac{e^{-s}s^y}{y!}, \qquad y = 0, 1, 2, \ldots$$

a. Verify that $\sum_{x=0}^{\infty} f_X(x) = 1$.
b. Let $Z = X + Y$. Find $f_Z(z)$. Do X and Y reproduce themselves?

Question 4.5.17 If X and Y have the joint pdf

$$f_{X,Y}(x, y) = 2(x + y), \qquad 0 \le x \le y \le 1$$

find $f_Z(z)$, where $Z = X + Y$.

Question 4.5.18 An urn contains five red chips, four white chips, and three blue chips. A sample of two chips is drawn without replacement. Let X denote the number of red chips in the sample and Y be the number of white chips. Let $Z = X + Y$. Find the cdf for Z.

Question 4.5.19 Let X and Y have the joint pdf

$$f_{X,Y}(x, y) = xe^{-x} \cdot e^{-y}, \qquad x > 0, y > 0$$

Find $f_Z(z)$ for $Z = X + Y$.

Question 4.5.20 If a random variable X is independent of two random variables Y and Z, prove that X is also independent of the *sum* of Y and Z.

Question 4.5.21 Let X and Y have the joint pdf

$$f_{X,Y}(x, y) = 4e^{-x-4y}$$

for $x > 0$ and $y > 0$. Define $Z = X + Y$. Find $f_Z(z)$.

Question 4.5.22 Let X_1, X_2, and X_3 each be independent random variables with densities

$$f_{X_i}(x) = e^{-x}, \qquad x > 0, i = 1, 2, 3$$

Let $Z = X_1 + X_2 + X_3$. Find $f_Z(z)$.

Sometimes arithmetic combinations of random variables other than sums have to be dealt with. What is $f_Z(z)$, for example, if $Z = X/Y$? Depending on the nature of the function, problems of this sort can be quite difficult. In certain simple cases, though, $f_Z(z)$ can be gotten quite readily by following the direct approach used earlier in this section: Find $F_Z(z)$ and take its derivative.

Example 4.5.7

Suppose X and Y have a joint uniform density over the unit square:

$$f_{X,Y}(x, y) = \begin{cases} 1 & 0 < x < 1, 0 < y < 1 \\ 0 & \text{elsewhere} \end{cases}$$

Find the pdf of their product—that is, find $f_Z(z)$, where $Z = XY$.

For $0 < z < 1$, $F_Z(z)$ is the volume under $f_{X,Y}(x, y)$ above the shaded region in Figure 4.5.4:

$$F_Z(z) = P(Z \le z) = P(XY \le z) = \iint_R f_{X,Y}(x, y) \, dy \, dx$$

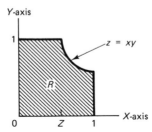

Figure 4.5.4

By inspection, we see that the double integral over R can be split up into *two* double integrals—one letting x range from 0 to z, the other with x-values extending from z to 1:

$$F_Z(z) = \int_0^z \left(\int_0^1 1 \, dy \right) dx + \int_z^1 \left(\int_0^{z/x} 1 \, dy \right) dx$$

But

$$\int_0^z \left(\int_0^1 1 \, dy \right) dx = \int_0^z 1 \, dx = z$$

and

$$\int_z^1 \left(\int_0^{z/x} 1 \, dy \right) dx = \int_z^1 \left(\frac{z}{x} \right) dx = z \ln x \Big|_z^1 = -z \ln z$$

It follows that

$$F_Z(z) = \begin{cases} 0 & z \le 0 \\ z - z \ln z & 0 < z < 1 \\ 1 & z \ge 1 \end{cases}$$

in which case

$$f_Z(z) = \begin{cases} -\ln z & 0 < z < 1 \\ 0 & \text{elsewhere} \end{cases}$$

Question 4.5.23 Suppose the random variables X and Y have the joint uniform density over the unit square. Find (a) the cdf and (b) the pdf for $Z = X/Y$.

Question 4.5.24 Let X_1 and X_2 be independent random variables with pdf's

$$f_{X_i}(x) = 2x, \qquad 0 < x < 1, i = 1, 2$$

Let $Z = X_1/X_2$. Find $F_Z(z)$.

Question 4.5.25 Let X and Y have the joint pdf

$$f_{X,Y}(x, y) = x + y, \qquad 0 < x < 1, 0 < y < 1$$

Define $Z = XY$. Find $f_Z(z)$.

Question 4.5.26 Suppose the random variables X and Y have the joint pdf

$$f_{X,Y}(x, y) = xe^{-x} \cdot (\tfrac{1}{3})e^{-y/3}, \qquad x > 0, y > 0$$

Find $P(X/Y > 1)$.

4.6 CONDITIONAL DENSITIES

We have already seen that many of the concepts defined in Chapter 2 relating to the probabilities of *events*—for example, independence—have their random-variable counterparts. Another of these carry-overs is conditional probability—it reappears here as a *conditional probability density function*. Applications of conditional pdf's are common. The height and girth of a tree, for instance, can be considered a pair of random variables. While it is easy to measure girth, it can be difficult to determine height; thus it might be of interest to a forester to know the probability that a Ponderosa pine attains certain heights given a known value for its girth. Or consider the plight of a factory owner trying to decide how much to invest in plant renovations. The task would be that much easier if the owner knew the conditional probability that x additional dollars invested would stimulate an average increase of y dollars in profit.

In the case of discrete random variables, a conditional pdf can be treated in the same way as a conditional probability. Note the similarity between Definitions 4.6.1 and 2.6.1.

> **Definition 4.6.1.** Let X and Y be two discrete random variables. The *conditional probability density function of Y given x*—that is, the probability that Y takes on the value y given that X is equal to x—is denoted $f_{Y|x}(y)$, where
>
> $$f_{Y|x}(y) = P(Y = y \mid X = x) = \frac{f_{X,Y}(x, y)}{f_X(x)}$$
>
> for $f_X(x) \neq 0$.

Example 4.6.1

Recall the coin-tossing experiment described in Example 4.3.4; Table 4.6.1 reproduces the joint pdf, $f_{X,Y}(x, y)$.

A simple application of Definition 4.6.1 gives

$$f_{Y|0}(0) = \frac{\frac{1}{8}}{\frac{1}{2}} = \tfrac{1}{4}, \qquad f_{Y|1}(0) = \frac{0}{\frac{1}{2}} = 0$$

TABLE 4.6.1

			Y			
		0	1	2	3	$f_X(x)$
	0	$\frac{1}{8}$	$\frac{1}{4}$	$\frac{1}{8}$	0	$\frac{1}{2}$
X	1	0	$\frac{1}{8}$	$\frac{1}{4}$	$\frac{1}{8}$	$\frac{1}{2}$
	$f_Y(y)$	$\frac{1}{8}$	$\frac{3}{8}$	$\frac{3}{8}$	$\frac{1}{8}$	

$$f_{Y|0}(1) = \frac{\frac{1}{4}}{\frac{1}{2}} = \frac{1}{2}, \qquad f_{Y|1}(1) = \frac{\frac{1}{8}}{\frac{1}{2}} = \frac{1}{4}$$

and so on.

Comment

The notion of a conditional pdf easily generalizes to situations involving more than two random variables. For example, if X, Y, and Z have the joint pdf $f_{X,Y,Z}(x, y, z)$, then the *joint conditional pdf* of, for instance, X and Y given that $Z = z$ would be the ratio

$$f_{X,Y|z}(x, y) = \frac{f_{X,Y,Z}(x, y, z)}{f_Z(z)}$$

(For random variables W, X, Y, and Z, how would $f_{W,X|y,z}(w, x)$ be defined?)

Question 4.6.1 Prove that $f_{Y|x}(y)$ as defined in Definition 4.6.1 is a legitimate probability density function.

Question 4.6.2 Five cards are dealt from a standard poker deck. Let X be the number of aces received and Y be the number of kings. Compute $P(X = 2 | Y = 2)$.

Question 4.6.3 Suppose X and Y have the joint pdf

$$f_{X,Y}(x, y) = \frac{xy^2}{39}$$

for the points $(1, 2)$, $(1, 3)$, $(2, 2)$, and $(2, 3)$ and has a value 0 otherwise. Find the conditional probability that X is 1 given that Y is 2.

Question 4.6.4 An urn contains eight red chips, six white chips, and four blue chips. A sample of size 3 is drawn without replacement. Let X denote the number of red chips in the sample and Y be the number of white chips. Find an expression for $f_{Y|x}(y)$.

Question 4.6.5 A fair coin is tossed five times. Let Y denote the total number of heads occurring in the five tosses and let X denote the number of heads occurring in the last two tosses. Find the conditional pdf, $f_{Y|x}(y)$.

Question 4.6.6 Let X and Y be discrete random variables. Show that

$$f_{X|y}(x) = \frac{f_{Y|x}(y) \cdot f_X(x)}{\sum_{\text{all } x} f_{Y|x}(y) f_X(x)}$$

Question 4.6.7 Suppose the random variables X, Y, and Z have a trivariate distribution described by the joint pdf

$$f_{X,Y,Z}(x, y, z) = \frac{xy}{9z}$$

defined for the points $(1, 1, 1)$, $(2, 1, 2)$, $(1, 2, 2)$, $(2, 2, 2)$, and $(2, 2, 1)$. Tabulate the joint conditional pdf of X and Y given each of the possible values for z.

Question 4.6.8 Let X have the pdf

$$f_X(x) = \binom{n}{x} p^x (1 - p)^{n-x}, \qquad x = 0, 1, \ldots, n$$

and let Y have the pdf

$$f_Y(y) = \binom{n}{y} p^y (1 - p)^{n-y}, \qquad y = 0, 1, \ldots, n$$

Define $Z = X + Y$. Show that the conditional pdf $f_{X|z}(x)$ is hypergeometric.

If the variables X and Y are continuous, we can still appeal to the quotient $f_{X,Y}(x, y)/f_X(x)$ as the definition of $f_{Y|x}(y)$ and argue its propriety by analogy. A more-satisfying approach, though, is to arrive at the same conclusion by taking the limit of Y's "conditional" *cdf*.

If X is continuous, a direct evaluation of $F_{Y|x}(y) = P(Y \le y | X = x)$ via Definition 2.6.1 is impossible, since the denominator would be 0. Alternatively, we can think of $P(Y \le y | X = x)$ as a limit:

$$P(Y \le y | X = x) = \lim_{h \to 0} P(Y \le y | x \le X \le x + h)$$

$$= \lim_{h \to 0} \frac{\int_x^{x+h} \int_{-\infty}^y f_{X,Y}(t, u) \, du \, dt}{\int_x^{x+h} f_X(t) \, dt}$$

The quotient of the limits is $0/0$, so l'Hôpital's rule is indicated:

$$P(Y \le y | X = x) = \lim_{h \to 0} \frac{\dfrac{d}{dh} \int_x^{x+h} \int_{-\infty}^y f_{X,Y}(t, u) \, du \, dt}{\dfrac{d}{dh} \int_x^{x+h} f_X(t) \, dt} \qquad (4.6.1)$$

By the fundamental theorem of calculus,

$$\frac{d}{dh} \int_x^{x+h} g(t) \, dt = g(x + h)$$

which simplifies Equation 4.6.1 to

$$P(Y \le y | X = x) = \lim_{h \to 0} \frac{\int_{-\infty}^y f_{X,Y}(x + h, u) \, du}{f_X(x + h)}$$

$$= \frac{\int_{-\infty}^{y} \lim_{h \to 0} f_{X,Y}(x + h, u) \, du}{\lim_{h \to 0} f_X(x + h)} = \int_{-\infty}^{y} \frac{f_{X,Y}(x, u)}{f_X(x)} \, du$$

provided the limit operation and the integration can be interchanged (see (3) for a discussion of when such an interchange is valid). It follows from this last expression that $f_{X,Y}(x, u)/f_X(x)$ behaves as a conditional probability density function should, meaning we are justified in extending Definition 4.6.1 to the continuous case.

The concept of a conditional pdf is also useful when one variable is discrete and the other is continuous, as Case Study 4.3 will show. For that setting, too, we get the same expression for $f_{Y|x}(y)$. All of these cases can be consolidated into a single result.

Definition 4.6.2. For random variables X and Y, the *conditional probability density of Y given x* is written $f_{Y|x}(y)$, where

$$f_{Y|x}(y) = \frac{f_{X,Y}(x, y)}{f_X(x)} \qquad \text{for} \qquad f_X(x) \neq 0$$

Example 4.6.2

A frequent assumption made in both Chapters 2 and 4 has been that the life of an electronic component can be modeled by an exponential pdf, $f_X(x) = \lambda e^{-\lambda x}$, $x > 0$, where λ is a positive constant. In some situations, though, λ is more properly thought of as being a random variable. This would be the case if the component being tested were selected from a large population of components, with the members of that population having a variety of operating characteristics. If λ *is* a random variable, what we have assumed about X is no longer its pdf but, rather, its *conditional* pdf:

$$f_{X|\Lambda}(x) = \lambda e^{-\lambda x}, \qquad x > 0$$

What can we say about the unconditional $f_X(x)$?

To begin, suppose the pdf for Λ is given by

$$f_{\Lambda}(\lambda) = \lambda e^{-\lambda}$$

Appealing to Definition 4.6.2, we can write the joint pdf for X and Λ as a product:

$$f_{X,\Lambda}(x, \lambda) = f_{X|\Lambda}(x) \cdot f_{\Lambda}(\lambda) = \lambda e^{-\lambda x} \cdot \lambda e^{-\lambda}$$
$$= \lambda^2 e^{-\lambda(x+1)} \tag{4.6.2}$$

From the latter it follows that the marginal (or unconditional) pdf for X is the integral,

$$f_X(x) = \int_0^{\infty} \lambda^2 e^{-\lambda(x+1)} \, d\lambda \tag{4.6.3}$$

But note that Equation 4.6.2 is the same joint pdf that appeared in Example 4.3.7, and it was shown there that the integral in Equation 4.6.3 reduces to

$$f_X(x) = \frac{2}{(x + 1)^3}$$

CASE STUDY 4.3

Experience has shown that the ash content of peat taken from a bog usually has a normal distribution (recall the pdf in Example 2.5.5). For the samples reported in Table 4.6.2, though, that was seemingly not the case (34). The histogram for the data (Figure 4.6.1) is sharply bimodal, with its two peaks occurring at widely separated points.

TABLE 4.6.2 DISTRIBUTION OF 430 PEAT SAMPLES BY ASH CONTENT

Ash content (%)	Observed number	Expected number
0–2.0	9	7.3
2.0–2.5	12	12.5
2.5–3.0	18	19.8
3.0–3.5	20	22.4
3.5–4.0	19	18.9
4.0–4.5	16	14.2
4.5–5.0	14	14.6
5.0–5.5	20	20.6
5.5–6.0	25	28.4
6.0–6.5	35	37.0
6.5–7.0	43	44.2
7.0–7.5	48	45.1
7.5–8.0	45	42.1
8.0–8.5	35	36.5
8.5–9.0	26	26.7
9.0–9.5	17	18.1
9.5–10.0	13	11.2
10.0–10.5	9	5.7
>10.5	6	4.8

Where does the shape of Figure 4.6.1 lead us, given that our objective is to find an $f_X(x)$ that describes the data? Must we immediately discard the normal as a potential model? Not necessarily. What is showing up as bimodality may be the result of *mixing* two unimodal distributions.

Suppose we assume the bog is actually made up of two different kinds of peat, each containing, on the average, a different concentration of ash. Let X denote a sample's measured ash content; let Y (1 or 2) indicate whether the peat being analyzed

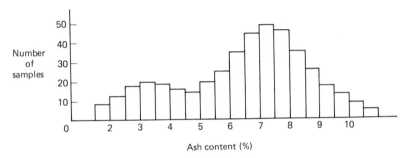

Figure 4.6.1

belongs to Type 1 or Type 2. We can express the unconditional pdf of X as a linear combination of pdf's conditioned on Y:

$$f_X(x) = \sum_{y=1}^{2} f_{X|y}(x) \cdot f_Y(y) \tag{4.6.4}$$

(Compare Equation 4.6.4 to the statement of Theorem 2.6.1.)

The next step in our model-building scenario is to make some specific assumptions about the density functions in the right-hand side of Equation 4.6.4 and see if the resulting $f_X(x)$ agrees with the data. Suppose it is known that 80% of the peat in the bog belongs to Type 1—that is, $f_Y(1) = 0.80$ and $f_Y(2) = 0.20$. If we further assume that the conditional pdf's of X given y are normal, we can write,

$$f_X(x) = (0.8)\,\frac{1}{\sqrt{2\pi}\sigma_1}\,e^{[-\frac{1}{2}((x-\mu_1)/\sigma_1)^2]} + (0.2)\,\frac{1}{\sqrt{2\pi}\sigma_2}\,e^{[-\frac{1}{2}((x-\mu_2)/\sigma_2)^2]}$$

Values for the four unknown parameters—μ_1, σ_1, μ_2, and σ_2—can be estimated from the data using standard statistical methods. Here, the numbers found were 7.2, 1.5, 3.1, and 0.8, respectively. The model being proposed, then, has the pdf,

$$f_X(x) = (0.8)\,\frac{1}{\sqrt{2\pi}(1.5)}\,e^{[-\frac{1}{2}((x-7.2)/1.5)^2]} + (0.2)\,\frac{1}{\sqrt{2\pi}(0.8)}\,e^{[-\frac{1}{2}((x-3.1)/0.8)^2]}$$

The probability of X falling in the interval from 6.0 to 6.5, for example, is given by

$$\int_{6.0}^{6.5} f_X(x)\,dx$$

We will show in Section 6.3 how to evaluate integrals of the sort just mentioned. For the problem at hand, $\int_{6.0}^{6.5} f_X(x)\,dx = 0.086$. Since there are 430 observations making up the sample, we would expect roughly 430×0.086, or 37, to fall between 6.0 and 6.5. The number actually observed was 35, so the model performed very well for that particular range. Column 3 in Table 4.6.2 lists the entire set of "expected" frequencies. The agreement between the second and third columns is excellent: Note, for instance, that the expected distribution has its two peaks located in the exact same intervals as the observed distribution.

Question 4.6.9 Let X and Y be continuous random variables with joint pdf

$$f_{X,Y}(x, y) = \begin{cases} (\frac{1}{8})(6 - x - y) & 0 < x < 2, 2 < y < 4 \\ 0 & \text{elsewhere} \end{cases}$$

Find $f_X(x)$ and $f_{Y|x}(y)$. Draw a diagram showing $f_{Y|1}(y)$.

Question 4.6.10 Let X be a nonnegative random variable. We say that X is *memoryless* if

$$P(X > s + t | X > t) = P(X > s) \qquad \text{for all } s, t \geq 0$$

Show that a random variable with pdf $f_X(x) = (1/\lambda)e^{-x/\lambda}$, $x > 0$, is memoryless.

Question 4.6.11 Given the joint pdf

$$f_{X,Y}(x, y) = 2e^{-(x+y)}, \qquad 0 < x < y, y > 0$$

evaluate the following expressions.
a. $P(Y < 1 | X < 1)$
b. $P(Y < 1 | X = 1)$
c. $f_{Y|x}(y)$

Question 4.6.12 Find the conditional pdf of Y given x if

$$f_{X,Y}(x, y) = x + y$$

for $0 \leq x \leq 1$ and $0 \leq y \leq 1$.

Question 4.6.13 If

$$f_{X,Y}(x, y) = 2, \qquad x \geq 0, y \geq 0, x + y \leq 1$$

show that the conditional pdf of Y given x is uniform.

Question 4.6.14 Suppose

$$f_{Y|x}(y) = \frac{2y + 4x}{1 + 4x} \quad \text{and} \quad f_X(x) = (\tfrac{1}{3})(1 + 4x)$$

for $0 < x < 1$ and $0 < y < 1$. Find the marginal pdf for Y.

Question 4.6.15 Suppose X and Y are jointly distributed according to the joint pdf

$$f_{X,Y}(x, y) = (\tfrac{2}{5})(2x + 3y), \qquad 0 \leq x \leq 1, 0 \leq y \leq 1$$

Find (a) $f_X(x)$, (b) $f_{Y|x}(y)$, and (c) $P(\tfrac{1}{4} \leq Y \leq \tfrac{3}{4} | X = \tfrac{1}{2})$.

Question 4.6.16 If X and Y have the joint pdf

$$f_{X,Y}(x, y) = 2, \qquad 0 < x < y < 1$$

find $P(0 < X < \tfrac{1}{2} | Y = \tfrac{3}{4})$.

Question 4.6.17 If

$$f_{X,Y}(x, y) = \frac{1}{2\pi\sqrt{1-\rho^2}} \cdot \exp\left[-\frac{1}{2(1-\rho^2)}(x^2 - 2\rho xy + y^2)\right]$$

for $-\infty < x < \infty$ and $-\infty < y < \infty$, show that

$$f_{Y|x}(y) = \frac{1}{\sqrt{2\pi(1-\rho^2)}} \cdot \exp\left[-\frac{(y-\rho x)^2}{2(1-\rho^2)}\right]$$

Question 4.6.18 Suppose X_1, X_2, X_3, X_4, and X_5 have the joint pdf $f_{X_1,X_2,X_3,X_4,X_5}(x_1, x_2, x_3, x_4, x_5) = 32x_1x_2x_3x_4x_5$ for $0 < x_i < 1$, $i = 1, 2, \ldots, 5$. Find the joint conditional pdf of X_1, X_2, and X_3 given that $X_4 = x_4$ and $X_5 = x_5$.

5

Expected Values

I am giddy, expectation whirls me around.
Shakespeare

5.1 INTRODUCTION

Probability density functions, as we have already seen, provide a global overview of a random variable's behavior. If X is discrete, $f_X(x)$ gives $P(X = x)$ for all x; if X is continuous and A is any interval or countable union of intervals, $P(X \in A) = \int_A f_X(x)\, dx$. Detail that explicit, though, is not always necessary—or even helpful. There are times when a more prudent strategy is to focus the information contained in $f_X(x)$ by summarizing certain of its features with single numbers. The search for these numbers—and the investigation, interpretation, and application of their properties—is the primary concern of this chapter.

The first feature of a pdf that we examine is *central tendency,* a term referring to the average value of a random variable. Consider the pdf's $f_X(x)$ and $g_Y(y)$ pictured in Figure 5.1.1. While we obviously cannot predict with certainty what values any

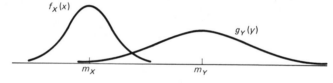

Figure 5.1.1

future X's and Y's will take on, it seems clear that X-values will tend to lie somewhere near m_X and Y-values will be somewhere near m_Y. In some sense, then, we can characterize $f_X(x)$ by m_X and $g_Y(y)$ by m_Y.

The most frequently used measure for describing central tendency—that is, for quantifying m_X and m_Y—is the *expected value*. Discussed at some length in Sections 5.2 and 5.3, the expected value of a random variable is a slightly more abstract formulation of what we are already familiar with in simple discrete settings as the arithmetic average.

Another characteristic of random variables that warrants our attention is *dispersion*, a term referring to how spread out its values are—or, equivalently, the extent to which its values are not all the same. Consider again the pdf's shown in Figure 5.1.1. Quite clearly, random variable Y is more dispersed than random variable X: For any $k > 0$, $P(|Y - m_Y| > k)$ is greater than $P(|X - m_X| > k)$. Section 5.4 addresses the problem of measuring dispersion by defining a random variable's *variance*.

The remainder of the chapter treats certain refinements and extensions of these basic ideas. *Higher moments* and *moment-generating functions* are discussed in Section 5.5, the *Bienaymé-Chebyshev inequality* in Section 5.6, and *conditional expectation* in Section 5.7.

5.2 EXPECTED VALUES

Gambling affords us a familiar illustration for the notion of an expected value. Consider the game of roulette. After all bets have been placed, the croupier spins the wheel, waits for the silver ball to come to rest, and then declares one of the 38 numbers 00, 0, 1, 2, . . . , 36 to be the winner. Disregarding what seems to be a perverse tendency of many roulette wheels to land on numbers for which no money has been wagered, we assume that each of these 38 outcomes is equally likely. Suppose our bet is \$1 on "odds" and we let X denote our winnings. Then X takes on the value 1 if an odd number occurs and -1 otherwise. By the uniformity assumption,

$$f_X(1) = P(X = 1) = P(\{1, 3, 5, \ldots, 35\}) = \tfrac{18}{38} = \tfrac{9}{19}$$

and

$$f_X(-1) = P(X = -1) = \tfrac{20}{38} = \tfrac{10}{19}$$

(0 and 00 are considered neither even nor odd). It follows that $\tfrac{9}{19}$ of the time we will win \$1 and $\tfrac{10}{19}$ of the time we will lose \$1. Intuitively, we stand to *lose*, on the average, a little more than 5¢ per bet:

$$\text{\textit{Expected} winnings} = \$1 \cdot \left(\tfrac{9}{19}\right) + (-\$1) \cdot \left(\tfrac{10}{19}\right) \qquad (5.2.1)$$

$$= -\$0.053 \doteq -5¢$$

The final weighted average, -0.053, is called the *expected value of X*.

Physically, an expected value can be thought of as the center of gravity for $f_X(x)$. Imagine two bars of height $\tfrac{10}{19}$ and $\tfrac{9}{19}$ positioned along a weightless X-axis at the points -1 and $+1$, respectively (see Figure 5.2.1). If a fulcrum were placed at the point

−0.053, the system would be in balance. It is in this sense that we can interpret the expected value of a random variable as being a measure of a pdf's "location."

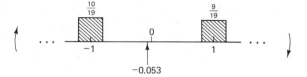

Figure 5.2.1

Extending our roulette example to the problem of finding expected values of arbitrary discrete random variables is straightforward. Suppose X takes on the values $x_1, x_2, \ldots$ with probabilities $f_X(x_1), f_X(x_2), \ldots$ By analogy with Equation 5.2.1, we define the expected value of X to be the weighted sum $\Sigma_i \, x_i \cdot f_X(x_i)$.

Note that if X is a discrete *uniform* random variable, what we are calling the expected value is simply the everyday notion of an arithmetic average:

$$\text{Expected value of } X = \sum_{i=1}^{n} x_i \cdot f_X(x_i) = \sum_{i=1}^{n} x_i \cdot \frac{1}{n}$$

$$= \frac{\sum_{i=1}^{n} x_i}{n}$$

Although the idea of an average predates Pythagoras, the probabilistic concept of an expected value made its debut in 1657 in Huygens' *De Ratiociniis in Alea Ludo*.

Definition 5.2.1. Let X be a random variable with probability density function $f_X(x)$. The *expected value of X* is denoted by $E(X)$, or μ, where:

1. $E(X) = \mu = \Sigma_{\text{all } x} \, x \cdot f_X(x)$ if X is discrete,
2. $E(X) = \mu = \int_{-\infty}^{\infty} x \cdot f_X(x) \, dx$ if X is continuous

provided the sum in (1) and the integral in (2) converge absolutely.

Comment

If it is not true that either

$$\sum_{\text{all } x} |x| f_X(x) < \infty \quad \text{or} \quad \int_{-\infty}^{\infty} |x| f_X(x) \, dx < \infty$$

we say that X has no finite expected value. *Absolute* convergence is a necessity here because the value of a sum that is convergent but not absolutely convergent depends on the order in which the terms are added. It makes no sense, though, for an expected value to have that sort of arbitrariness, so we must insist on the stronger form of convergence. Instances of random variables not having finite means will be discussed in Example 5.2.5 and in several of the questions throughout this section.

We begin our discussion of expected values by looking at the computation of $E(X)$ for several discrete random variables. Particularly important are Examples 5.2.3 and 5.2.4 because they derive $E(X)$ for two of the most useful of all discrete pdf's, the binomial and the geometric. We refer to these two results many times in the chapters ahead. Also of special interest here is Example 5.2.5, which describes a famous paradox in probability: What appears to be a very simple coin-tossing game turns out to have a very unlikely expected value.

Example 5.2.1

An urn contains nine chips, five red and four white (see Figure 5.2.2). Three are drawn out at random without replacement. Let X denote the number of red chips in the sample. Find $E(X)$, the expected number of red chips selected.

Draw 3 (without replacement).

Let X = number of ●'s in sample.

Figure 5.2.2

From Section 3.6, we recognize X to be a hypergeometric random variable, where

$$P(X = x) = f_X(x) = \frac{\binom{5}{x}\binom{4}{3-x}}{\binom{9}{3}}, \qquad x = 0, 1, 2, 3$$

By Definition 5.2.1,

$$E(X) = \sum_{x=0}^{3} x \cdot \frac{\binom{5}{x}\binom{4}{3-x}}{\binom{9}{3}}$$

$$= (0)\left(\frac{4}{84}\right) + (1)\left(\frac{30}{84}\right) + (2)\left(\frac{40}{84}\right) + (3)\left(\frac{10}{84}\right)$$

$$= \frac{5}{3}$$

(What would $E(X)$ be if the three chips were drawn *with* replacement?)

Example 5.2.2

Among the more common versions of the "numbers" racket is a game called D.J., its name deriving from the fact that the winning ticket is determined from Dow Jones averages. Three sets of stocks are used, Industrials, Transportations, and Utilities. Traditionally, the three are quoted at two different times, 11 A.M. and noon. The last digits of the earlier quotation are arranged to form a three-digit number; the noon quotation generates a second three-digit number, formed the same way. Those two numbers are then added together and the last three digits of that sum become the winning pick. Figure 5.2.3 shows a set of quotations for which *906* would be declared the winner.

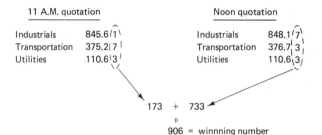

Figure 5.2.3

The payoff in D.J. is 700 to 1. Suppose we bet $5. How much do we stand to win or lose, *on the average?*

Let p denote the probability of our number being the winner and let X denote our earnings. Then

$$X = \begin{cases} \$3500 & \text{with probability } p \\ -\$5 & \text{with probability } 1 - p \end{cases}$$

and

$$E(X) = \$3500 \cdot p - \$5 \cdot (1 - p)$$

Our intuition would suggest (and this time it would be correct!) that each of the possible winning numbers, 000 through 999, is equally likely. That being the case, $p = 1/1000$ (see Question 5.2.16) and

$$E(X) = \$3500 \cdot \left(\tfrac{1}{1000}\right) - \$5 \cdot \left(\tfrac{999}{1000}\right) = -\$1.50$$

On the average, we lose $1.50 on a $5 bet. (There are actually two types of bets that can be placed in D.J.—we just described a *straight bet;* a *box bet* is a wager on a specific number—such as 906—as well as on all of its permutations. The payoff on the latter is $\tfrac{700}{6}$ to 1. In terms of their expected values, is there any difference between a straight bet and a box bet?)

Example 5.2.3

Let X be a binomial random variable defined on n trials, where $p = P(\text{success})$. Find $E(X)$.

Applying Definition 5.2.1 to the pdf for a binomial, we can write

$$E(X) = \sum_{x=0}^{n} x \cdot f_X(x) = \sum_{x=0}^{n} x \binom{n}{x} p^x (1 - p)^{n-x}$$

$$= \sum_{x=0}^{n} \frac{x \cdot n!}{x! \, (n - x)!} p^x (1 - p)^{n-x}$$

$$= \sum_{x=1}^{n} \frac{n!}{(x - 1)! \, (n - x)!} p^x (1 - p)^{n-x} \qquad (5.2.2)$$

To evaluate the sum in Equation 5.2.2 we need to use a summation technique similar to one we first saw in Chapter 2. If $E(X) = \sum_x g(x)$ can be factored in such a way that $E(X) = h \sum_x f_Y(x)$, where $f_Y(x)$ is the pdf for some random variable Y, then $E(X) = h$, since the sum of a pdf over its entire range is 1. Here, suppose np is factored out of the right-hand side of Equation 5.2.2. Then

$$E(X) = np \sum_{x=1}^{n} \frac{(n-1)!}{(x-1)!\,(n-x)!} p^{x-1}(1-p)^{n-x}$$

$$= np \sum_{x=1}^{n} \binom{n-1}{x-1} p^{x-1}(1-p)^{n-x}$$

If y is substituted for $x - 1$, the sum just written becomes

$$E(X) = np \sum_{y=0}^{n-1} \binom{n-1}{y} p^{y}(1-p)^{n-y-1}$$

Finally, letting $m = n - 1$ gives

$$E(X) = np \sum_{y=0}^{m} \binom{m}{y} p^{y}(1-p)^{m-y}$$

Since the value of this latter sum is 1 (why?),

$$E(X) = np \qquad (5.2.3)$$

Equation 5.2.3 should come as no surprise. If a multiple choice test, for example, has 100 questions, each with 5 possible answers, we would expect to get 20 correct if we were just guessing. But $20 = E(X) = 100(\frac{1}{5}) = np$. (Compare Equation 5.2.3 with your answer to the question posed at the end of Example 5.2.1.)

Example 5.2.4

In some of our earlier problems, we dealt with the probability function for X, the number of trials needed to obtain the *first* success in a series of independent Bernoulli trials (recall Example 2.4.2). With p defined to be the chance of success at any given trial, the pdf for any such problem can be written

$$f_X(x) = pq^{x-1}, \qquad x = 1, 2, \ldots$$

What is the corresponding $E(X)$? (Any X having this particular pdf is said to be a *geometric* random variable.)

Substituting $f_X(x)$ into the definition for $E(X)$ gives

$$E(X) = \sum_{x=1}^{\infty} xpq^{x-1} = p \sum_{x=1}^{\infty} xq^{x-1} \qquad (5.2.4)$$

but carrying out the summation here is not a simple matter (try it for yourself!). The "trick" in getting a closed-form expression for $E(X)$ is to use the basic notions of derivative and antiderivative. Let

$$h(q) = \sum_{x=1}^{\infty} xq^{x-1}$$

Recalling the formula for the sum of an infinite series, we find that the indefinite integral of $h(q)$ is $q/(1 - q)$:

$$\int h(q) \, dq = \sum_{x=1}^{\infty} \int xq^{x-1} \, dq = \sum_{x=1}^{\infty} q^x$$

$$= \sum_{x=0}^{\infty} q^x - q^0 = \frac{1}{1 - q} - 1$$

$$= \frac{q}{1 - q}$$

Now, write $h(q)$ as the derivative of its indefinite integral:

$$h(q) = \frac{d}{dq} \int h(q) \, dq$$

$$= \frac{d}{dq}[q(1 - q)^{-1}] = \frac{q}{(1 - q)^2} + \frac{1}{1 - q}$$

$$= \frac{1}{p^2}$$

Putting this last result together with Equation 5.2.4 gives

$$E(X) = p \cdot \frac{1}{p^2} = \frac{1}{p}$$

(Is this result intuitively reasonable? Explain.)

Example 5.2.5

Consider the following game. A fair coin is flipped until a *tail* appears; we win $2 if it appears on the first toss, $4 if it first appears on the second toss, and, in general, 2^k if the first tail occurs on the kth toss. Let the random variable X denote our winnings. How much should we have to pay in order for this to be a fair game? (*Note:* A fair game is one where the difference between the ante and $E(X)$ is 0.)

Known as the St. Petersburg paradox, this problem has an astonishing answer—in order for the game to be fair, we would have to ante an infinite amount of money! Perhaps even more surprising, the derivation is trivial. We note, first of all, that X is restricted to the values 2^1, 2^2, 2^3, ... Also, by inspection,

$$f_X(2^k) = P(X = 2^k) = \frac{1}{2^k}, \quad k = 1, 2, \ldots$$

That $E(X)$ is not finite follows immediately from Definition 5.2.1:

$$E(X) = \sum_{\text{all } x} xf_X(x) = \sum_{k=1}^{\infty} 2^k \cdot \frac{1}{2^k} = 1 + 1 + 1 + \cdots$$

Comment

Mathematicians have been trying to explain the St. Petersburg paradox for almost 200 years. The answer clearly seems absurd—no gambler would consider paying even $25 to play such a game, much less an infinite amount—yet the computations involved in showing that X has no finite expected value are unassailably correct. Where the difficulty lies, according to one common theory, is with our inability to put in perspective the very small probabilities of winning very large payoffs. Furthermore, the problem assumes that our opponent has infinite capital, which is an impossible state of affairs. We get a much more reasonable answer for $E(X)$ if the stipulation is added that our winnings can be no more than a certain amount (see Question 5.2.8), or if the payoffs are assigned according to some formula other than 2^k (see Question 5.2.9).

Comment

There are two important lessons to be learned from the St. Petersburg paradox. First is the realization that $E(X)$ is not necessarily a meaningful characterization of the "location" of a distribution. Questions 5.2.10 and 5.2.17 show two other situations where the formal computations of $E(X)$ give similarly inappropriate answers. Second, we need to be aware that the notion of expected value is not necessarily synonymous with the concept of worth. Just because a game, for example, has a positive expected value—even a very *large* positive expected value—does not imply that someone would want to play it. Suppose, for example, you had the opportunity to spend your last $10,000 on a sweepstakes ticket where the prize was $1 billion but the probability of winning was only 1 in 10,000. The expected value of such a bet would be over $90,000,

$$E(X) = 1,000,000,000\left(\frac{1}{10,000}\right) + (-10,000)\left(\frac{9999}{10,000}\right)$$

$$= \$90,001$$

but it is doubtful that many individuals would rush out to buy a ticket. (Economists have long recognized the distinction between a payoff's numerical value and its perceived desirability. They refer to the latter as *utility*.)

As a final example showing the calculation of an expected value for a discrete $f_X(x)$, we consider an application taken from the medical sciences. The random variable here is an unusual one: It has positive probability only for the two values 1 and $k + 1$.

Example 5.2.6

Suppose N people are to be given a blood test to see which of them, if any, has a certain rare disease. The obvious approach is to examine each person's blood individually, meaning that a total of N tests will be run. But that may be inefficient. An alternate strategy is to pool the N blood samples into groups of k and run the test on each group. If a given test proves negative, all k in that group are free from the disease. If the result is positive, each person's blood in that group must be rerun, ultimately resulting in $k + 1$ tests being done on those particular k individuals. Suppose the probability of any one person showing a positive result is p. What will be the expected number of tests under the "pooling" plan, and how does it compare to N?

Let X be the random variable counting the number of tests done on a pooled sample. Then X takes on only two values, either 1 or $k + 1$, where

$$f_X(1) = P(\text{none of the } k \text{ gives a positive test})$$
$$= (1 - p)^k$$

and

$$f_X(k + 1) = 1 - (1 - p)^k$$

By Definition 5.2.1,

$$E(X) = 1 \cdot (1 - p)^k + (k + 1)[1 - (1 - p)^k]$$
$$= (k + 1) - k(1 - p)^k$$

Since there are approximately N/k groups of size k, the total expected number of tests is roughly

$$\left(\frac{N}{k}\right) \cdot E(X) = \left(\frac{N}{k}\right)[(k + 1) - k(1 - p)^k] = N\left[1 - (1 - p)^k + \left(\frac{1}{k}\right)\right]$$

Table 5.2.1 gives $(N/k) \cdot E(X)$ for p values of $\frac{1}{100}$ and $\frac{1}{1000}$ and for $k = 2, 5, 10, 20, 40,$ and 100. Notice that for small p, especially, pooling can considerably reduce the number of tests that need to be performed.

TABLE 5.2.1

p	k	Expected number of tests
$\frac{1}{100}$	2	$0.52N$
$\frac{1}{100}$	5	$0.25N$
$\frac{1}{100}$	10	$0.20N$
$\frac{1}{100}$	20	$0.23N$
$\frac{1}{100}$	40	$0.36N$
$\frac{1}{100}$	100	$0.64N$
$\frac{1}{1000}$	2	$0.50N$
$\frac{1}{1000}$	5	$0.20N$
$\frac{1}{1000}$	10	$0.11N$
$\frac{1}{1000}$	20	$0.07N$
$\frac{1}{1000}$	40	$0.06N$
$\frac{1}{1000}$	100	$0.10N$

(Find the optimal pooling size k if $p = \frac{1}{50}$.)

Question 5.2.1 Nick has a total of $700 he is willing to bet on a series of college football games. On the opening week of the season, he picks one game listed by his bookie and bets $100. If he wins, he quits for the year; if he loses, he picks a game the following week and bets $200. If he wins that second week, he quits for the year; if he loses, he picks a game played in the third week and bets his last $400. If the point spreads for each game are entirely appropriate, making each contest a 50-50 proposition, what is Nick's expected gain for the season?

Question 5.2.2 In Example 2.4.4, the Poisson distribution

$$f_S(s) = \frac{e^{-0.82}(0.82)^s}{s!}, \qquad s = 0, 1, 2, \ldots$$

was shown to provide a good model for the daily number of fatalities among London senior citizens. Find $E(S)$. (*Hint:* Follow the same approach used in Example 5.2.3.)

Question 5.2.3 Suppose two evenly matched teams are playing in the World Series. On the average, how many games will be played? (The winner is the first team to get four victories.) Assume each game is an independent trial. How long would we expect the series to last if the National League team had a 60% chance of winning each game?

Question 5.2.4 Below are the last five lines of Shelley's poem *Ozymandias:*

> "My name is Ozymandias, king of kings:
> Look on my works, ye Mighty, and despair!"
> Nothing beside remains. Round the decay
> Of that colossal wreck, boundless and bare
> The lone and level sands stretch far away.

Suppose a word is selected at random from those lines. What is its expected length?

Question 5.2.5 a. If X is a nonnegative, integer-valued random variable, show that

$$E(X) = \sum_{k=1}^{\infty} P(X \geq k)$$

b. Use the result given in part (a) to show that if $f_X(x) = q^{x-1}p$, $x = 1, 2, \ldots$, then $E(X) = 1/p$.

Question 5.2.6 Stan Musial retired from baseball with a lifetime batting average of .331. On the average, how many hits did he get in a game where he had five official at bats?

Question 5.2.7 A fair die is rolled three times. Let X denote the number of different faces showing, $X = 1, 2, 3$. Find $E(X)$.

Question 5.2.8 How much would you have to ante to make the St. Petersburg game "fair" (recall Example 5.2.5) if the most you could win was $1000? That is, the payoffs are 2^k for $1 \leq k \leq 9$ and $1000 for $k \geq 10$.

Question 5.2.9 a. Find the expected payoff for the St. Petersburg problem if the amounts won are c^k instead of 2^k, where $0 < c < 2$.

b. Find the expected payoff if the amounts won are $\log 2^k$. (This was a modification suggested by D. Bernoulli (a nephew of James Bernoulli) to take into account the decreasing marginal utility of money—the more you have, the less useful a bit more is.)

Question 5.2.10 An urn contains one white chip and one black chip. A chip is drawn at random. If it is white, the game is over; if it is black, that chip and another black one are put into the urn.

Then another chip is drawn at random from the new urn and the same rules for ending or continuing the game are followed (if the chip is white, the game is over; if the chip is black, it is replaced in the urn, together with another chip of the same color). The drawings continue until a white chip is selected. Show that the expected number of drawings necessary to get a white chip is not finite.

Question 5.2.11 For Pascal's distribution,

$$f_X(x) = \left(\frac{1}{1+\mu}\right)\left(\frac{\mu}{1+\mu}\right)^x, \quad x = 0, 1, \ldots, \mu > 0$$

show that $E(X) = \mu$. (*Hint:* Let $y = \mu/(1+\mu)$. The identity

$$\sum_{x=0}^{\infty} xy^x = y\frac{d}{dy}\sum_{x=0}^{\infty} y^x$$

can be used.)(Pascal's distribution is a reparameterization of the geometric model described in Example 5.2.4.)

Question 5.2.12 It can be proved (see (18)) that

$$\sum_{k=1}^{\infty} \frac{1}{k^2} = \frac{\pi^2}{6}$$

which implies that a discrete pdf can be defined by

$$f_X(k) = P(X = k) = \frac{3}{\pi^2 k^2}, \quad k = \pm 1, \pm 2, \ldots$$

Show that even though $f_X(k)$ is symmetric around 0, it does not have a finite expected value. What does this say about the center-of-gravity analogy referred to earlier in this section?

Question 5.2.13 A young couple plan to continue having children until they have their first boy. Suppose the probability that each child is a boy is $\frac{1}{2}$ and the outcome of each birth is an independent event. How many children can they expect to have?

Question 5.2.14 A single-elimination tennis tournament has eight players. Assume the players can be ranked from 1 to 8, with player 1 always being able to defeat player 2, player 2 always superior to player 3, and so on. If the initial draw for the tournament is done at random, what is the expected rank of the runner-up? (See (98) for a proof showing that as the number of players increases, the expected rank of the runner-up converges to 3.)

Question 5.2.15 A random variable X is said to have a *negative binomial distribution* if

$$f_X(x) = \binom{x-1}{r-1}p^r q^{x-r}, \quad x = r, r+1, r+2, \ldots$$

(We prove in Chapter 6 that $\sum_{x=r}^{\infty} f_X(x) = 1$.) Conceptually, the negative binomial is a generalization of the model described in Example 5.2.4: Given a series of independent Bernoulli trials, X can be thought of as the trial number at which the rth success occurs. Show that $E(X) = r/p$.

Question 5.2.16 To validate the calculation of $E(X)$ done for the numbers racket described in Example 5.2.2, prove that the last three digits in the sum of two random three digit numbers is random. (*Hint:* Consider the analogous problem of adding two independent uniform random variables, X_1 and X_2, drawn from (0, 1). Show that the decimal part of that sum is uniform. Recall that the pdf of the sum of two uniform variables is triangular.)

Question 5.2.17 A sequence of random independent observations, $x_0, x_1, x_2, \ldots$ is to be drawn from a continuous pdf with mean μ. Suppose the first observation, x_0, is selected and it turns out to be greater than μ. Show that the expected number of samples that need to be drawn before finding one that exceeds x_0 is not finite. (*Hint:* If N denotes the index of the first observation that exceeds x_0, write $P(N = n)$ as $P(N > n - 1) - P(N > n)$.)

The next three examples illustrate the computation of $E(X)$ for continuous random variables. We do further calculations of this sort in Chapter 6 when we profile several of the continuous models that are considered especially important.

Example 5.2.7

A criminal court judge has heard many cases where the defendant was charged with grand theft, auto, and the jury returned a guilty verdict. But the outcome was not always the same. Mitigating circumstances of various kinds led the judge over the years to impose unequal jail terms for what was basically the same offense. Looking back over his court records, he sees that X, the imposed sentence length (in years), has a distribution that can be described quite well by a quadratic pdf,

$$f_X(x) = \tfrac{1}{9}x^2, \qquad 0 < x < 3$$

Compute the average sentence length the judge recommended for grand theft, auto.

From Definition 5.2.1,

$$E(X) = \int_0^3 x \cdot \left(\frac{1}{9}\right)x^2 \, dx = \frac{x^4}{36} \Big|_0^3 = 2\tfrac{1}{4} \text{ years}$$

(Find the *median* sentence length imposed by the judge. By definition, half the sentence lengths will be less than the median and half will be greater. For what sorts of pdf's will the median and the mean be the same?)

Example 5.2.8

One of the continuous pdf's that has many applications in physics is the Rayleigh distribution,

$$f_X(x) = \frac{x}{a^2}e^{-x^2/2a^2}, \qquad a > 0, 0 < x < \infty \qquad (5.2.5)$$

Find its expected value.

Appealing again to Definition 5.2.1, we can write

$$E(X) = \int_0^\infty x \cdot \frac{x}{a^2}e^{-x^2/2a^2} \, dx$$

Let $y = x/(a\sqrt{2})$. Then

$$E(X) = 2\sqrt{2}a \int_0^\infty y^2 e^{-y^2}\, dy$$

The integrand here is a special case of the general form $y^{2k}e^{-y^2}$, where

$$\int_0^\infty y^{2k} e^{-y^2}\, dy = \frac{1}{2}\Gamma\left(k + \frac{1}{2}\right)$$

Therefore, with k set equal to 1, the expected value reduces to $a\sqrt{\pi/2}$:

$$E(X) = 2\sqrt{2}a \cdot \frac{1}{2}\Gamma\left(\frac{3}{2}\right)$$

$$= 2\sqrt{2}a \cdot \frac{1}{2} \cdot \frac{1}{2}\Gamma\left(\frac{1}{2}\right)$$

$$= a\sqrt{\frac{\pi}{2}}$$

(Recall that $\Gamma(r + 1) = r \cdot \Gamma(r)$ and $\Gamma(\tfrac{1}{2}) = \sqrt{\pi}$.)

Comment

The distribution of Example 5.2.8 is named for John William Strutt, Baron Rayleigh, the nineteenth- and twentieth-century British physicist who showed that that particular pdf is the solution to a problem arising in the study of wave motion. If two waves are superimposed, it is well known that the height of the resultant at any time t is simply the algebraic sum of the corresponding heights of the waves being added (see Figure 5.2.4). Seeking to extend that notion, Rayleigh posed the following question: If n waves, each having the same amplitude h and the same wavelength, are superimposed randomly with respect to phase, what can we say about the amplitude R of the resultant? Clearly, R is a random variable, its value depending on the particular collection of phase angles represented by the sample. What Rayleigh was able to show in his 1880 paper (84) is that when n is large, the probabilistic behavior of R is described by the pdf

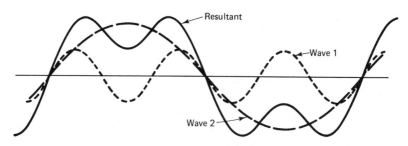

Figure 5.2.4

$$f_R(r) = \frac{2r}{nh^2} \cdot e^{-r^2/nh^2}, \qquad r > 0$$

which is just a special case of Equation 5.2.5 with $a = \sqrt{2/nh^2}$.

Example 5.2.9

As part of a reliability study, a total of n items are put "on test." Suppose each has the same exponential failure time distribution—that is, if T_i denotes the time when the ith item fails,

$$f_{T_i}(t) = \lambda e^{-\lambda t}, \qquad t > 0$$

where λ is a constant greater than 0. On the average, how long will we have to wait for the *first* failure?

Let T denote the first failure time. Note that T exceeds some particular value t only if all the T_i's exceed t. But that occurs with probability $e^{-n\lambda t}$:

$$P(T > t) = P(T_1 > t, T_2 > t, \ldots, T_n > t)$$
$$= P(T_1 > t) \cdot P(T_2 > t) \cdots P(T_n > t)$$
$$= \int_t^\infty \lambda e^{-\lambda t} \, dt_1 \cdot \int_t^\infty \lambda e^{-\lambda t} \, dt_2 \cdots \int_t^\infty \lambda e^{-\lambda t} \, dt_n$$
$$= e^{-\lambda t} \cdot e^{-\lambda t} \cdots e^{-\lambda t} = e^{-n\lambda t}$$

Therefore, the cdf for T is $1 - e^{-n\lambda t}$, from which it follows that

$$f_T(t) = F_T'(t) = \frac{d}{dt}(1 - e^{-n\lambda t}) = n\lambda e^{-n\lambda t}, \qquad t > 0$$

Having found $f_T(t)$, we can easily determine $E(T)$:

$$E(T) = \int_0^\infty t \cdot n\lambda e^{-n\lambda t} \, dt$$
$$= \frac{1}{n\lambda} \int_0^\infty y e^{-y} \, dy = \frac{1}{n\lambda} \cdot 1 = \frac{1}{n\lambda}$$

Thus, while the average waiting time for any particular item to fail is $1/\lambda$, that figure reduces to $1/n\lambda$ if we are simply waiting for whichever of the n items fails first.

Comment

The relationship between $E(T_i)$ and $E(T)$—specifically, the latter's dependence on n—reflects an everyday phenomenon with which we all are familiar. As our necessities of life become more and more complicated, whether that means owning a car with all the latest options or relying on a comprehensive word-processing system in the office, the waiting time (T) for *something* to malfunction gets shorter and shorter. In a sense, we find ourselves in a no-win situation: As technology advances, the failure parameters (λ_i) of individual manufactured goods decreases, but those same improvements encourage the multiplicity of components and complexity of objectives that can have the opposite effect of *increasing* the failure parameter ($n\lambda$) for the overall system.

Question 5.2.18 Find $E(X)$ for the following pdf's.
  a. $f_X(x) = \frac{1}{3}$, $0 \le x \le 3$
  b. $f_X(x) = 3(1 - x)^2$, $0 \le x \le 1$
  c. $f_X(x) = 4xe^{-2x}$, $x > 0$
  d. $f_X(x) = \begin{cases} \frac{3}{4} & 0 \le x \le 1 \\ \frac{1}{4} & 2 \le x \le 3 \\ 0 & \text{elsewhere} \end{cases}$
  e. $f_X(x) = \sin x$, $0 \le x \le \pi/2$

Question 5.2.19 Show that

$$f_X(x) = \frac{1}{x^2}, \qquad x \ge 1$$

is a valid pdf but does not have a finite expected value.

Question 5.2.20 Like Rayleigh's distribution, another continuous pdf that has frequent applications in physics is Cauchy's distribution,

$$f_X(x) = \frac{1}{\pi\alpha} \cdot \frac{1}{1 + (x - \mu)^2/\alpha^2}, \qquad -\infty < x < \infty;\ \alpha > 0;\ -\infty < \mu < \infty$$

Show that the associated $E(X)$ is not finite.

Question 5.2.21 Show that if X is a random variable with cdf $F_X(x)$, then

$$E(X) = \int_0^\infty [1 - F_X(x)]\, dx - \int_{-\infty}^0 F_X(x)\, dx$$

Draw a diagram illustrating the relationship between $E(X)$ and $F_X(x)$.

5.3 PROPERTIES OF EXPECTED VALUES

We defined $E(X)$ in Section 5.2—the obvious next step is to investigate some of its mathematical properties. Three of the most important will be taken up in this section. Specifically, we will show that (1) the expected value of a linear combination is equal to the linear combination of expected values ($E(a_1 X_1 + \cdots + a_n X_n) = a_1 E(X_1) + \cdots + a_n E(X_n)$), (2) the expected value of a *function* of X, $Y = g(X)$, can be found by simply summing (or integrating) $g(x) \cdot f_X(x)$ (we do not have to determine $f_Y(y)$ and evaluate the sum (or integral) of $y \cdot f_Y(y)$), and (3) if X and Y are independent, the expected value of their product is equal to the product of their expected values.

We begin by considering the problem of finding the expected value of a sum. What we derive proves to be a very useful result: The examples that follow Theorem

5.3.1 and its corollary show very clearly that if a random variable X can be written in the form $X = \sum_{i=1}^{n} a_i X_i$, it may be much easier to evaluate $\sum_{i=1}^{n} a_i E(X_i)$ than it is to work with $E(X)$ directly.

Theorem 5.3.1. For any random variables X and Y and any numbers a and b,

$$E(aX + bY) = aE(X) + bE(Y)$$

Proof. We will prove the theorem by establishing that $E(aX) = aE(X)$ and $E(X + Y) = E(X) + E(Y)$. Only the continuous case will be considered.

Let $f_{aX}(t)$ be the probability density function for aX. Then

$$E(aX) = \int_{-\infty}^{\infty} t f_{aX}(t)\, dt = \int_{-\infty}^{\infty} t \frac{1}{|a|} f_X\!\left(\frac{t}{a}\right) dt$$

the last equality being a consequence of Theorem 4.5.1. Assume $a > 0$ so $a = |a|$. Making the substitutions $x = t/a$ and $dx = (1/a)\, dt$, we can write

$$E(aX) = \int_{-\infty}^{\infty} ax\!\left(\frac{1}{a}\right) f_X(x) a\, dx = a \int_{-\infty}^{\infty} x f_X(x)\, dx = aE(X)$$

The derivation for $a < 0$ follows similarly.

The crux of the second part of the proof is to find an expression for $E(X + Y)$. One approach would be to derive a general formula for $f_{X+Y}(t)$ and then apply Definition 5.2.1. A simpler and more illuminating mode of attack, though, is to return to the basic concept of an expected value. The random variable $X + Y$ takes on values of the form $x + y$, with the likelihood of such a value being $f_{X,Y}(x, y)$. It follows that $E(X + Y)$ should equal

$$\sum_{\text{all } x} \sum_{\text{all } y} (x + y) f_{X,Y}(x, y) \quad \text{or} \quad \int_{-\infty}^{\infty} \int_{-\infty}^{\infty} (x + y) f_{X,Y}(x, y)\, dy\, dx$$

depending on the nature of X and Y.

Once we have written $E(X + Y)$ in terms of $f_{X,Y}(x, y)$, verifying that $E(X + Y) = E(X) + E(Y)$ becomes a simple exercise in the definition of marginal distributions. Consider the case where X and Y are continuous. By splitting the integrand into two parts, we find that

$$
\begin{aligned}
E(X + Y) &= \int_{-\infty}^{\infty} \int_{-\infty}^{\infty} (x + y) f_{X,Y}(x, y)\, dy\, dx \\[1mm]
&= \int_{-\infty}^{\infty} \int_{-\infty}^{\infty} x f_{X,Y}(x, y)\, dy\, dx + \int_{-\infty}^{\infty} \int_{-\infty}^{\infty} y f_{X,Y}(x, y)\, dy\, dx \\[1mm]
&= \int_{-\infty}^{\infty} x\!\left(\int_{-\infty}^{\infty} f_{X,Y}(x, y)\, dy\right) dx + \int_{-\infty}^{\infty} y\!\left(\int_{-\infty}^{\infty} f_{X,Y}(x, y)\, dx\right) dy \\[1mm]
&= \int_{-\infty}^{\infty} x f_X(x)\, dx + \int_{-\infty}^{\infty} y f_Y(y)\, dy
\end{aligned}
$$

But the latter two integrals are simply $E(X)$ and $E(Y)$, respectively, so the result is proved.

An easy induction argument extends Theorem 5.3.1 to the case of n variables. We omit the details.

Corollary. Given n random variables $X_1, X_2, \ldots, X_n$ and a set of n constants $a_1, a_2, \ldots, a_n$,

$$E(a_1 X_1 + a_2 X_2 + \cdots + a_n X_n) = a_1 E(X_1) + a_2 E(X_2) + \cdots + a_n E(X_n)$$

Comment

An equivalent way of writing the statement of the corollary is to say that E is a linear transformation.

A good example for demonstrating the usefulness of the corollary is the problem of finding the expected value of a binomial random variable. Contrast the simplicity of the approach given below with the intricacies of the direct summation described in Example 5.2.3. An even more difficult problem that is much simplified by the corollary is described in Example 5.3.2.

Example 5.3.1

Let X be a binomial random variable defined on n independent trials, each trial resulting in success with probability p. Find $E(X)$.

Note, first of all, that X can be thought of as a sum, $X = X_1 + X_2 + \cdots + X_n$, where X_i is a *Bernoulli* random variable denoting the number of successes at the ith trial:

$$X_i = \begin{cases} 1 & \text{if the ith trial produces a success} \\ 0 & \text{if the ith trial produces a failure} \end{cases}$$

By assumption, $f_{X_i}(1) = p$ and $f_{X_i}(0) = q$, $i = 1, 2, \ldots, n$. Using the corollary,

$$E(X) = E(X_1) + E(X_2) + \cdots + E(X_n)$$
$$= n \cdot E(X_1)$$

The last step is a consequence of the X_i's having identical distributions. But

$$E(X_1) = 1 \cdot p + 0 \cdot q = p$$

so $E(X) = np$, which is what we found before (recall Equation 5.2.3).

Example 5.3.2

A disgruntled secretary is upset about having to stuff envelopes. Handed a box of n letters and n envelopes, he vents his frustration by putting the letters into the envelopes *at random*. How many people, on the average, will receive the correct mail?

If X denotes the number of envelopes properly stuffed, what we are seeking is $E(X)$. However, applying Definition 5.2.1 here would prove formidable because of the difficulty in getting a workable expression for $f_X(x)$ (see Question 5.3.7). By using the corollary to Theorem 5.3.1, though, we can solve the problem easily. Let X_i denote a random variable equal to the number of correct letters put into the ith envelope, $i = 1, 2, \ldots, n$. Then X_i equals 0 or 1, and

$$f_{X_i}(k) = P(X_i = k) = \begin{cases} \dfrac{1}{n} & \text{for } k = 1 \\[2mm] \dfrac{n-1}{n} & \text{for } k = 0 \end{cases}$$

But $X = X_1 + X_2 + \cdots + X_n$ and $E(X) = E(X_1) + E(X_2) + \cdots + E(X_n)$. Furthermore, each of the X_i's has the same expected value, $1/n$:

$$E(X_i) = \sum_{k=0}^{1} k \cdot P(X_i = k) = 0 \cdot \frac{n-1}{n} + 1 \cdot \frac{1}{n} = \frac{1}{n}$$

It follows that

$$E(X) = \sum_{i=1}^{n} E(X_i) = n\left(\frac{1}{n}\right) = 1$$

showing that, *regardless of n*, the expected number of properly stuffed envelopes is 1. (Are the X_i's independent? Does it matter?)

In the next several examples, we explore some variations in the use of the corollary to Theorem 5.3.1. In Example 5.3.3 we face a reverse-type situation where $E(X)$ is known at the outset, and what we need to find is $E(X_i)$. Next come two problems where X is a linear combination but the component X_i's are not identically distributed. Case Study 5.1 follows, with an expected-value derivation in a combinatorial setting where the answer is integral to a statistical argument. As a final example, we look at the expected outcome of a rather unusual baseball strategy.

Example 5.3.3

The honor count in a (13-card) bridge hand can vary from 0 to 37 according to the formula

Honor count $= 4 \cdot$ (number of aces) $+ 3 \cdot$ (number of kings)

$$+ 2 \cdot \text{(number of queens)} + 1 \cdot \text{(number of jacks)}$$

What is the expected honor count of North's hand?

If X_i, $i = 1, 2, 3, 4$, denotes the honor count for North, South, East, and West, respectively, and if X denotes the analogous sum for the entire deck, we can write

$$X = X_1 + X_2 + X_3 + X_4$$

But

$$X = E(X) = 4 \cdot 4 + 3 \cdot 4 + 2 \cdot 4 + 1 \cdot 4 = 40$$

By symmetry, $E(X_i) = E(X_j)$, $i \neq j$, so it follows that $40 = 4 \cdot E(X_1)$, which implies that 10 is the expected honor count of North's hand. (Try doing this problem directly, without making use of the deck's honor count being 40.)

Example 5.3.4

A grocery store is sponsoring a sales promotion where the cashiers give away one of the letters A, E, L, S, U, and V for each purchase. If a customer collects all six (spelling *VALUES*), he or she gets $10 worth of groceries free. What is the expected number of trips to the store a customer needs to make in order to get a complete set? Assume the different letters are given away randomly.

Let X_i denote the number of purchases necessary to get the ith different letter, $i = 1, 2, \ldots, 6$, and let X denote the number of purchases necessary to qualify for the $10. Then $X = X_1 + X_2 + \cdots + X_6$ (see Figure 5.3.1). Clearly, X_1 equals 1 with probability 1, so $E(X_1) = 1$. Having received the first letter, the chances of getting a different one are $\frac{5}{6}$ for each subsequent trip to the store. Therefore,

$$f_{X_2}(k) = P(X_2 = k) = (\tfrac{1}{6})^{k-1}\tfrac{5}{6}, \qquad k = 1, 2, \ldots$$

Since a random variable of this type has expected value $1/p$ (recall Example 5.2.4), $E(X_2) = \frac{6}{5}$. Similarly, the chances of getting a *third* different letter are $\frac{4}{6}$ (for each purchase), so

$$f_{X_3}(k) = P(X_3 = k) = (\tfrac{2}{6})^{k-1}\tfrac{4}{6}, \qquad k = 1, 2, \ldots$$

and $E(X_3) = \frac{6}{4}$. Continuing in this fashion, we can find the remaining $E(X_i)$'s. It follows that a customer will have to make 14.7 trips to the store, on the average, to collect a complete set of six letters:

$$E(X) = \sum_{i=1}^{6} E(X_i)$$

$$= 1 + \tfrac{6}{5} + \tfrac{6}{4} + \tfrac{6}{3} + \tfrac{6}{2} + \tfrac{6}{1}$$

$$= 14.7$$

(What would $E(X)$ be if the customer had to get the six letters *in order*—first the V, then the A, and so on?)

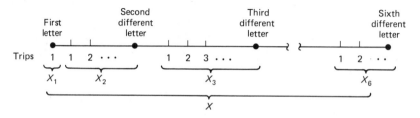

Figure 5.3.1

Example 5.3.5

A friend proposes the following game. She will select at random some number between 0 and 100 and offer you that amount in cash. You may accept the money or reject it. If you reject it, she will choose another number at random from the same interval (0 to 100) and now offer you *that* amount. As before, you may either accept this second offer or reject it. If you reject it, she will make a third and final offer (again choosing a number at random from 0 to 100); you must accept this one. Should you play the game if your friend insists the ante be $65, and if you *do* elect to play, what should your strategy be?

Let X denote the amount we ultimately receive playing the game. We need to find $E(X)$ *under the optimal strategy* and compare that figure to 65. If the former exceeds the latter, it would make sense mathematically to play the game (recall the comment following Example 5.2.5.)

Intuitively, the *form* of the optimal strategy seems clear: We should accept any initial offer exceeding some cutoff c_1 and any second offer exceeding a cutoff c_2. (Will c_1 be less than, greater than, or equal to c_2?) Let X_i denote the amount received from the ith offer (after we decide whether or not to accept it) and let X denote the amount we ultimately receive at the game's conclusion. Then $X = X_1 + X_2 + X_3$ and $E(X) = E(X_1) + E(X_2) + E(X_3)$. Let Y_i, $i = 1$, $2, 3$, denote the random number chosen on each occasion. Then

$$X_1 = \begin{cases} 0 & \text{if } 0 \le y_1 < c_1 \\ Y_1 & \text{if } y_1 \ge c_1 \end{cases}$$

and

$$E(X_1) = \int_0^{c_1} 0 \cdot \frac{1}{100} \, dy_1 + \int_{c_1}^{100} y_1 \cdot \frac{1}{100} \, dy_1$$

$$= \frac{(100 - c_1)(100 + c_1)}{2 \cdot 100}$$

Similarly,

$$X_2 = \begin{cases} 0 & \text{if } 0 \le y_2 < c_2 \text{ and } y_1 < c_1 \\ Y_2 & \text{if } y_2 \ge c_2 \text{ and } y_1 < c_1 \end{cases}$$

and

$$E(X_2) = \int_0^{c_2} \int_0^{c_1} 0 \cdot f_{Y_1, Y_2}(y_1, y_2) \, dy_1 \, dy_2 + \int_{c_2}^{100} \int_0^{c_1} y_2 \cdot f_{Y_1, Y_2}(y_1, y_2) \, dy_1 \, dy_2$$

$$= 0 + \int_{c_2}^{100} \int_0^{c_1} y_2 \cdot \frac{1}{100} \cdot \frac{1}{100} \, dy_1 \, dy_2 \qquad \text{(Why?)}$$

$$= \frac{c_1(100 - c_2)(100 + c_2)}{2(100)^2}$$

Finally,

$$X_3 = \begin{cases} 0 & \text{if } y_1 \geq c_1 \text{ or } y_1 < c_1 \cap y_2 \geq c_2 \\ Y_3 & \text{if } y_1 \leq c_1 \cap y_2 \leq c_2 \end{cases}$$

which gives

$$E(X_3) = \int_0^{100} \int_0^{c_2} \int_0^{c_1} y_3 \cdot f_{Y_1, Y_2, Y_3}(y_1, y_2, y_3) \, dy_1 \, dy_2 \, dy_3$$

$$= \frac{1}{(100)^3} \, c_1 c_2 \cdot \frac{y_3^2}{2} \bigg|_0^{100} = \frac{c_1 c_2}{2 \cdot 100}$$

Thus our expression for $E(X)$—in terms of c_1 and c_2—becomes

$$E(X) = \frac{(100 - c_1)(100 + c_1)}{200} + \frac{c_1(100 - c_2)(100 + c_2)}{2(100)^2} + \frac{c_1 c_2}{200}$$

$$= 50 - \frac{c_1^2}{200} + \frac{c_1}{2} - \frac{c_1 c_2^2}{20,000} + \frac{c_1 c_2}{200}$$

To determine the optimal strategy, we need to solve

$$\frac{\partial E(X)}{\partial c_1} = 0$$

and

$$\frac{\partial E(X)}{\partial c_2} = 0$$

simultaneously. But

$$\frac{\partial E(X)}{\partial c_1} = -\frac{c_1}{100} + \frac{1}{2} - \frac{c_2^2}{20,000} + \frac{c_2}{200}$$

and

$$\frac{\partial E(X)}{\partial c_2} = -\frac{c_1 c_2}{10,000} + \frac{c_1}{200}$$

Setting these two equations equal to 0 gives

$$c_1 = \$62.50 \quad \text{and} \quad c_2 = \$50.00$$

When the c's are put back into the expression for $E(X)$, we find that

$$E(X) = \frac{(100 - 62.50)(100 + 62.50)}{200} + \frac{62.50(100 - 50.00)(100 + 50.00)}{2(100)^2}$$

$$+ \frac{(62.50)(50.00)}{200}$$

$$= \$69.53$$

Our optimal strategy, then, is to take the first offer if it exceeds (or equals)

$62.50; if it happens to be less than that, we decline it and take the second offer if *it* exceeds (or equals) $50.00. Should we proceed in that fashion, our expected value is $69.53. Therefore, if our decision to play hinges on the relative sizes of $E(X)$ and the $65 ante, we should go ahead and put up the money—we stand to make a profit of $4.53 per game, on the average. (What would be the break-even ante if the same cutoff had to be used for both offers?)

In certain situations it becomes of interest to investigate whether or not a sequence of observations is random. Among the standard statistical procedures for doing this is a technique based on the *number of runs*. By definition, a run is an uninterrupted series of observations of a given kind. For example, in the sequence of 4 *A*'s and 3 *B*'s,

$$\underbrace{A\ A}\ \underbrace{B}\ \underbrace{A}\ \underbrace{B\ B}\ \underbrace{A}$$

there are *five* runs—an initial run of two *A*'s, followed by a run of one *B*, a run of one *A*, and so on. Too few runs in a sequence (for example, *AAAABBB*) or too many (such as *ABABABA*) would suggest the *A*'s and *B*'s are not positioned randomly. In Case Study 5.1, we see how the properties of expected values can help us interpret the number of runs found in a sequence of *n* observations.

CASE STUDY 5.1

The first widespread labor dispute in the United States occurred in 1877. Railroads were the target and workers were idled from Pittsburgh to San Francisco. That first dispute may have been a long time coming but workers were quick to recognize what a powerful weapon the strike really was—between 1881 and 1905 they called 36,757 more.

We look here at those 36,757 strikes, year by year (13). Specifically, our intention is to focus on the proportion each year that were considered successful. (A successful strike, by definition, was one where most or all of the workers' demands were met.) An obvious question is whether these proportions exhibit any patterns or trends. We might, for example, expect them gradually to increase, as unions acquired more and more power. Or, we might think that years or high success rates would tend to alternate with years of low success rates, indicating a give-and-take relationship on the part of labor and management. Still another hypothesis would be that the numbers show *no* patterns and behave like a random sequence.

Figures for the 25 years in question are shown in Table 5.3.1. To generate a series of runs, we categorize each year's percentage of successful strikes as being either above or below the median percentage of successful strikes. Here the median value—the 13th largest entry in the third column of Table 5.3.1—is the 1893 figure of 50%. The last column of the table indicates whether that year's percentage was above the median (*A*) or below the median (*B*). Scanning down that column, we see that the 24 observations (omitting the year when the median occurred) gave rise to a total of 8 runs. What should we infer from that total? Is it compatible with the hypothesis that the sequence was generated in a random fashion?

To answer fully any question about the randomness of a sequence requires a statistical argument beyond our present intentions. Still, we can derive the cornerstone of that argument by finding an expression for the *expected* number of runs.

TABLE 5.3.1 STRIKE STATISTICS (1881–1905)

Year	Number of strikes	Percent successful		A or B
1881	451	61		A
1882	454	53		A
1883	478	58		A
1884	443	51		A
1885	645	52		A
1886	1432	34		B
1887	1436	45		B
1888	906	52		A
1889	1075	46		B
1890	1833	52		A
1891	1717	37		B
1892	1298	39	8	B
1893	1305	50	runs	***
1894	1349	38		B
1895	1215	55		A
1896	1026	59		A
1897	1078	57		A
1898	1056	64		A
1899	1797	73		A
1900	1779	46		B
1901	2924	48		B
1902	3161	47		B
1903	3494	40		B
1904	2307	35		B
1905	2077	40		B

Suppose a two-symbol sequence of length n is being randomly generated by a process where

$$P(A \text{ appears in } i\text{th position}) = p$$

and

$$P(B \text{ appears in } i\text{th position}) = q = 1 - p$$

Let X_i denote the symbol in position i, $i = 1, 2, \ldots, n$. We can express the number of runs in the sequence in terms of the $n - 1$ "transitions" from X_i to X_{i+1}, $i = 1, 2, \ldots, n - 1$. Specifically, let

$$Q(X_i, X_{i+1}) = \begin{cases} 0 & X_i = X_{i+1} \\ 1 & X_i \neq X_{i+1} \end{cases}$$

If R denotes the total number of runs in the sequence,

$$R = 1 + Q(X_1, X_2) + Q(X_2, X_3) + \cdots + Q(X_{n-1}, X_n)$$

and

$$E(R) = 1 + \sum_{i=1}^{n-1} E[Q(X_i, X_{i+1})]$$

But

$$
\begin{aligned}
E[Q(X_i, X_{i+1})] &= 0 \cdot P(X_i = X_{i+1}) + 1 \cdot P(X_i \neq X_{i+1}) \\
&= P(X_i \neq X_{i+1}) \\
&= P(X_i = A \cap X_{i+1} = B) + P(X_i = B \cap X_{i+1} = A) \\
&= pq + qp = 2pq
\end{aligned}
$$

Therefore,

$$E(R) = 1 + (n - 1)(2pq)$$

In our situation, categorizing the observations as being above or below their median is equivalent to setting $p = q = \frac{1}{2}$. This implies that on the average we would have gotten 12.5 runs $(1 + (24 - 1)(2 \cdot \frac{1}{2} \cdot \frac{1}{2}))$. Whether the difference between $E(R)$ and R (12.5 versus 8) is great enough to justify rejecting the hypothesis of randomness is a question for statistics (see, for example, (12) or (94)). (Would it be possible for a sequence to have R *equal* to $E(R)$ and still be clearly nonrandom? Explain.)

While the baseball strategy described in the next example is not likely to get a manager (or the players) enshrined in Cooperstown, it does showcase a number of the expected-value results that we have developed in Sections 5.2 and 5.3.

Example 5.3.6

The California Mellows are a semipro baseball team. Eschewing all forms of violence, the laid-back Mellow batters never swing at a pitch; should they be fortunate enough to reach base on a walk, they never try to steal. On the average, how many runs will the Mellows score in a nine-inning road game, assuming the opposing pitcher has a 50% chance of throwing a strike on any given pitch (39)?

First, note that the probability of any given Mellow batter being called out on strikes—that is, of getting three strikes before four balls—is $\frac{21}{32}$:

$$
\begin{aligned}
P(\text{batter strikes out}) &= \sum_{k=3}^{6} P(\text{batter strikes out on } k \text{ pitches}) \\
&= (\tfrac{1}{2})^3 + \binom{3}{2}(\tfrac{1}{2})^4 + \binom{4}{2}(\tfrac{1}{2})^5 + \binom{5}{2}(\tfrac{1}{2})^6 \\
&= \tfrac{21}{32}
\end{aligned}
$$

The probability of a Mellow batter drawing a walk, then, is $1 - \frac{21}{32}$, or $\frac{11}{32}$.

Let W denote the number of walks the Mellows get in a given inning. In order for W to take on the value w, exactly two of the first $w + 2$ batters must strike out, as must the $(w + 3)$rd (see Figure 5.3.2).

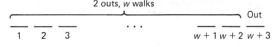

2 outs, w walks

Out

| 1 | 2 | 3 | . . . | w + 1 | w + 2 | w + 3 |

Batters

Figure 5.3.2

Therefore,

$$P(W = w) = \binom{w+2}{2}(\tfrac{21}{32})^3(\tfrac{11}{32})^w, \qquad w = 0, 1, 2, \ldots$$

The relationship between W and R_i, the number of runs the Mellows score in inning i, is obvious:

$$R_i = \begin{cases} 0 & w \le 3 \\ w - 3 & w > 3 \end{cases}$$

Multiplying R_i by $P(W = w)$ and summing over $w \ge 4$ gives an expression for the number of runs the Mellows can expect to score in the ith inning:

$$E(R_i) = \sum_{w=4}^{\infty} (w - 3)P(W = w)$$

$$= \sum_{w=0}^{\infty} (w - 3)P(W = w) - \sum_{w=0}^{3} (w - 3)P(W = w)$$

$$= E(W) - 3 + \sum_{w=0}^{3} (3 - w)P(W = w) \qquad (5.3.1)$$

To find $E(W)$ we make a simple transformation that allows us to use the result given in Question 5.2.15. Set $X = W + 3$. Then

$$P(X = x) = P(W = x - 3)$$

$$= \binom{x - 1}{2}\left(\frac{21}{32}\right)^3\left(\frac{11}{32}\right)^{x-3}, \qquad x = 3, 4, \ldots$$

implying that X has a negative binomial distribution with r equal to 3 and p equal to $\frac{21}{32}$. From Question 5.2.15,

$$E(X) = \frac{r}{p} = \frac{3}{\frac{21}{32}} = \frac{32}{7}$$

so

$$E(W) = E(X) - 3 = \frac{32}{7} - 3 = \frac{11}{7}$$

With $E(W)$ calculated, the expected number of runs scored by the Mellows per inning is easily determined. From Equation 5.3.1,

$$E(R_i) = \frac{11}{7} - 3 + 3 \cdot \binom{2}{2}\left(\frac{21}{32}\right)^3\left(\frac{11}{32}\right)^0 + 2 \cdot \binom{3}{2}\left(\frac{21}{32}\right)^3\left(\frac{11}{32}\right)^1$$

$$+ 1 \cdot \binom{4}{2}\left(\frac{21}{32}\right)^3\left(\frac{11}{32}\right)^2$$

$$= 0.202$$

Now, let R denote the number of runs the Mellows score throughout the game. Then $R = R_1 + R_2 + \cdots + R_9$, and by the corollary to Theorem 5.3.1, $E(R)$ is 1.82:

$$E(R) = \sum_{i=1}^{9} E(R_i) = 9(0.202) = 1.82$$

(Would the Mellows necessarily score an average of 1.82 runs per game if they were playing at home?)

Question 5.3.1 A standard result in analysis for estimating the sum of the first n terms in a harmonic series is Euler's approximation:

$$1 + \frac{1}{2} + \frac{1}{3} + \cdots + \frac{1}{n} \doteq \ln n + \frac{1}{2n} + 0.57721 \ldots$$

where the last term is known as *Euler's constant*. Use this result to estimate $E(X)$ for a letter-collecting game where the customer needs to get an $A, C, E, I, L, T, U, V,$ and Y (to spell *VALUE CITY*). Also, compare the exact $E(X)$ derived in Example 5.3.4 with its estimate based on Euler's approximation.

Question 5.3.2 Cards are dealt one by one from the top of a shuffled poker deck until the first ace appears. On the average, how many cards are above that first ace?

Question 5.3.3 Suppose X_i is a random variable for which $E(X_i) = \mu$, $i = 1, 2, \ldots, n$. Under what conditions will

$$E\left(\sum_{i=1}^{n} a_i X_i\right) = \mu$$

Question 5.3.4 Let $X_1, X_2, \ldots, X_n$ be a random sample of size n drawn from a population with mean μ. The *sample mean*, denoted by $\overline{X}$, is the arithmetic average of the X_i's:

$$\overline{X} = \frac{1}{n} \sum_{i=1}^{n} X_i$$

Find $E(\overline{X})$.

Question 5.3.5 Suppose the daily closing price of a stock goes up an eighth of a point with probability p and down an eighth of a point with probability q, where $p > q$. After n days how much gain can we expect the stock to have achieved? Assume the daily price fluctuations are independent events.

Question 5.3.6 Eight cards are drawn from a poker deck. What is the expected number of clubs?

Question 5.3.7 Show directly that the expected value for the letter-stuffing problem (Example 5.3.2) is 1. (*Hint:* Let X = total number of matches. Then $X = 0, 1, 2, \ldots, n - 2, n$. Recall from Example 3.7.2 that $P(X = 0) = (1/n!) \cdot$ number of derangements of n objects. Also, $P(X = k) = \binom{n}{k} P$ (first k letters all match and none of the remaining $n - k$ match).)

Question 5.3.8 If $X = 3X_1 + 4X_2 + 2X_3$, where

$$f_{X_1}(x) = \binom{6}{x}\left(\frac{1}{3}\right)^x \left(\frac{2}{3}\right)^{6-x}, \qquad x = 0, 1, \ldots, 6$$

$$f_{X_2}(x) = \frac{e^{-4} 4^x}{x!}, \qquad\qquad x = 0, 1, 2, \ldots$$

and

$$f_{X_3}(x) = 5e^{-5x}, \qquad\qquad x > 0$$

find $E(5X)$.

Question 5.3.9 An urn contains n chips numbered 1 through n. A sample of size r (less than n) is drawn. What is the expected value of the sum of the chips in the sample? Does it matter if the selections are made with replacement or without replacement?

Question 5.3.10 Sharpshooting competition at a certain level requires each contestant to take 10 shots with each of two different hand guns. Final scores are computed by taking a weighted average of four times the number of bulls-eyes made with the first gun plus six times the number gotten with the second. If Cathie has a 30% chance of hitting the bulls-eye with each shot from the first gun and a 40% chance with each shot from the second gun, what is her expected score?

Question 5.3.11 Ten fair dice are thrown. What is the expected value of the sum of the faces showing?

Question 5.3.12 A die has two of its sides colored red, one white, and three blue. On the average, how many times does the die have to be tossed in order for each color to appear at least once?

Question 5.3.13 Two college football teams play each other the last game of the season to decide the mythical state championship. The winner gets possession of a silver trophy. Suppose that during the last 20 years, State has won 12 times and Tech has won 8 times. How often should we expect the trophy to have changed hands? (What assumption are you making?)

The utility of linear combinations notwithstanding, there are many situations that call for the expected value of a *nonlinear* function of one or more random variables. Extending what we have already done to cover this more general case is not particularly difficult: The technique introduced in the proof of Theorem 5.3.1 for finding the expected value of $X + Y$ also works for arbitrary functions.

Theorem 5.3.2. If X is a discrete random variable with pdf $f_X(x)$ and if $g(x)$ is any function of X, then

$$E[g(X)] = \sum_{\text{all } x} g(x) f_X(x)$$

provided

$$\sum_{\text{all } x} |g(x)| f_X(x) < \infty$$

If X is a continuous random variable,

$$E[g(X)] = \int_{-\infty}^{\infty} g(x) f_X(x) \, dx$$

provided

$$\int_{-\infty}^{\infty} |g(x)| \cdot f_X(x) \, dx < \infty$$

Proof. We will prove the result for the discrete case. See (75) for details showing how the argument is modified when $f_X(x)$ is continuous.

Let $Y = g(X)$. The set of all possible x-values, $x_1, x_2, \ldots$, will give rise to a set of y-values, $y_1, y_2, \ldots$, where, in general, more than one x may be associated with a given y. Let S_j be the set of x's for which $g(x) = y_j$ (so $\cup_j S_j$ is the entire set of x-values for which $f_X(x)$ is defined). Obviously, $P(Y = y_j) = P(X \in S_j)$, and we can write

$$E(Y) = \sum_j y_j \cdot P(Y = y_j) = \sum_j y_j \cdot P(X \in S_j)$$

$$= \sum_j y_j \sum_{x \in S_j} f_X(x)$$

$$= \sum_j \sum_{x \in S_j} y_j \cdot f_X(x)$$

$$= \sum_j \sum_{x \in S_j} g(x) f_X(x) \qquad \text{(Why?)}$$

$$= \sum_{\text{all } x} g(x) f_X(x)$$

Since it is assumed that $\sum_{\text{all } x} |g(x)| f_X(x) < \infty$, the statement of the theorem holds.

Example 5.3.7

Suppose X is a random variable whose pdf is nonzero only for the three values -2, 1, and $+2$:

x	$f_X(x)$
-2	$\frac{5}{8}$
1	$\frac{1}{8}$
2	$\frac{2}{8}$
	1

Let $Y = g(X) = X^2$. Verify the statement of Theorem 5.3.2 by computing $E(Y)$ two ways—first, by finding $f_Y(y)$ and summing $y \cdot f_Y(y)$ over y and, second, by summing $g(x) \cdot f_X(x)$ over x.

By inspection, the pdf for Y is defined for only two values, 1 and 4:

$y \ (= x^2)$	$f_Y(y)$
1	$\frac{1}{8}$
4	$\frac{7}{8}$
	1

Following the first approach just mentioned gives

$$E(Y) = \sum_y y \cdot f_Y(y) = 1 \cdot (\tfrac{1}{8}) + 4 \cdot (\tfrac{7}{8})$$

$$= \tfrac{29}{8}$$

To find the expected value using Theorem 5.3.2, we take

$$E[g(X)] = \sum_x x^2 \cdot f_X(x) = (-2)^2 \cdot \tfrac{5}{8} + (1)^2 \cdot \tfrac{1}{8} + (2)^2 \cdot \tfrac{2}{8}$$

with the sum here reducing to the answer we already found, $\tfrac{29}{8}$.

For this particular situation, neither approach was easier than the other. In general, that will not be the case. Finding $f_Y(y)$ is often quite difficult, and on those occasions Theorem 5.3.2 can be of great benefit.

Example 5.3.8

In one of the early applications of probability to physics, James Clerk Maxwell (1831–1879) showed that the speed S of a molecule in a perfect gas has a density function given by

$$f_S(s) = 4\sqrt{\frac{a^3}{\pi}} \, s^2 e^{-as^2}, \qquad s > 0$$

where a is a constant depending on the temperature of the gas and the mass of the particle. What is the average *energy* of a molecule in a perfect gas?

Let m denote the molecule's mass. Recall from physics that energy (W), mass (m), and speed (S) are related through the equation

$$W = \frac{1}{2}mS^2 = g(S)$$

To find $E(W)$ we appeal to the second part of Theorem 5.3.2:

$$E(W) = \int_0^\infty g(s)f_S(s) \, ds$$

$$= \int_0^\infty \frac{1}{2}ms^2 \cdot 4\sqrt{\frac{a^3}{\pi}} \, s^2 e^{-as^2} \, ds$$

$$= 2m\sqrt{\frac{a^3}{\pi}} \int_0^\infty s^4 e^{-as^2} \, ds$$

Make the substitution $t = as^2$. Then

$$E(W) = \frac{m}{a\sqrt{\pi}} \int_0^\infty t^{3/2} e^{-t} \, dt$$

$$= \frac{m}{a\sqrt{\pi}} \Gamma\left(\frac{5}{2}\right) \qquad \text{(See Theorem 6.6.1.)}$$

$$= \frac{m}{a\sqrt{\pi}} \left(\frac{3}{2}\right)\left(\frac{1}{2}\right)\sqrt{\pi}$$

$$= \frac{3m}{4a}$$

(Find $E(W)$ the direct way: Differentiate $F_W(w) = P(W \le w)$ to get $f_W(w)$; then integrate $w \cdot f_W(w)$.)

Example 5.3.9

A point is chosen at random along a line of length l, dividing the line into two segments. What is the expected value of the ratio of the shorter segment to the longer segment?

Without loss of generality, we can think of the original line as being the unit interval. Then the ratio

$$\frac{\text{shorter segment}}{\text{longer segment}}$$

has two expressions, depending on the location of the random point (see Figure 5.3.3):

$$Q = \frac{\text{shorter segment}}{\text{longer segment}} = \begin{cases} \dfrac{x}{1-x} & 0 \le x \le \dfrac{1}{2} \\[2mm] \dfrac{1-x}{x} & \dfrac{1}{2} < x \le 1 \end{cases}$$

Since $f_X(x) = 1$ for $0 \le x \le 1$,

$$E(Q) = \int_0^{\frac{1}{2}} \frac{x}{1-x} \cdot 1 \, dx + \int_{\frac{1}{2}}^1 \frac{1-x}{x} \cdot 1 \, dx$$

Writing the second integrand as $(1/x - 1)$ gives

$$\int_{\frac{1}{2}}^1 \frac{1-x}{x} \cdot 1 \, dx = \int_{\frac{1}{2}}^1 \left(\frac{1}{x} - 1\right) dx = (\ln x - x) \Big|_{\frac{1}{2}}^1$$

$$= \ln 2 - \frac{1}{2}$$

By symmetry, the two integrals are the same, so

$$E(Q) = 2 \ln 2 - 1 \doteq 0.39$$

On the average, the longer segment will be a little more than $2\frac{1}{2}$ times the length of the shorter segment.

0 $\frac{1}{2}$ 1

Figure 5.3.3

Example 5.3.10

Consolidated Industries is planning to market a new product and they are trying to decide how many to manufacture. They estimate that each item sold will return a profit of m dollars; each one not sold represents a loss of n dollars. Furthermore, they suspect the demand for the product, V, will have an exponential distribution,

$$f_V(v) = \lambda e^{-\lambda v}, \qquad v > 0$$

How many items should the company produce if they want to maximize their expected profit? (Assume n, m, and λ are known.)

If a total of x items are made, the company's profit function, Q, is given by

$$Q = Q(v) = \begin{cases} mv - n(x-v) & v < x \\ mx & v \geq x \end{cases}$$

It follows that their *expected* profit is

$$E(Q) = \int_0^\infty Q \cdot f_V(v) \, dv$$

$$= \int_0^x [(m+n)v - nx]\lambda e^{-\lambda v} \, dv + \int_x^\infty mx \cdot \lambda e^{-\lambda v} \, dv \qquad (5.3.2)$$

The integration here is straightforward, though a bit tedious. Equation 5.3.2 eventually simplifies to

$$E(Q) = \frac{m+n}{\lambda} - \left(\frac{m+n}{\lambda}\right)e^{-\lambda x} - nx$$

To find the optimal production level, we need to solve $dE(Q)/dx = 0$ for x. But

$$\frac{dE(Q)}{dx} = (m+n)e^{-\lambda x} - n$$

and the latter equals 0 when

$$x = -\frac{1}{\lambda} \ln\left(\frac{n}{m+n}\right)$$

(How many items should the company manufacture if the demand function is

$$f_V(v) = 0.0001e^{-0.0001v}, \qquad v > 0$$

and m and n are fixed at 50¢ and \$2, respectively?)

Question 5.3.14 Let X have the probability density function

$$f_X(x) = \begin{cases} 2(1-x) & 0 < x < 1 \\ 0 & \text{elsewhere} \end{cases}$$

Suppose $Y = g(X) = X^3$. Find $E(Y)$ two different ways.

Question 5.3.15 A tool-and-die company makes castings for steel stress-monitoring gauges. Their annual profit Q, in hundreds of thousands of dollars, can be expressed as a function of product demand, x:

$$Q = Q(x) = 2(1 - e^{-2x})$$

Suppose the demand (in thousands) for their castings follows an exponential pdf, $f_X(x) = 6e^{-6x}$, $x > 0$. Find the company's expected profit.

Theorem 5.3.2 can be extended to include functions of any number of random variables. We formally state the generalization only for the bivariate case, but the examples that follow involve functions of two, three, and four random variables. For a proof of Theorem 5.3.3, see (64).

Theorem 5.3.3. Suppose X and Y are discrete random variables with joint density $f_{X,Y}(x, y)$. Let $g(x, y)$ be any function of X and Y. Then

$$E[g(X, Y)] = \sum_{\text{all } x} \sum_{\text{all } y} g(x, y) \cdot f_{X,Y}(x, y)$$

provided

$$\sum_{\text{all } x} \sum_{\text{all } y} |g(x, y)| \cdot f_{X,Y}(x, y) < \infty$$

If X and Y are continuous random variables,

$$E[g(X, Y)] = \int_{-\infty}^{\infty} \int_{-\infty}^{\infty} g(x, y) \cdot f_{X,Y}(x, y)\, dy\, dx$$

provided

$$\int_{-\infty}^{\infty} \int_{-\infty}^{\infty} |g(x, y)| \cdot f_{X,Y}(x, y)\, dx\, dy < \infty$$

Example 5.3.11

Consider a triangle ABC with a point P being chosen at random along AB and a point Q chosen at random along AC (see Figure 5.3.4). Show that the expected value of the area of the quadrilateral $PBCQ$ is three-fourths of the area of the triangle ABC.

To simplify notation, we introduce the symbols x, y, b, and c to denote the lengths of the line segments shown in Figure 5.3.4. Note that the area of $PBCQ$ can be written as the difference of the areas of two triangles:

$$\text{Area of } PBCQ = \text{area of } ABC - \text{area of } APQ$$

$$= \frac{1}{2} bc \sin \theta - \frac{1}{2} xy \sin \theta$$

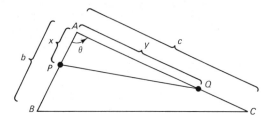

Figure 5.3.4

Since the joint pdf of the random variables X and Y is

$$f_{X,Y}(x, y) = \begin{cases} \dfrac{1}{bc} & 0 \le x \le b,\, 0 \le y \le c \\ 0 & \text{elsewhere} \end{cases}$$

it follows that

$$E(\text{area of } PBCQ) = \int_0^c \int_0^b \left(\frac{1}{2} bc \sin \theta - \frac{1}{2} xy \sin \theta \right) \frac{1}{bc}\, dx\, dy$$

which reduces to

$$E(\text{area of } PBCQ) = \left(\frac{3}{8} \right) bc \sin \theta$$

Therefore,

$$\frac{E(\text{area of } PBCQ)}{\text{area of } ABC} = \frac{\left(\frac{3}{8}\right) bc \sin \theta}{\left(\frac{1}{2}\right) bc \sin \theta} = \frac{3}{4}$$

the latter being the ratio we set out to establish.

Example 5.3.12

An electrical circuit has three resistors, R_X, R_Y, and R_Z, wired in parallel (see Figure 5.3.5). The nominal resistance of each is 15 Ω, but their *actual* resistances, X, Y, and Z, vary between 10 Ω and 20 Ω according to the joint pdf,

$$f_{X,Y,Z}(x, y, z) = \frac{1}{675{,}000} (xy + xz + yz),$$

$$10 \le x \le 20,\ 10 \le y \le 20,\ 10 \le z \le 20$$

What is the expected resistance for the circuit?

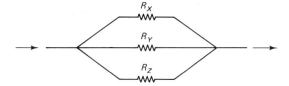

Figure 5.3.5

Let R denote the circuit's resistance. A well-known result in physics is that

$$\frac{1}{R} = \frac{1}{X} + \frac{1}{Y} + \frac{1}{Z}$$

or, equivalently,

$$R = \frac{XYZ}{XY + XZ + YZ} = R(X, Y, Z)$$

Integrating $R(x, y, z) \cdot f_{X,Y,Z}(x, y, z)$ shows that the expected resistance is 5.0:

$$E(R) = \int_{10}^{20} \int_{10}^{20} \int_{10}^{20} \frac{xyz}{xy + xz + yz} \cdot \frac{1}{675,000}(xy + xz + yz) \, dx \, dy \, dz$$

$$= \frac{1}{675,000} \int_{10}^{20} \int_{10}^{20} \int_{10}^{20} xyz \, dx \, dy \, dz$$

$$= 5.0$$

(What would $E(R)$ be if X, Y, and Z were independent and each had a uniform pdf over the interval $(10, 20)$?)

Example 5.3.13

Two points, $Q = (x, y)$ and $Q' = (x', y')$, are chosen at random inside a square having sides of length s. What is the expected value of D^2, the square of the distance between Q and Q' (see Figure 5.3.6)?

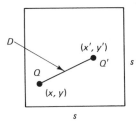

Figure 5.3.6

Written in terms of the coordinates of Q and Q', D^2 is a function of *four* random variables,

$$D^2 = g(X, X', Y, Y') = (X' - X)^2 + (Y' - Y)^2$$

Therefore,

$$E(D^2) = \int_0^s \int_0^s \int_0^s \int_0^s [(x' - x)^2$$

$$+ (y' - y)^2] \cdot f_{X, X', Y, Y'}(x, x', y, y') \, dx \, dx' \, dy \, dy'$$

Since X, X', Y, and Y' are all to be chosen at random over the interval $(0, s)$,

$$f_{X, X', Y, Y'}(x, x', y, y') = f_X(x) \cdot f_{X'}(x') \cdot f_Y(y) \cdot f_{Y'}(y') = \frac{1}{s^4}$$

so the expression for $E(D^2)$ simplifies to

$$E(D^2) = \frac{1}{s^4} \int_0^s \int_0^s \int_0^s \int_0^s [(x' - x)^2 + (y' - y)^2] \, dx \, dx' \, dy \, dy'$$

The integrations here are lengthy but straightforward; this finally reduces to

$$E(D^2) = \frac{a^2}{3}$$

(Is $E(D) = a/\sqrt{3}$?)

Question 5.3.16 Two points, X and Y, are chosen at random along perpendicular sides of the unit square. What is the expected value of the area of the rectangle whose sides are of length X and Y?

Question 5.3.17 Suppose $f_{X, Y}(x, y) = e^{-x-y}$, $x > 0$, $y > 0$. Find $E(X + Y)$.

Question 5.3.18 An urn contains n chips numbered 1 through n. A sample of size 2 is drawn without replacement. Show that the expected value of the product of the two drawn is

$$(\tfrac{1}{12})(n + 1)(3n + 2)$$

The final result in this section is a special case of Theorem 5.3.3 dealing with the expected value of a *product* of random variables. Unlike the opening theorem in this section that dealt with expected values of sums, the statement we are about to see for products is true *only if the variables are independent*.

Theorem 5.3.4. If X and Y are independent random variables,

$$E(XY) = E(X) \cdot E(Y)$$

Proof. Let X and Y both be discrete. Using Theorems 4.4.1 and 5.3.3, we can easily establish the desired factorization:

$$E(XY) = \sum_{\text{all } x} \sum_{\text{all } y} xy f_{X, Y}(x, y)$$

$$= \sum_{\text{all } x} \sum_{\text{all } y} xy f_X(x) \cdot f_Y(y)$$

$$= \sum_{\text{all } x} x f_X(x) \cdot \left[\sum_{\text{all } y} y f_Y(y) \right]$$

$$= E(X) \cdot E(Y)$$

The proof for continuous X and Y is left as an exercise.

Comment

Theorem 5.3.4 is not an if and only if statement: Just because $E(XY)$ equals $E(X) \cdot E(Y)$, it does not follow that X and Y are independent. As an illustration, suppose X and Y are defined over the sample space of four (equally likely) points 1, 2, 3, and 4 according to the equations

$$X(1) = X(2) = 1; \qquad\qquad X(3) = X(4) = 0$$
$$Y(1) = 1, \quad Y(2) = -1; \quad Y(3) = Y(4) = 0 \qquad (5.3.3)$$

Then $f_X(0) = \frac{1}{2}, f_X(1) = \frac{1}{2}, f_Y(-1) = \frac{1}{4}, f_Y(0) = \frac{1}{2}$, and $f_Y(1) = \frac{1}{4}$. Notice, here, that X and Y are *not* independent—for example,

$$P(Y = 1 \,|\, X = 1) = \tfrac{1}{2} \neq P(Y = 1) = \tfrac{1}{4}$$

Looking at the two *expected* values, we see that $E(X) = \frac{1}{2}$ and $E(Y) = 0$, making $E(X) \cdot E(Y) = 0$. Is $E(XY)$ similarly equal to 0? Yes. Consider the joint pdf of X and Y. Table 5.3.2 shows the values of $f_{X,Y}(x, y)$ that derive from Equation 5.3.3.

TABLE 5.3.2 THE JOINT PDF OF X AND Y

		Y		
		−1	0	1
X	0	0	$\frac{1}{2}$	0
	1	$\frac{1}{4}$	0	$\frac{1}{4}$

By Theorem 5.3.3,

$$E(XY) = -1(\tfrac{1}{4}) + 1(\tfrac{1}{4}) = 0$$

so $E(XY)$ factors into $E(X) \cdot E(Y)$, even though X and Y are not independent.

Question 5.3.19 Let $X_1, X_2, \ldots, X_n$ be a set of mutually independent, continuous random variables. Show that

$$E(X_1 X_2 \cdots X_n) = E(X_1) \cdot E(X_2) \cdots E(X_n)$$

Question 5.3.20 Two fair dice are tossed. What is the expected value of the product of the faces showing?

5.4 THE VARIANCE

The expected value is an effective measure of central tendency, but it fails to address a second important aspect of the behavior of a random variable—its dispersion. Unless we are also provided with some indication of how *spread out* a pdf is, the expected value by itself can be misleading, and we are prey to absurdities such as: A person whose head is in the freezer and feet are in the oven is *on the average* quite comfortable. Imagine two football teams—both with interior lines averaging 200 pounds—but one having all five players weighing close to that figure and the other lining up four 150-pounders next to a 400-lb behemoth at left tackle. One would expect the opposition to recognize the tactical significance of such a difference in weight dispersion rather quickly. Or, consider two games of chance—one, where $1 is won or lost with equal probability; the other, where the stake is $100,000. Both wagers have the same expected value of 0, but these are certainly very different games (or, at the very least, played by decidedly different people).

Given, then, that the assessment of a random variable's dispersion, or variability, is a worthwhile pursuit, how should we proceed? One seemingly reasonable approach would be to average, in the generalized sense of Definition 5.2.1, the deviations of the values of X from their mathematical expectation. But that proves futile, since the numerical value of such an average will necessarily be 0:

$$E(X - \mu) = E(X) - \mu = \mu - \mu = 0 \tag{5.4.1}$$

Another possibility is to modify the left-hand side of Equation 5.4.1 by making all deviations positive—that is, replace $E(X - \mu)$ with $E(|X - \mu|)$. This does work, and $E(|X - \mu|)$ *is* occasionally used to measure dispersion, but the absolute value is troublesome mathematically. It does not have a simple arithmetic formula, nor is it a differentiable function. *Squaring* the deviations proves to be a much more attractive solution.

Definition 5.4.1. The *variance* of a random variable X, denoted by Var(X), is the expected value of its squared deviations from μ:

$$\text{Var}(X) = E((X - \mu)^2)$$

where $\mu = E(X)$. (If $E(X^2)$ is not finite, the variance is not defined.) The symbol σ^2 is often used in place of Var(X).

Comment

One unfortunate consequence of Definition 5.4.1 is that the units for the variance are the square of the units for X: If X is measured in inches, the units for Var(X) are square inches. This causes obvious problems in relating variances back to sample values. Applied statisticians sidestep the difficulty by expressing dispersion in terms of the *standard deviation,* defined to be the square root of the variance:

$$\text{Standard deviation} = \sigma = \sqrt{\text{Var}(X)}$$

Comment

The analogy between the expected value of a random variable and the center of gravity of a physical system has already been pointed out. A similar duality holds between the variance and what physicists call a *moment of inertia*. If a set of k weights having masses $m_1, m_2, \ldots, m_k$ are positioned along a (weightless) rigid bar at distances $r_1, r_2, \ldots, r_k$ from an axis of rotation (see Figure 5.4.1), the *moment of inertia* of the system is defined to be the sum $\sum_{i=1}^{k} m_i r_i^2$. If the masses were values of a discrete pdf at points $x_1, x_2, \ldots, x_k$ and if the axis of rotation were set at μ, the numerical value of $\sum_{i=1}^{k} m_i r_i^2$ would be the same as what Definition 5.4.1 gives for the variance of X.

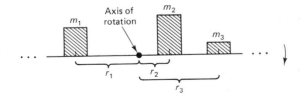

Figure 5.4.1

For some random variables, Definition 5.4.1 does not provide a particularly convenient computing formula for Var(X). Theorem 5.4.1 gives an equivalent expression that is often much easier to use. Its proof follows from a simple application of the distributive property of expected values.

Theorem 5.4.1. Let X be a random variable having mean μ and for which $E(X^2)$ is finite. Then

$$\mathrm{Var}(X) = E(X^2) - \mu^2$$

Proof. By Definition 5.4.1 and the corollary to Theorem 5.3.1,

$$\mathrm{Var}(X) = E((X - \mu)^2) = E(X^2 - 2\mu X + \mu^2)$$
$$= E(X^2) - 2\mu E(X) + \mu^2 = E(X^2) - 2\mu^2 + \mu^2$$
$$= E(X^2) - \mu^2$$

Example 5.4.1

Two red and three white chips are placed in an urn. Suppose two are drawn out at random, *without replacement*. Let X denote the number of red chips in the sample. Find Var(X).

Regardless of which formula we elect to use, Definition 5.4.1 or Theorem 5.4.1, we first need to find μ. Here, since X is hypergeometric,

$$\mu = E(X) = \sum_{x=0}^{2} x \cdot \frac{\binom{2}{x}\binom{3}{2-x}}{\binom{5}{2}} = 0.8$$

Using Definition 5.4.1 the variance becomes

$$\mathrm{Var}(X) = E[(X - \mu)^2] = \sum_{\mathrm{all}\,x} (x - \mu)^2 \cdot f_X(x)$$

$$= (0 - 0.8)^2 \cdot \frac{\binom{2}{0}\binom{3}{2}}{\binom{5}{2}} + (1 - 0.8)^2 \cdot \frac{\binom{2}{1}\binom{3}{1}}{\binom{5}{2}} + (2 - 0.8)^2 \cdot \frac{\binom{2}{2}\binom{3}{0}}{\binom{5}{2}}$$

$$= 0.36$$

To find $\text{Var}(X)$ using Theorem 5.4.1, we begin by computing $E(X^2)$. From Theorem 5.3.2,

$$E(X^2) = \sum_{\text{all } x} x^2 \cdot f_X(x) = 0^2 \cdot \frac{\binom{2}{0}\binom{3}{2}}{\binom{5}{2}} + 1^2 \cdot \frac{\binom{2}{1}\binom{3}{1}}{\binom{5}{2}} + 2^2 \cdot \frac{\binom{2}{2}\binom{3}{0}}{\binom{5}{2}}$$

$$= 1.00$$

Therefore, according to our second formula,

$$\text{Var}(X) = E(X^2) - \mu^2 = 1.00 - (0.8)^2$$

$$= 0.36$$

confirming what we calculated earlier. (We derive a general expression for the variance of a hypergeometric random variable in Example 5.4.7.)

Question 5.4.1 Find $\text{Var}(X)$ for the urn problem of Example 5.4.1 if the sampling is done *with* replacement.

Question 5.4.2 Listed below is the length distribution of World Series competition for the 50 years from 1926 to 1975.

WORLD SERIES LENGTHS

Number of games	Number of years
4	9
5	11
6	8
7	22

Let X denote the length of the World Series in a year picked at random from 1926 to 1975. Find the variance of X.

Question 5.4.3 Find the variance of X if

$$f_X(x) = \begin{cases} \frac{3}{4} & 0 \leq x \leq 1 \\ \frac{1}{4} & 2 \leq x \leq 3 \\ 0 & \text{elsewhere} \end{cases}$$

Question 5.4.4 Ten equally qualified applicants, six men and four women, apply for three lab technician positions. Unable to justify choosing any of the applicants over all the others, the personnel director decides to select the three at random. Let Y denote the number of men hired. Compute the standard deviation of Y.

Question 5.4.5 Compute the variance for a uniform random variable defined on the unit interval.

Question 5.4.6 Find the standard deviation of the sentence-length random variable described in Example 5.2.7.

Question 5.4.7 Use Theorem 5.4.1 to find the variance of the random variable X, where

$$f_X(x) = 3(1 - x)^2, \qquad 0 < x < 1$$

Question 5.4.8 Consider the pdf defined by

$$f_X(x) = \frac{2}{x^3}, \qquad x \geq 1$$

Show that (a) $\int_1^\infty f_X(x) \, dx = 1$, (b) $E(X) = 2$, and (c) $\text{Var}(X)$ is not finite.

Question 5.4.9 Semester grades (expressed as percentages) for students in a Physics for Poets class are known to be described fairly well by the pdf

$$f_X(x) = 6x(1 - x), \qquad 0 < x < 1$$

Anyone earning a grade of less than 60% fails. Five suitemates are among the students enrolled in the course. Let Y denote the number of failures that will be recorded among that group. Find the standard deviation of Y.

Question 5.4.10 Frankie and Johnny play the following game. Frankie selects a number at random from the interval $[a, b]$. Johnny, not knowing Frankie's number, is to pick a second number from that same interval and pay Frankie an amount W equal to the squared difference between the two (so $0 \leq W \leq (b - a)^2$). What should Johnny's strategy be if he wants to minimize his expected loss?

Certain random variables have pdf's whose particular mathematical formulation would make the computation of $\text{Var}(X)$ very difficult no matter which of the previous two approaches we elected to try. (Density functions having combinatorial factors typically fall into this category.) Fortunately, a third option is available—as an intermediate step we can compute *factorial moments*. From those, we can "recover" the variance.

By definition, the kth factorial moment, μ_k^*, is given by

$$\mu_k^* = E[X(X - 1)(X - 2) \cdots (X - k + 1)]$$

Written in terms of μ_k^*, the variance of X becomes

$$\text{Var}(X) = \mu_2^* + \mu_1^* - (\mu_1^*)^2 \qquad (5.4.2)$$

Among the random variables for which Equation 5.4.2 is preferable to either Definition 5.4.1 or Theorem 5.4.1 is the Pascal. Example 5.4.2 shows why.

Example 5.4.2

The geometric pdf that has been discussed several times—most recently in Example 5.2.4—is sometimes reparameterized and referred to as *Pascal's distribution*. Specifically, X is a Pascal random variable if

$$f_X(x) = \left(\frac{1}{1 + \mu} \right) \left(\frac{\mu}{1 + \mu} \right)^x, \qquad x = 0, 1, 2, \ldots, \mu > 0$$

Find the variance of X.

From Question 5.2.11, we know that $E(X) = \mu$, so to find $\text{Var}(X)$ via Equation 5.4.2, all we need to compute is μ_2^*, the second factorial moment. By definition,

$$\mu_2^* = E[X(X - 1)] = \sum_{i=0}^{\infty} i(i - 1)\left(\frac{1}{1 + \mu}\right)\left(\frac{\mu}{1 + \mu}\right)^i$$

Let $r = \mu/(1 + \mu)$. Then

$$\mu_2^* = \frac{r^2}{1 + \mu} \cdot \frac{d^2}{dr^2}\left(\sum_{i=0}^{\infty} r^i\right)$$

$$= \frac{r^2}{1 + \mu} \cdot \frac{d^2}{dr^2}\left(\frac{1}{1 - r}\right)$$

$$= \frac{r^2}{1 + \mu} \cdot \frac{2}{(1 - r)^3} = \frac{r^3}{\mu} \cdot \frac{1}{(1 - r)^3}$$

$$= 2\mu^2$$

It follows that

$$\text{Var}(X) = 2\mu^2 + \mu - \mu^2$$

$$= \mu^2 + \mu$$

(If the mean for a geometric random variable is $1/p$, why is the mean for a Pascal random variable *not* equal to $1/[1/(1 + \mu)] = \mu + 1$?)

Question 5.4.11 Find the variance for a geometric random variable

$$f_X(x) = pq^{x-1}, \qquad x = 1, 2, \ldots$$

and compare your answer to the variance found for Pascal's distribution.

Question 5.4.12 Use factorial moments to show that the variance of a binomial random variable is npq. (We will derive this result a much easier way in Example 5.4.5.)

Examined in Section 5.3 were some of the properties associated with expected values. Analogous results for the variance are motivated by asking two simple questions: (1) how is the variance affected by a linear transformation, and (2) if $Y = X_1 + X_2 + \cdots + X_n$, how is the variance of Y related to the variances of the X_i's?

Theorem 5.4.2. Let X and Y be random variables and a and b be constants such that $Y = aX + b$. Then

$$\text{Var}(Y) = a^2\text{Var}(X)$$

Proof. Let $Y = aX + b$. Since $E(aX + b) = a\mu + b$,

$$\text{Var}(Y) = E([(aX + b) - (a\mu + b)]^2)$$
$$= E(a^2(X - \mu)^2)$$
$$= a^2 E((X - \mu)^2)$$
$$= a^2 \text{Var}(X)$$

Example 5.4.3

Over a recent 90-year period, temperatures recorded in Bismarck, North Dakota, in December ranged from $-36°F$ to $+72°F$ (95). Assume the standard deviation of the distribution of daily extremes is approximately $18°F$. What is the corresponding standard deviation in degrees *Celsius?*

Let X denote a temperature recorded in degrees Fahrenheit. The well-known temperature conversion formula says that if Y is the same temperature expressed in degrees Celsius, then

$$Y = \tfrac{5}{9}(X - 32) = \tfrac{5}{9}X - \tfrac{160}{9}$$

An application of Theorem 5.4.2 shows the *variance* of Y to be 100:

$$\text{Var}(Y) = (\tfrac{5}{9})^2 \text{Var}(X) = \tfrac{25}{81}(18)^2 = 100$$

The standard deviation of (Celsius) temperature extremes recorded in December in Bismarck is therefore $\sqrt{100}$, or $10°C$.

Example 5.4.4

Theorem 5.4.2 allows us to simplify the computations required to find $\text{Var}(Y)$ if the units of Y are inconvenient. For example, suppose Y were a simple equally likely random variable, but its values were 10,000, 10,002, and 10,010 (each with probability $\tfrac{1}{3}$). Computing $\text{Var}(Y)$ would be straightforward, but the calculations cumbersome: From Theorem 5.4.1,

$$\text{Var}(Y) = (10,000)^2(\tfrac{1}{3}) + (10,002)^2(\tfrac{1}{3}) + (10,010)^2(\tfrac{1}{3}) - (10,004)^2$$
$$= 18.67$$

The same answer can be found with much less work by considering the linear transformation $Y = 2X + 10,002$. The original Y-values then correspond to X-values of $-1, 0$, and 4, respectively, and *their* variance is computed easily:

$$\text{Var}(X) = (-1)^2(\tfrac{1}{3}) + (0)^2(\tfrac{1}{3}) + (4)^2(\tfrac{1}{3}) - (1)^2$$
$$= 4.67$$

Applying Theorem 5.4.2 confirms the answer found the long way:

$$\text{Var}(Y) = (2)^2 \text{Var}(X) = 4(4.67) = 18.67$$

Prior to the ready availability of electronic calculators, the use of linear transformations to rescale random variables was a widely used computational shortcut.

Question 5.4.13 If $E(X) = \mu$ and $\text{Var}(X) = \sigma^2$, show that

$$E\left(\frac{X - \mu}{\sigma}\right) = 0$$

and

$$\text{Var}\left(\frac{X - \mu}{\sigma}\right) = 1$$

Question 5.4.14 Advertising copy for the performance capabilities of a tachometer on a sports car claims it has a variance of 100 rev^2/min^2 when the car is traveling 55 mi/h. What is its standard deviation in revolutions per hour?

Question 5.4.15 Let Y be a uniform random variable defined on the interval $(1000, 3000)$. Use Theorem 5.4.2 to find the variance of Y by considering the transformation $X = (Y - 2000)/1000$.

If $Y = X_1 + X_2 + \cdots + X_n$, the expected value of Y is the sum of the corresponding $E(X_i)$'s, *whether or not the X_i's are independent*. What would be the parallel result for variances is not true: The variance of Y is not necessarily equal to the sum of the $\text{Var}(X_i)$'s. Equality *does* hold, though, if the X_i's are independent.

> **Theorem 5.4.3.** Let $X_1, X_2, \ldots, X_n$ be independent random variables and let $Y = X_1 + X_2 + \cdots + X_n$. Then
>
> $$\text{Var}(Y) = \text{Var}(X_1) + \text{Var}(X_2) + \cdots + \text{Var}(X_n)$$

Proof. We give a proof for $Y = X_1 + X_2$; an easy induction completes the argument for general n. From Theorems 5.3.1 and 5.4.1,

$$\text{Var}(Y) = E[(X_1 + X_2)^2] - [E(X_1) + E(X_2)]^2$$

After expanding the indicated squares, we can write

$$\text{Var}(Y) = E(X_1^2 + 2X_1X_2 + X_2^2) - [E(X_1)]^2 - 2E(X_1)E(X_2) - [E(X_2)]^2$$
$$= E(X_1^2) - [E(X_1)]^2 + E(X_2^2) - [E(X_2)]^2 + 2[E(X_1X_2) - E(X_1)E(X_2)]$$

$$(5.4.3)$$

By the independence of X_1 and X_2, $E(X_1X_2) = E(X_1)E(X_2)$, making the last term in Equation 5.4.3 vanish. The remaining terms combine to give the desired result: $\text{Var}(Y) = \text{Var}(X_1) + \text{Var}(X_2)$.

Example 5.4.5

The binomial random variable, being a sum of n independent Bernoulli variables, is an obvious candidate for Theorem 5.4.3. Let X_i denote the number of successes occurring on the ith trial. Then

$$X_i = \begin{cases} 1 & \text{with probability } p \\ 0 & \text{with probability } 1 - p \end{cases}$$

Write $Y = X_1 + X_2 + \cdots + X_n$. But

$$E(X_i) = 1 \cdot p + 0 \cdot q = p$$

and

$$E(X_i^2) = (1)^2 \cdot p + (0)^2 \cdot q = p$$

so

$$\text{Var}(X_i) = E(X_i^2) - [E(X_i)]^2 = p - p^2$$

$$= pq$$

It follows that the *variance of a binomial random variable is npq:*

$$\text{Var}(Y) = \sum_{i=1}^{n} \text{Var}(X_i) = npq$$

(Contrast the simplicity of this approach with the detailed summation argument needed to derive the result directly, as asked for in Question 5.4.12.)

Example 5.4.6

We have seen how Var(X), or σ^2, serves as a numerical measure of the dispersion of a probability distribution. In statistics, an analogous quantity is defined to reflect the dispersion of a random sample. Let $X_1, X_2, \ldots, X_n$ denote a set of n independent observations made on a random variable X having pdf $f_X(x)$. Let $\sigma^2 = E[(X - \mu)^2]$ denote the variance of X. The *sample variance* of the X_i's, denoted by S^2, is defined as

$$S^2 = \frac{1}{n-1} \sum_{i=1}^{n} (X_i - \bar{X})^2$$

(recall Question 5.3.4). Show that $E(S^2) = \sigma^2$.

To begin, we rewrite S^2 in a form that enables us to apply Theorem 5.4.1:

$$E(S^2) = E\left[\frac{1}{n-1} \sum_{i=1}^{n} (X_i - \bar{X})^2 \right]$$

$$= E\left[\frac{1}{n-1} \sum_{i=1}^{n} (X_i^2 - 2X_i\bar{X} + \bar{X}^2) \right]$$

$$= E\left[\frac{1}{n-1} \left(\sum_{i=1}^{n} X_i^2 - 2\bar{X} \sum_{i=1}^{n} X_i + n\bar{X}^2 \right) \right]$$

$$= E\left[\frac{1}{n-1} \left(\sum_{i=1}^{n} X_i^2 - n\bar{X}^2 \right) \right]$$

$$= \frac{1}{n-1} \sum_{i=1}^{n} E(X_i^2) - \frac{n}{n-1} E(\bar{X}^2)$$

By Theorem 5.4.1, $E(X_i^2) = \sigma^2 + \mu^2$. Also, $E(\bar{X}^2) = \sigma^2/n + \mu^2$ (see Question 5.4.19). Therefore,

$$E(S^2) = \frac{1}{n-1} \sum_{i=1}^{n} (\sigma^2 + \mu^2) - \frac{n}{n-1}\left(\frac{\sigma^2}{n} + \mu^2\right)$$

$$= \frac{1}{n-1}(n\sigma^2 + n\mu^2) - \frac{n}{n-1}\left(\frac{\sigma^2}{n} + \mu^2\right)$$

$$= \sigma^2$$

(In the language of mathematical statistics, what we have just shown is that S^2 *is an unbiased estimator for* σ^2.)

Question 5.4.16 A mason is contracted to build a short retaining wall along a patio. Plans call for the base of the wall to be a row of 50 bricks 10 in. long, each separated by $\frac{1}{2}$-in.-thick mortar. Suppose the bricks used are randomly chosen from a population of bricks whose mean length is 10 in. and whose standard deviation is $\frac{1}{32}$ in. Also, suppose the mason, on the average, will make the mortar $\frac{1}{2}$ in. thick, but the actual dimension varies from brick to brick, the standard deviation of the thicknesses being $\frac{1}{16}$ in. What is the standard deviation of L, the length of the first row of the wall? What assumptions are you making?

Question 5.4.17 A man enters a sharpshooter competition that requires participants to fire 10 shots with a handgun and 10 with a rifle. The winner is determined on the basis of a weighted sum Y, where

$$Y = 4X_1 + 6X_2$$

with X_1 and X_2 being the numbers of bulls-eyes recorded with the handgun and with the rifle, respectively. If the man has a 20% chance of hitting the bulls-eye with each shot he fires from the handgun and a corresponding 30% success rate with the rifle, what is the standard deviation associated with his final score? Assume X_1 and X_2 are independent.

Question 5.4.18 A gambler plays n hands of poker. If she wins the kth hand, she collects k dollars; if she loses the kth hand, she collects nothing. Let T denote the total amount she wins in n hands. Assuming her chances of winning each hand are constant (equal to p) and are independent of her success or failure at any other hand, find $E(T)$ and $\text{Var}(T)$.

Question 5.4.19 Let $X_1, X_2, \ldots, X_n$ be a random sample from a pdf whose variance is σ^2. Show that the variance of $\bar{X}$, the sample mean, is σ^2/n.

Question 5.4.20 Carry out the induction argument to complete the proof of Theorem 5.4.3.

Question 5.4.21 An electric circuit has six resistors wired in series, each nominally being 5 Ω. What is the maximum standard deviation that can be allowed in the manufacture of these resistors if the combined circuit resistance is to have a standard deviation no greater than 0.4 Ω?

Generalizing the statement of Theorem 5.4.3 to cover sums of *arbitrary* random variables (independent or dependent) requires the introduction of a new term, the *covariance of X and Y*.

> **Definition 5.4.2.** Let X and Y be random variables with means μ_X and μ_Y, respectively. The *covariance of X and Y*, written $\text{Cov}(X, Y)$, is given by
>
> $$\text{Cov}(X, Y) = E((X - \mu_X)(Y - \mu_Y))$$

An expression for the covariance that is often more convenient is found by taking the expected value of $(X - \mu_X)(Y - \mu_Y)$ term by term. We leave the details as an exercise.

> **Theorem 5.4.4.** For any random variables X and Y with means μ_X and μ_Y,
>
> $$\text{Cov}(X, Y) = E(XY) - \mu_X\mu_Y$$

Having defined the covariance, we can now restate Theorem 5.4.3 in complete generality.

> **Theorem 5.4.5.** Let $X_1, X_2, \ldots, X_n$ be any set of random variables. Let $Y = X_1 + X_2 + \cdots + X_n$. Then
>
> $$\text{Var}(Y) = \sum_{i=1}^{n} \text{Var}(X_i) + 2 \sum_{j<k} \text{Cov}(X_j, X_k)$$

Proof. The proof is a simple exercise in algebra:

$$\text{Var}(Y) = E(Y^2) - [E(Y)]^2$$

$$= E\left[\left(\sum_{i=1}^{n} X_i\right)^2\right] - \left[\sum_{i=1}^{n} E(X_i)\right]^2$$

$$= E\left[\sum_{i=1}^{n} X_i^2 + 2 \sum_{j<k} X_j X_k\right] - \left(\sum_{i=1}^{n} [E(X_i)]^2 + 2 \sum_{j<k} E(X_j)E(X_k)\right)$$

$$= \sum_{i=1}^{n} [E(X_i^2) - [E(X_i)]^2] + 2 \sum_{j<k} [E(X_j X_k) - E(X_j)E(X_k)]$$

$$= \sum_{i=1}^{n} \text{Var}(X_i) + 2 \sum_{j<k} \text{Cov}(X_j, X_k)$$

(Note that when the X_i's are independent, $\text{Cov}(X_j, X_k) = E(X_j X_k) - \mu_j \mu_k = E(X_j) \cdot E(X_k) - \mu_j \mu_k = 0$ for all j and k, and the statement of Theorem 5.4.5 reduces to the statement of Theorem 5.4.3.)

Example 5.4.7

We can use Theorem 5.4.5 to calculate the variance of a hypergeometric random variable. As a physical model, consider an urn containing N chips, r red and w white ($r + w = N$). A sample of n is to be selected *without replacement*. Let Y denote the number of red chips drawn. From Theorem 3.6.3,

$$P(Y = y) = \frac{\binom{r}{y}\binom{w}{n-y}}{\binom{N}{n}}, \qquad y = 0, 1, \ldots, \min(r, n) \qquad (5.4.4)$$

Although the expected value and variance of Y can be determined directly from their fundamental definitions, it is more instructive (and much easier!) to find $E(Y)$ and $\text{Var}(Y)$ by introducing auxiliary variables, as we did previously when dealing with the binomial. Define

$$X_i = \begin{cases} 1 & \text{if the } i\text{th chip drawn is red} \\ 0 & \text{otherwise} \end{cases}$$

Then $Y = X_1 + X_2 + \cdots + X_n$. We see immediately that $E(X_i) = r/N$, and the latter makes $E(Y) = n(r/N) = np$, where $p = r/N$. What we have shown, then, is that the expected value of Y, the number of red chips drawn, remains the same whether the sampling is done *with* replacement or *without* replacement (recall Example 5.3.1).

For the variance, though, the mode of sampling *does* make a difference. Note that the X_i's are not independent, but, by symmetry, they *are* identically distributed. Also, any pair $(X_j, X_k), j \neq k$, has the same distribution as (X_1, X_2), again by symmetry. (There is no reason to suppose that we are more (or less) likely to get red chips on, say, the third and fifth draws than on the first and second.) Since $X_i^2 = X_i$, $E(X_i^2) = E(X_i) = r/N$, and $\text{Var}(X_i) = r/N - (r/N)^2 = (r/N)(1 - r/N) = p(1 - p)$. For any $j \neq k$, $\text{Cov}(X_j, X_k) = \text{Cov}(X_1, X_2) = E(X_1 X_2) - E(X_1)E(X_2)$. But $E(X_1 X_2) = 1 \cdot P(X_1 X_2 = 1) = 1 \cdot P(\text{first chip is red and second chip is red}) = (r/N) \cdot [(r - 1)/(N - 1)]$. Therefore,

$$\text{Cov}(X_j, X_k) = \frac{r}{N} \cdot \frac{r-1}{N-1} - \frac{r}{N} \cdot \frac{r}{N} = \frac{r}{N}\left(\frac{r-1}{N-1} - \frac{r}{N}\right)$$

$$= -\frac{r}{N} \cdot \frac{N-r}{N} \cdot \frac{1}{N-1} = -\frac{p(1-p)}{N-1}$$

Applying Theorem 5.4.5, we get an expression for $\text{Var}(Y)$:

$$\text{Var}(Y) = \sum_{i=1}^{n} \text{Var}(X_i) + 2 \sum_{j<k} \text{Cov}(X_j, X_k)$$

$$= np(1-p) - 2\binom{n}{2}p(1-p) \cdot \frac{1}{N-1}$$

$$= np(1-p) \cdot \frac{N-n}{N-1}$$

Comment

The difference between the $\text{Var}(Y)$ expression for a binomial $[np(1-p)]$—recall Example 5.4.5—and its counterpart for a hypergeometric $[np(1-p) \cdot (N-n)/(N-1)]$ is important when viewed in a statistical context. Suppose p, the *true* proportion of red chips in the urn, is unknown, and our intention is to estimate it by computing the *observed* red chip proportion, Y/n, found in a sample of size n. Should we draw the n out *with replacement* or *without replacement?* Without replacement: Since $(N - n)/(N - 1) < 1$ (for $n > 1$), we see that $\text{Var}(Y/n) = (1/n^2) \cdot \text{Var}(Y)$ will be smaller—meaning p will be estimated more precisely—if we sample without replacement, rather than with replacement. The improvement afforded by the former is particularly pronounced when n is relatively large compared to N.

Question 5.4.22 Recall the letter-stuffing problem described in Example 5.3.2. Follow the method used in Example 5.4.7 to show that the variance of the number of letters put into their proper envelopes is also 1.

Question 5.4.23 Show that

$$\text{Cov}(aX + b, cY + d) = ac\,\text{Cov}(X, Y)$$

for any constants a, b, c, and d.

Question 5.4.24 Find the variance of a hypergeometric random variable (Equation 5.4.4) *directly*. (*Hint:* Begin with the factorial moment, $\mu_2^* = E(Y(Y - 1))$.)

Question 5.4.25 Let X and Y have the joint pdf

$$f_{X,Y}(x, y) = \begin{cases} \dfrac{x + 2y}{22} & (x, y) = \{(1, 1), (1, 3), (2, 1), (2, 3)\} \\ 0 & \text{elsewhere} \end{cases}$$

Find $\text{Cov}(X, Y)$.

Question 5.4.26 An urn contains n chips, numbered 1 through n. A sample of k chips is drawn out without replacement. Find the variance of the sum of the numbers on the chips drawn.

Theorems 5.4.3 and 5.4.5 are special cases of what is commonly called the *propagation-of-error* problem. Specifically, how do the *individual* variances of a set of random variables $X_1, X_2, \ldots, X_n$ affect the variance of some *function* of those variables? For example, consider a physics student trying to determine the acceleration due to gravity, g. As a theoretical exercise, it is a simple-enough calculus problem to show that the distance d traveled by a freely falling body in time t is related to g by the equation

$$d = \tfrac{1}{2}gt^2$$

(assuming the body is initially at rest). Equivalently,

$$g = \frac{2d}{t^2}$$

To estimate g empirically, the student needs to make a distance measurement and a corresponding time measurement. It must be remembered, of course, that the values actually obtained, d and t, will not be exactly correct—they are simply realizations of two random variables D and T, the latter having expected values μ_D and μ_T and variances $Var(D)$ and $Var(T)$. From past experience, the precision of the procedures giving rise to the distance and time measurements may be known. If they are, an obvious question suggests itself: Knowing $Var(D)$ and $Var(T)$, what can we say about $Var(G)$?

In general, if $Y = g(X_1, X_2, \ldots, X_n)$ and $g(X_1, X_2, \ldots, X_n)$ is some function other than a simple sum, it is very difficult to find $Var(Y)$ knowing only the variances of the X_i's. However, it *is* possible—if the X_i's are independent—to approximate $Var(Y)$ by taking a Taylor expansion of $g(X_1, X_2, \ldots, X_n)$.

Assume that each X_i has mean μ_i and variance $Var(X_i)$. By expanding $g(X_1, X_2, \ldots, X_n)$ around the point $(\mu_1, \mu_2, \ldots, \mu_n)$ and ignoring all but the first-order terms, we can write

$$Y \doteq g(\mu_1, \mu_2, \ldots, \mu_n) + (X_1 - \mu_1)\left[\frac{\partial g}{\partial X_1}\bigg|_{(\mu_1, \ldots, \mu_n)}\right]$$

$$+ (X_2 - \mu_2)\left[\frac{\partial g}{\partial X_2}\bigg|_{(\mu_1, \ldots, \mu_n)}\right] + \cdots + (X_n - \mu_n)\left[\frac{\partial g}{\partial X_n}\bigg|_{(\mu_1, \ldots, \mu_n)}\right]$$

Since the evaluated partials are constants, Y is a simple linear combination of the X_i's, and its variance derives from Theorem 5.4.3:

$$Var(Y) \doteq \left[\frac{\partial g}{\partial X_1}\bigg|_{(\mu_1, \mu_2, \ldots, \mu_n)}\right]^2 Var(X_1)$$

$$+ \left[\frac{\partial g}{\partial X_2}\bigg|_{(\mu_1, \mu_2, \ldots, \mu_n)}\right]^2 Var(X_2) + \cdots + \left[\frac{\partial g}{\partial X_n}\bigg|_{(\mu_1, \mu_2, \ldots, \mu_n)}\right]^2 Var(X_n)$$

$$(5.4.5)$$

(If the μ_i's are unknown, which will often be the case, the partial derivatives are evaluated at the vector of observations, $(x_1, x_2, \ldots, x_n)$, the latter being our best estimate of $(\mu_1, \mu_2, \ldots, \mu_n)$.)

CASE STUDY 5.2

In the typical dental X-ray machine, electrons from the cathode of an X-ray tube are decelerated by nuclei in the anode, producing Bremsstrahlung radiation (X rays) in the process. These emissions, when collimated by a lead-lined tube, effect the desired image on a sheet of film.

Tennessee state regulations (83) require that the distance Q from the focal spot on the anode of the X-ray tube to the patient's skin be at least 18 cm. On some equipment, particularly older units, that distance cannot be measured directly because the exact location of the focal spot cannot be determined just by looking at the tube's outer housing. On confronting such a tube, state inspectors resort to an indirect measuring procedure. Two films are exposed, one at the unknown distance Q and a second at a distance $Q + Z$. The two diameters X and Y of the resulting circular images are then measured (see Figure 5.4.2). By similar triangles,

$$\frac{X}{Q} = \frac{Y}{Q + Z}$$

meaning that

$$Q = \frac{XZ}{Y - X}$$

Or, phrased in the context of our previous notation,

$$Q = g(X, Y, Z) = XZ(Y - X)^{-1} \qquad (5.4.6)$$

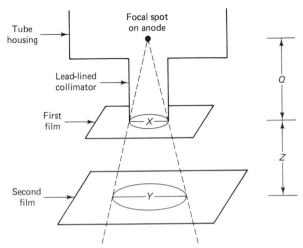

Focal spot on anode

Tube housing

Lead-lined collimator

First film — X

Q

Z

Second film — Y

Figure 5.4.2

During the course of one such inspection (47), values measured for the two diameters X and Y and the film separation distance Z were 6.4 cm, 9.7 cm, and 10.2 cm, respectively. From Equation 5.4.6, the anode-to-patient distance is estimated to be

$$q = \frac{(6.4)(10.2)}{9.7 - 6.4} = 19.8 \text{ cm}$$

indicating the unit is in compliance. If the "error" in q, though, were sufficiently large, there might still be a sizeable probability that the *true* Q was less than 18 cm, an event meaning that—despite appearances—the unit was out of compliance. It is not unreasonable, therefore, to inquire about the magnitude of Var(Q).

To apply Equation 5.4.5, we first need to compute the partial derivatives of $g(X, Y, Z)$. Here,

$$\frac{\partial g}{\partial X} = \frac{XZ}{(Y - X)^2} + \frac{Z}{(Y - X)}$$

$$\frac{\partial g}{\partial Y} = \frac{-XZ}{(Y - X)^2}$$

and

$$\frac{\partial g}{\partial Z} = \frac{X}{(Y - X)}$$

From past experience, inspectors feel that the standard deviation in any of their measurements is on the order of 0.08 cm, so $\mathrm{Var}(X) = \mathrm{Var}(Y) = \mathrm{Var}(Z) = (0.08)^2$. Substituting the variance estimates and the partial derivatives—evaluated at the point $(x, y, z) = (6.4, 9.7, 10.2)$—into Equation 5.4.5 gives

$$\mathrm{Var}(Q) \doteq \left[\frac{(6.4)(10.2)}{(9.7 - 6.4)^2} + \frac{10.2}{(9.7 - 6.4)} \right]^2 (0.08)^2$$

$$+ \left[\frac{-(6.4)(10.2)}{(9.7 - 6.4)^2} \right]^2 (0.08)^2 + \left[\frac{6.4}{(9.7 - 6.4)} \right]^2 (0.08)^2$$

$$= 0.782$$

The estimated *standard deviation* associated with the calculated value of Q is $\sqrt{0.782}$, or 0.88 cm.

We can now restate the inspector's initial concern in more-precise terms: What is the probability that the X-ray unit is out of compliance if the observed Q is 19.8 cm (1.8 cm greater than the "critical" distance) but has an associated error of 0.88 cm? Since the probability distribution of the estimate for Q is unknown, there is no way to answer such a question exactly. Approximations are available, though, and in Section 5.6 we use a result known as the *Bienaymé-Chebyshev inequality* to show that the probability that the X-ray unit is out of compliance—despite the value of q to the contrary—can be as large as 0.24.

Question 5.4.27 Recall the equation defining the gravitational constant in terms of distance and time:

$$g = \frac{2d}{t^2}$$

Suppose that the standard deviation of the measurement errors in D is 0.0025 ft and in T is 0.045 s. If the experimental apparatus is set up so that D is 4 ft, then T is approximately $\frac{1}{2}$ s. If D is set at 16 ft, T is close to 1 s. Which of these two sets of values for D and T will give a smaller variance for the calculated g?

Question 5.4.28 If d, r, and t represent distance, rate, and time, respectively, then

$$d = rt$$

Suppose an experiment is performed where both r and t are measured, the particular values observed being $r = 2.1$ ft/s and $t = 6.0$ s. Also, suppose the standard deviations associated with the measurement of r and t are 0.1 ft/s and 0.2 s, respectively. Use a first-order Taylor approximation to estimate the standard deviation associated with the calculated d.

5.5 HIGHER MOMENTS

The quantities we have identified as the mean and the variance are actually special cases of what are referred to more generally as the *moments* of a random variable. More precisely, $E(X)$ is the *first moment about the origin* and σ^2 is the *second moment about the mean*. As the terminology suggests, we will have occasion to define higher moments of X. Just as $E(X)$ and σ^2 reflect a random variable's location and dispersion, so is it possible to characterize other aspects of a distribution in terms of other moments. We will see, for example, that the skewness of a distribution—that is, the extent to which it is not symmetric around μ—can be effectively measured in terms of a *third* moment.

We begin by investigating some of the basic results associated with higher moments. The section concludes with a discussion of the moment-generating function, an infinite series that gives us a particularly powerful technique for calculating higher moments, as well as a way of deducing the distribution of sums of independent random variables.

Definition 5.5.1. Let X be any random variable with pdf $f_X(x)$. For any positive integer r, the following hold:

1. The *rth moment of X about the origin*, μ_r, is given by

$$\mu_r = E(X^r)$$

provided $\int_{-\infty}^{\infty} |x|^r \cdot f_X(x)\,dx < \infty$ (or provided the analogous condition on the *summation* of $|x|^r$ holds, if X is discrete). When $r = 1$, we usually delete the subscript and write $E(X)$ as μ rather than μ_1.

2. The *rth moment of X about the mean*, μ_r', is given by

$$\mu_r' = E[(X - \mu)^r]$$

provided the finiteness conditions of part 1 hold.

Comment

We can express μ_r' in terms of μ_j, $j = 1, 2, \ldots, r$, by simply writing out the binomial expansion of $(X - \mu)^r$:

$$\mu_r' = E[(X - \mu)^r] = \sum_{j=0}^{r} \binom{r}{j} E(X^j)(-\mu)^{r-j}$$

Thus,

$$\mu_2' = E[(X - \mu)^2] = \sigma^2 = \mu_2 - \mu_1^2$$
$$\mu_3' = E[(X - \mu)^3] = \mu_3 - 3\mu_1\mu_2 + 2\mu_1^3$$
$$\mu_4' = E[(X - \mu)^4] = \mu_4 - 4\mu_1\mu_3 + 6\mu_1^2\mu_2 - 3\mu_1^4$$

and so on.

Example 5.5.1

The normal distribution was first mentioned in Example 2.5.5. Indexed by two parameters, μ and σ^2, its pdf is given by

$$f_X(x) = \frac{1}{\sqrt{2\pi}\sigma} \cdot e^{-\frac{1}{2}[(X-\mu)/\sigma]^2}, \qquad -\infty < x < \infty$$

Consider the special case where $\mu = 0$:

$$f_X(x) = \frac{1}{\sqrt{2\pi}\sigma} \cdot e^{-(x^2/2\sigma^2)}, \qquad -\infty < x < \infty \qquad (5.5.1)$$

The symmetry (about 0) of $f_X(x)$ clearly forces every *odd* moment of X to be zero:

$$E(X^{2k+1}) = \int_{-\infty}^{\infty} x^{2k+1} \cdot \frac{1}{\sqrt{2\pi}\sigma} e^{-(x^2/2\sigma^2)} \, dx = 0, \qquad k = 0, 1, 2, \ldots$$

The *even* moments of X do *not* vanish, but a general formula for the $(2k)$th moment can be derived fairly readily. Let

$$I(t) = \int_{-\infty}^{\infty} e^{-tx^2/2\sigma^2} \, dx$$

From Equation 5.5.1,

$$I(t) = \int_{-\infty}^{\infty} e^{-tx^2/2\sigma^2} \, dx = \sqrt{2\pi}\sigma t^{-1/2} \qquad \text{(Why?)}$$

Therefore,

$$\frac{d}{dt}[I(t)] = \int_{-\infty}^{\infty} \frac{-x^2}{2\sigma^2} \cdot e^{-tx^2/2\sigma^2} \, dx = -\frac{1}{2}\sqrt{2\pi}\sigma t^{-3/2}$$

$$\frac{d^2}{dt^2}[I(t)] = \int_{-\infty}^{\infty} \left(\frac{-x^2}{2\sigma^2}\right)\left(\frac{-x^2}{2\sigma^2}\right) e^{-tx^2/2\sigma^2} \, dx = \left(-\frac{1}{2}\right)\left(-\frac{3}{2}\right)\sqrt{2\pi}\sigma t^{-5/2}$$

and, in general,

$$\frac{d^k}{dt^k}[I(t)] = \int_{-\infty}^{\infty} \left(\frac{-x^2}{2\sigma^2}\right)^k e^{-tx^2/2\sigma^2} \, dx$$

$$= \left(-\frac{1}{2}\right)\left(-\frac{3}{2}\right) \cdots \left(-\frac{2k-1}{2}\right) \cdot t^{-(2k+1)/2}\sqrt{2\pi}\sigma$$

Set $t = 1$. Then the formula for the $(2k)$th moment of a normal random variable having $\mu = 0$ follows immediately:

$$\frac{1}{\sqrt{2\pi}\sigma} \int_{-\infty}^{\infty} x^{2k} e^{-x^2/2\sigma^2} \, dx = E(X^{2k}) = (1)(3) \cdots (2k-1)\sigma^{2k} \qquad (5.5.2)$$

(Note that Equation 5.5.2 verifies that the parameter we called σ^2 in Equation 5.5.1 is actually what its name suggests, the variance of X: $\text{Var}(X) = E(X^2) - \mu^2 = (1)\sigma^{2(1)} - 0^2 = \sigma^2$.)

Question 5.5.1 Let X be a uniform random variable defined over (a, b). Find an expression for the rth moment of X about the origin. Also, use the binomial expansion as described in the comment to find $E[(X - \mu)^6]$.

Question 5.5.2 Let X have an exponential pdf with parameter λ:

$$f_X(x) = \left(\frac{1}{\lambda}\right) \cdot e^{-x/\lambda}, \qquad x > 0$$

Integrate by parts to find $E(X^3)$.

Question 5.5.3 A random variable X is said to have a *beta distribution* with parameters α and β if

$$f_X(x) = \frac{\Gamma(\alpha + \beta + 2)}{\Gamma(\alpha + 1)\Gamma(\beta + 1)} \cdot x^\alpha (1 - x)^\beta, \qquad 0 < x < 1$$

Show that

$$E(X^r) = \frac{(2\alpha + 2)(2\alpha + 4) \cdots (2\alpha + 2r)}{w(w + 2) \cdots (w + 2r - 2)}$$

where $w = 2\alpha + 2\beta + 4$.

Earlier in this chapter we encountered random variables whose means did not exist—recall, for example, the St. Petersburg paradox. More generally, there are random variables for which certain of their higher moments are finite and certain others are not finite. The following existence theorem addresses the question of whether or not a given $E(X^j)$ is finite.

Theorem 5.5.1. If the kth moment of a random variable exists, all moments of order less than k exist.

Proof. Let $f_X(x)$ be the pdf of X. By Definition 5.5.1, $E(X^k)$ exists if and only if

$$\int_{-\infty}^{\infty} |x|^k \cdot f_X(x) \, dx < \infty \qquad (5.5.3)$$

Let $1 \le j < k$. To prove the theorem we must show that

$$\int_{-\infty}^{\infty} |x|^j \cdot f_X(x) \, dx < \infty$$

is implied by Inequality 5.5.3. But

$$\int_{-\infty}^{\infty} |x|^j \cdot f_X(x) \, dx = \int_{|x| \le 1} |x|^j \cdot f_X(x) \, dx + \int_{|x| > 1} |x|^j \cdot f_X(x) \, dx$$

$$\le \int_{|x| \le 1} f_X(x) \, dx + \int_{|x| > 1} |x|^j \cdot f_X(x) \, dx$$

$$\leq 1 + \int_{|x|>1} |x|^j \cdot f_X(x)\, dx$$

$$\leq 1 + \int_{|x|>1} |x|^k \cdot f_X(x)\, dx < \infty$$

Therefore, $E(X^j)$ exists, $j = 1, 2, \ldots, k - 1$.

Example 5.5.2

Many of the random variables enjoying frequent application as statistical models have moments existing for *all* k, as does, for instance, the normal pdf described in Example 5.5.1. Still, it is not difficult to find well-known models for which this is *not* true. A case in point is the *student t*, a one-parameter model widely used in statistical inference. The pdf for a student t random variable X is given by

$$f_X(x) = \frac{\Gamma[(n + 1)/2]}{\sqrt{n\pi}\,\Gamma(n/2)(1 + x^2/n)^{(n+1)/2}}, \qquad -\infty < x < \infty,\ n \geq 1$$

(The parameter n is referred to as the number of *degrees of freedom*.) By definition, the $(2k)$th moment of a student t random variable with n degrees of freedom is the integral

$$E(X^{2k}) = \int_{-\infty}^{\infty} \frac{\Gamma[(n + 1)/2]}{\sqrt{n\pi}\,\Gamma(n/2)} \cdot \frac{x^{2k}}{(1 + x^2/n)^{(n+1)/2}}\, dx \qquad (5.5.4)$$

Is $E(X^{2k})$ finite?

Not necessarily. Recall from calculus that an integral of the form

$$\int_{-\infty}^{\infty} \frac{1}{x^\alpha}\, dx$$

will converge only if $\alpha > 1$. Therefore, if $E(X^{2k})$ is to be finite, we must have

$$2\left(\frac{n + 1}{2}\right) - 2k > 1$$

or, equivalently, $2k < n$. Thus, a student t random variable with, for instance, $n = 9$ degrees of freedom has $E(X^8) < \infty$, but no moment of order higher than eight exists.

Question 5.5.4 A random variable X is said to have a *gamma distribution* with parameters b and c if

$$f_X(x) = \frac{c^{b+1}}{\Gamma(b + 1)} \cdot x^b e^{-cx}, \qquad x > 0$$

Show that

$$E(X^r) = \frac{c^{b+1}\Gamma(b + r + 1)}{c^{b+r+1}\Gamma(b + 1)} = (b + 1)(b + 2) \cdots (b + r)c^{-r}$$

Question 5.5.5 Make the substitution $z = 1/(1 + x^2/n)$ and find an expression for the $(2k)$th moment of a student t random variable (where $2k < n$).

Question 5.5.6 If $r > 0$ and $E|X_i|^r < \infty$ for all i, show that $E|X_1 + \cdots + X_n|^r < \infty$.

Recall that the variance was introduced in Section 5.4 because information provided by the expected value is incomplete in terms of how it summarizes $f_X(x)$. We saw that by coupling a measure of dispersion (σ^2) with a measure of location (μ), a much clearer picture of a random variable's probabilistic behavior emerges. Pursuing that train of thought, it seems reasonable to ask whether any additional information of value can be gleaned by computing even higher moments of X.

The answer in some cases is yes. What μ and σ^2 say nothing about is the *shape* of $f_X(x)$. If that sort of information is needed, we can turn to the third and fourth moments.

First, consider the problem of characterizing the symmetry, or lack of symmetry, of $f_X(x)$. For any symmetric pdf, every odd moment about the mean will be 0. It makes sense, then, to use the magnitude of an odd moment—that is, the extent to which $E[(X - \mu)^{2k+1}]$ is not zero—as a measure of the extent to which a pdf is not symmetric. The *first* odd moment, of course, is unacceptable since $E[(X - \mu)] = 0$, regardless of the shape of $f_X(x)$. This leaves the third moment, $E[(X - \mu)^3]$, as the obvious (simplest) choice. To make μ_3' easier to interpret, we divide it by σ^3.

Definition 5.5.2. Let X be a random variable whose third moment exists. The symmetry of $f_X(x)$ is measured by the ratio,

$$\gamma_1 = \frac{E[(X - \mu)^3]}{\sigma^3}$$

Dimensionless, γ_1, is called the *coefficient of skewness*.

A second shape parameter in common use, the *coefficient of kurtosis,* involves the fourth moment. References to the coefficient of kurtosis often associate it with the "peakedness" of a probability distribution: Certain values of the coefficient presumably represent relatively flat (or *platykurtic*) pdf's, while other values are thought to indicate more peaked (or *leptokurtic*) pdf's. Sometimes—for example, in the family of pdf's examined in Example 5.5.3—the coefficient of kurtosis *does* coincide with our intuitive notion of what peakedness means. For certain other distributions, though, that interpretation proves to be a bit oversimplistic and somewhat misleading (see Questions 5.5.8 and 5.5.9). Nevertheless, the value of the fourth moment plays an important role in a number of approximation results, one of which we will see in Chapter 7.

Definition 5.5.3. Let X be a random variable whose fourth moment exists. The *coefficient of kurtosis,* γ_2, of $f_X(x)$ is given by

$$\gamma_2 = \frac{E[(X - \mu)^4]}{\sigma^4} - 3$$

Comment

The number 3 that appears in the definition for γ_2 is the value of $E[(X - \mu)^4]/\sigma^4$ for any normal distribution (recall Example 5.5.1). Thus the value γ_2 is sometimes referred to as a distribution's *excess of kurtosis*, the extent of its peakedness beyond what is characteristic of curves having the equation

$$y = \frac{1}{\sqrt{2\pi}\sigma} \cdot e^{-\frac{1}{2}[(x-\mu)/\sigma]^2}, \qquad -\infty < x < \infty$$

Example 5.5.3

The one-parameter gamma pdf can be written in the form

$$f_X(x) = \frac{1}{\Gamma(t)} \cdot x^{t-1} e^{-x}, \qquad x > 0, t > 0$$

What is its coefficient of kurtosis?

Using the result given in Question 5.5.4, we can set $c = 1$ and $b = t - 1$, to give

$$E(X^r) = t(t + 1) \cdots (t + r - 1) \qquad (5.5.5)$$

Also, from the comment following Definition 5.5.1,

$$E[(X - \mu)^4] = E(X^4) - 4E(X)E(X^3) + 6[E(X)]^2 E(X^2) - 3[E(X)]^4 \qquad (5.5.6)$$

Substitute Equation 5.5.5 into Equation 5.5.6. After simplifying,

$$E[(X - \mu)^4] = 3t^2 + 6t$$

Similarly,

$$\sigma^4 = (\sigma^2)^2 = [E(X^2) - \mu^2]^2$$
$$= [t^2 + t - t^2]^2$$
$$= t^2$$

From Definition 5.5.3, the coefficient of kurtosis is $6/t$:

$$\gamma_2 = \frac{3t^2 + 6t}{t^2} - 3$$

$$= \frac{6}{t}$$

Figure 5.5.1 shows four members of the one-parameter gamma family. Values chosen for t are $\frac{1}{2}$, 1, 2, and 4, giving coefficients of kurtosis equal to 12, 6, 3, and 1.5, respectively. Here the relationship between kurtosis and shape is obvious: As γ_2 increases, the peakedness of the corresponding pdf increases.

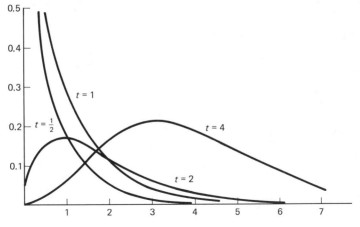

Figure 5.5.1

Example 5.5.4

We showed in Example 5.4.6 that the expected value of the sample variance—$E(S^2)$—is σ^2, regardless of what pdf the X_i's come from. A more difficult problem is finding the *variance* of the sample variance. It can be shown that if the X_i's come from a normal distribution,

$$\text{Var}(S^2) = \frac{2\sigma^4}{n-1}$$

(The derivation follows by considering the distribution of $(n-1)S^2/\sigma^2$ (see (48)).) More generally, if the sample comes from a nonnormal distribution, the variance of S^2 is a function of the kurtosis of $f_X(x)$:

$$\text{Var}(S^2) = \frac{2\sigma^4}{n-1} \cdot \left[1 + \frac{n-1}{n} \cdot \frac{\gamma_2}{2}\right]$$

(see (97)).

Comment

Having seen that certain higher moments of a random variable reflect various aspects of its pdf, we might well inquire whether knowing *all* the moments of X would uniquely characterize $f_X(x)$. That is, if we were given values for $E(X)$, $E(X^2)$, $E(X^3)$, . . . , could we theoretically deduce $f_X(x)$? For all practical purposes, the answer is yes.[1] More significantly, if we know only a very few moments—say, the first three or four—it is often possible to approximate $f_X(x)$ quite well. Anyone familiar with the theory of statistical estimation will recognize that the implicit relationship between a set of $E(X^j)$'s and an $f_X(x)$ is the rationale behind the *method of moments*.

[1] It *is* possible, though, to construct situations, admittedly "pathological" ones, where the same set of moments can be associated with two different distributions. See (45) for a discussion of when the $E(X^j)$'s, $j = 1, 2, \ldots$, *do* uniquely determine $f_X(x)$.

Question 5.5.7 Find the coefficient of skewness for an exponential random variable having pdf $f_X(x) = e^{-x}$, $x > 0$.

Question 5.5.8 If $Y = aX + b$, show that Y has the same coefficients of skewness and kurtosis as X.

Question 5.5.9 Draw a diagram of the following pdf's and compute γ_2 for each one. Comment on the "reliability" of γ_2 as a measure of peakedness.

a. $f_X(x) = \begin{cases} \dfrac{60 + x}{3600} & -60 < x < 0 \\[2mm] \dfrac{60 - x}{3600} & 0 \le x < 60 \end{cases}$

b. $f_X(x) = \begin{cases} 1 + x & -1 < x < 0 \\ 1 - x & 0 \le x < 1 \end{cases}$

c. $f_X(x) = \dfrac{1}{2\theta}$, $\quad -\theta < x < \theta$

d. $f_X(x) = |x|$, $\quad -1 \le x \le 1$

e. $f_X(x) = \dfrac{3x^2}{2}$, $\quad -1 \le x \le 1$

Question 5.5.10 Consider a family of curves of the form

$$f_X(x) = c \cdot e^{-|x|^n}, \quad -\infty < x < \infty$$

Find an expression for the coefficient of kurtosis. Show that it reduces to $\gamma_2 = 3$ when $n = 2$.

Finding moments of random variables, particularly certain higher moments, is conceptually straightforward, but the actual calculations may be quite difficult to carry out: Depending on the nature of $f_X(x)$, integrals of the form $\int_{-\infty}^{\infty} x^r f_X(x)\, dx$ are not always easy to evaluate. Fortunately, an alternate method is available. For some densities, we can find a *moment-generating function* (or *mgf*), $M_X(t)$, one of whose properties is that its rth derivative evaluated at 0 is equal to $E(X^r)$. If $M_X(t)$ can be found, $M_X^{(r)}(t)$ will often be easier to evaluate than $\int_{-\infty}^{\infty} x^r f_X(x)\, dx$.

Moment-generating functions can also be extremely useful in deriving the distribution of a *sum* of independent random variables. Such problems are important in statistics and typically difficult: Recall, for instance, Example 4.5.5, where it took considerable effort to prove that the sum of two binomials is also binomial. Using moment-generating functions, that particular problem is trivial.

> **Definition 5.5.4.** Let X be a random variable. The *moment-generating function for X*, denoted $M_X(t)$, is given by
>
> $$M_X(t) = E(e^{tX})$$
>
> at all values of t for which the expected value exists.

Comment

In Chapter 7 we generalize this idea somewhat by introducing the notion of a *characteristic function*. Denoted $\phi_X(t)$, the characteristic function of X is the expected value of e^{itX}, where $i = \sqrt{-1}$. The two functions $M_X(t)$ and $\phi_X(t)$ have similar applications, but the latter is superior in the sense that it always exists, whereas there are some random variables for which $M_X(t)$ is undefined for any point other than $t = 0$. Still, we choose to discuss the moment-generating function here in some detail because $M_X(t)$ is easier than $\phi_X(t)$ to manipulate and it proves to be quite adequate for most of the problems with which we will need to deal.

Before investigating the properties of moment-generating functions, we first simply *find* $M_X(t)$ for two of the densities already encountered, the binomial and the exponential. The computations here are straightforward applications of Theorem 5.3.2.

Example 5.5.5

Let X be a random variable with pdf $f_X(x) = \binom{n}{x}p^x q^{n-x}$, $x = 0, 1, \ldots, n$. Derive the moment-generating function for X.

From Definition 5.5.4, and with the help of the formula for a binomial expansion, we find that $M_X(t) = (pe^t + q)^n$:

$$M_X(t) = E(e^{tX}) = \sum_{x=0}^{n} e^{tX}\binom{n}{x}p^x q^{n-x} = \sum_{x=0}^{n} \binom{n}{x}(pe^t)^x q^{n-x}$$

$$= (pe^t + q)^n$$

In this case, $M_X(t)$ is defined for all values of t.

Example 5.5.6

Suppose X has an exponential pdf, $f_X(x) = \lambda e^{-\lambda x}$, $x > 0$. Find $M_X(t)$.

Since the exponential is a continuous random variable, $M_X(t)$ is an integral:

$$M_X(t) = E(e^{tX}) = \int_0^{\infty} e^{tX}\lambda e^{-\lambda x}\, dx$$

$$= \int_0^{\infty} \lambda e^{(t-\lambda)x}\, dx$$

$$= \frac{\lambda}{t - \lambda} e^{(t-\lambda)x}\,\Big|_0^{\infty}$$

$$= \frac{\lambda}{t - \lambda}\left[\lim_{x \to \infty} e^{(t-\lambda)x} - 1\right]$$

The limit here exists (and equals 0) only if $t - \lambda < 0$. Thus, for $t < \lambda$, the moment-generating function of X is given by

$$M_X(t) = \frac{\lambda}{\lambda - t}$$

For $t \geq \lambda$, $M_X(t)$ fails to exist.

Question 5.5.11 Let X be a discrete random variable with

$$f_X(x) = \begin{cases} pq^{x-1} & x = 1, 2, \ldots \\ 0 & \text{elsewhere} \end{cases}$$

Show that $M_X(t) = (pe^t)/(1 - qe^t)$. (*Hint:* Recall the formula for the sum of a geometric series,

$$\sum_{t=0}^{\infty} r^t = \frac{1}{1 - r}$$

for $0 < r < 1$.)

Question 5.5.12 Show that the moment-generating function of the random variable X having pdf $f_X(x) = \frac{1}{3}, -1 < x < 2$, is

$$M_X(t) = \begin{cases} \dfrac{e^{2t} - e^{-t}}{3t} & t \neq 0 \\ 1 & t = 0 \end{cases}$$

Question 5.5.13 Let X have pdf

$$f_X(x) = \begin{cases} x & 0 \leq x \leq 1 \\ 2 - x & 1 \leq x \leq 2 \\ 0 & \text{elsewhere} \end{cases}$$

Find $M_X(t)$.

Question 5.5.14 Let $C_X(t) = \ln M_X(t)$. Show that $C_X'(0) = \mu$ and $C_X''(0) = \sigma^2$.

Question 5.5.15 The factorial moment-generating function for a random variable X is defined as $\psi_X(t) = E(t^X)$. What is the relationship between $\psi_X(t)$ and $M_X(t)$?

Question 5.5.16 For a *Laplace* pdf,

$$f_X(x) = \frac{1}{2\lambda} \cdot e^{-|x - \mu|/\lambda}, \qquad -\infty < x < \infty;\ \lambda > 0;\ -\infty < \mu < \infty$$

show that

$$M_X(t) = (1 - \lambda^2 t^2)^{-1} e^{\mu t}, \qquad t < \frac{1}{\lambda}$$

The next theorem shows that $M_X(t)$ does, indeed, generate moments. We give a sketch of the proof, leaving its details as an exercise.

Theorem 5.5.2. Let X be a random variable with probability density function $f_X(x)$. (If X is continuous, f_X must be sufficiently smooth to allow the order of differentiation and integration to be interchanged.) Let $M_X(t)$ be the moment-generating function for X. Then

$$M_X^{(r)}(0) = E(X^r)$$

provided the rth moment exists.

Proof. We verify the theorem for the continuous case, where r is either 1 or 2. The extensions to discrete random variables and to an arbitrary positive integer r are straightforward.

For $r = 1$,

$$M_X^{(1)}(0) = \frac{d}{dt} \int_{-\infty}^{\infty} e^{tx} f_X(x)\, dx \bigg|_{t=0} = \int_{-\infty}^{\infty} \frac{d}{dt} e^{tx} f_X(x)\, dx \bigg|_{t=0}$$

$$= \int_{-\infty}^{\infty} xe^{tx} f_X(x)\, dx \bigg|_{t=0} = \int_{-\infty}^{\infty} xe^{0 \cdot x} f_X(x)\, dx$$

$$= \int_{-\infty}^{\infty} xf_X(x)\, dx = E(X)$$

Similarly, for $r = 2$,

$$M_X^{(2)}(0) = \frac{d^2}{dt^2} \int_{-\infty}^{\infty} e^{tx} f_X(x)\, dx \bigg|_{t=0} = \int_{-\infty}^{\infty} \frac{d^2}{dt^2} e^{tx} f_X(x)\, dx \bigg|_{t=0}$$

$$= \int_{-\infty}^{\infty} x^2 e^{tx} f_X(x)\, dx \bigg|_{t=0} = \int_{-\infty}^{\infty} x^2 e^{0 \cdot x} f_X(x)\, dx$$

$$= \int_{-\infty}^{\infty} x^2 f_X(x)\, dx = E(X^2)$$

That the theorem is true for any positive integer r should be readily apparent.

Example 5.5.7

Let X be a binomial random variable defined over a series of n independent trials. Find $E(X)$ using $M_X(t)$.

From Example 5.5.5,

$$M_X(t) = (pe^t + q)^n$$

Differentiating (once) with respect to t gives

$$M_X^{(1)}(t) = n(pe^t + q)^{n-1} pe^t$$

implying that

$$E(X) = M_X^{(1)}(0) = n(p + q)^{n-1} p$$

$$= np$$

(Does this agree with the value we found earlier for $E(X)$ when our approach was more direct?)

Question 5.5.17 Use Theorem 5.5.2 to find $E(X^2)$ for a binomial random variable. Then find Var(X) using Theorem 5.4.1.

Question 5.5.18 Show that the moment-generating function for a Poisson random variable having

$$f_X(x) = \frac{e^{-\lambda}\lambda^x}{x!}, \qquad x = 0, 1, \ldots$$

is $M_X(t) = e^{-\lambda + \lambda e^t}$. Use $M_X(t)$ to verify that $E(X) = \lambda$ and Var$(X) = \lambda$.

Question 5.5.19 Use the result in Question 5.5.11 to find $E(X^3)$ for the discrete random variable having pdf $f_X(x) = pq^{x-1}$, $x = 1, 2, \ldots$.

Question 5.5.20 Find a function which, when differentiated r times and appropriately evaluated, gives $E[(X - \mu)^r]$.

In some situations, we can determine the rth moment for a random variable without doing *any* differentiation: If we know the Taylor series representation for $M_X(t)$, $E(X^r)$ is just the coefficient of $t^r/r!$. To see why, just expand $M_X(t)$ about the point $t = 0$: From a theorem in elementary calculus,

$$M_X(t) = \sum_{r=0}^{\infty} \frac{M_X^{(r)}(0)}{r!} t^r \tag{5.5.7}$$

The next example applies Equation 5.5.7 to the problem of finding moments for an exponential random variable. Notice how much we gain in generality if we can write $M_X(t)$ as a power series.

Example 5.5.8

Let X have the exponential pdf, $f_X(x) = \lambda e^{-\lambda x}$, $x > 0$. Find an expression for $E(X^r)$.

From Example 5.5.6,

$$M_X(t) = \frac{\lambda}{\lambda - t} = \frac{1}{1 - (t/\lambda)}$$

$$= \sum_{r=0}^{\infty} (t/\lambda)^r = \sum_{r=0}^{\infty} \frac{1}{\lambda^r} \cdot t^r$$

A comparison of $M_X(t)$ with the right-hand side of Equation 5.5.7 gives

$$\frac{M_X^{(r)}(0)}{r!} = \frac{1}{\lambda^r}$$

and the latter implies that

$$E(X^r) = \frac{r!}{\lambda^r}$$

Question 5.5.21 Suppose X is a random variable for which

$$M_X(t) = e^{t^2/2}$$

Write the series expansion for $e^{t^2/2}$, compare it to the Taylor series for $M_X(t)$, and deduce a general formula for the rth moment of X.

Recall that moment-generating functions do more than just generate moments—they also characterize probability density functions and can often facilitate the derivation of $f_Z(z)$, where $Z = X_1 + X_2 + \cdots + X_n$ (and the X_i's are independent). The next two theorems state three important results. The first gives a uniqueness property of moment-generating functions: If X and Y have the same mgf's, then they must necessarily have the same pdf's. The second provides formulas for (1) the moment-generating function of a linear transformation and (2) the moment-generating function of a sum. We defer the proof of Theorem 5.5.3 until Chapter 7, when the result reappears as a special case of a property satisfied by characteristic functions. The proof of Theorem 5.5.4 is left as an exercise.

Theorem 5.5.3. Suppose X and Y are random variables such that $M_X(t) = M_Y(t)$ for some interval of t's containing 0. Then $f_X = f_Y$.

Theorem 5.5.4.

1. Let X be a random variable with moment-generating function $M_X(t)$. Suppose $Y = aX + b$. Then

$$M_Y(t) = e^{bt}M_X(at)$$

2. Let $X_1, X_2, \ldots, X_n$ be independent random variables with moment-generating functions $M_{X_1}(t)$, $M_{X_2}(t)$, $\ldots$, and $M_{X_n}(t)$, respectively. Let $Y = X_1 + X_2 + \cdots + X_n$. Then

$$M_Y(t) = M_{X_1}(t) \cdot M_{X_2}(t) \cdots M_{X_n}(t)$$

For the final example in this section, we apply Theorems 5.5.3 and 5.5.4 to the transformation problem described in Example 4.5.5. The solution is much simplified when we use moment-generating functions.

Example 5.5.9

Suppose X and Y are independent binomial random variables defined on m and n trials, respectively. Let the success probability p be the same for both sets of trials. Find the pdf for Z, where $Z = X + Y$.

From the second part of Theorem 5.5.4,

$$M_Z(t) = M_X(t) \cdot M_Y(t) = (pe^t + q)^m (pe^t + q)^n$$

$$= (pe^t + q)^{m+n} \qquad (5.5.8)$$

Does the right-hand side of Equation 5.5.8 look familiar? It should. The moment-generating function for a binomial random variable defined on $m + n$ trials would be $(pe^t + q)^{m+n}$. By Theorem 5.5.3 it follows that the pdf for Z is binomial—specifically,

$$f_Z(z) = \binom{m+n}{z} p^z q^{m+n-z}, \qquad z = 0, 1, \ldots, m + n$$

Question 5.5.22 Let X and Y be two independent exponential random variables:

$$f_X(x) = \lambda e^{-\lambda x}, \qquad x > 0$$
$$f_Y(y) = \lambda e^{-\lambda y}, \qquad y > 0$$

Let $Z = X + Y$. Does Z have an exponential distribution?

Question 5.5.23 Let X and Y be two independent Poisson random variables, each with the same parameter λ (see Question 5.5.18). Is their sum, $X + Y$, also Poisson?

Question 5.5.24 Prove Theorem 5.5.4.

5.6 THE BIENAYMÉ-CHEBYSHEV INEQUALITY

If X is a continuous random variable with pdf $f_X(x)$ and if we *know* $f_X(x)$, then

$$P(a \le X \le b) = \int_a^b f_X(x) \, dx \qquad (5.6.1)$$

for any interval (a, b). If, on the other hand, X is a random variable and we know nothing about its pdf, then obviously the only probability statement we can make about X lying between a and b is the trivial one,

$$0 \le P(a \le X \le b) \le 1 \qquad (5.6.2)$$

Equation 5.6.1 and Inequality 5.6.2 represent the extremes in making pronouncements about the behavior of a random variable; in this section we introduce a famous result that addresses a compromise.

Suppose $f_X(x)$ is unknown *but we do know the variance of* X. In light of the comment following Example 5.5.4, we might suspect that having a value for σ^2 would enable us to formulate some sort of approximation for $P(a \le X \le b)$. In point of fact, what can be derived is an upper bound for the probability that X lies outside an ϵ-neighborhood of μ. That is, we can find a nontrivial constant c, where $c = c(\sigma^2, \epsilon)$, such that

$$P(|X - \mu| \ge \epsilon) \le c(\sigma^2, \epsilon) \qquad (5.6.3)$$

for *any* $f_X(x)$ having mean μ and variance σ^2.

Inequality 5.6.3 is actually a special case of an even more powerful result that gives an upper bound for the probability that a nonnegative function of X exceeds ϵ. We will begin by proving the latter.

Theorem 5.6.1 (Bienaymé-Chebyshev Inequality). Let X be a random variable with pdf $f_X(x)$. Let $g(X)$ be any nonnegative integrable function of X. For any $\epsilon > 0$,

$$P\{g(X) \geq \epsilon\} \leq \frac{E[g(X)]}{\epsilon}$$

Proof. Let A and A^C be the sets of x-values for which $g(x) \geq \epsilon$ and $g(x) < \epsilon$, respectively. By Theorem 5.3.2,

$$E[g(X)] = \int_{-\infty}^{\infty} g(x) \cdot f_X(x) \, dx \tag{5.6.4}$$

$$= \int_A g(x) \cdot f_X(x) \, dx + \int_{A^C} g(x) \cdot f_X(x) \, dx$$

$$\geq \int_A g(x) \cdot f_X(x) \, dx$$

The inequality holds because $g(x) \cdot f_X(x) \geq 0$ for all x. Furthermore, since $g(x) \geq \epsilon$ over the set A,

$$\int_A g(x) \cdot f_X(x) \, dx \geq \epsilon \cdot \int_A f_X(x) \, dx$$

$$= \epsilon \cdot P(X \in A)$$

$$= \epsilon \cdot P\{g(X) \geq \epsilon\} \tag{5.6.5}$$

Dividing the inequality implied by Equations 5.6.4 and 5.6.5 by ϵ gives the result:

$$P\{g(X) \geq \epsilon\} \leq \frac{E[g(X)]}{\epsilon}$$

Comment

There are several special cases of the Bienaymé-Chebyshev inequality that should be singled out. These represent particular assignments for the function $g(X)$ that prove to be especially useful. We list the three as corollaries, leaving the details of their proofs as exercises. (Corollary 2 is the version described in the introduction to this section.)

Corollary 1. Let X be a nonnegative random variable with a finite mean. Let $\epsilon > 0$. Then

$$P(X \geq \epsilon) \leq \frac{E(X)}{\epsilon}$$

Example 5.6.1

As our intuition would suggest, the sweeping generality of the Bienaymé-Chebyshev inequality often makes the numerical value of its upper bound quite conservative given any particular $f_X(x)$. For instance, suppose X is an exponential random variable having $f_X(x) = e^{-x}$, $x > 0$. What is the probability X is more than two standard deviations away from its mean? Because $E(X) = 1$ and Var$(X) = 1$, what we are looking for is $P(|X - 1| \geq 2)$. Using Corollary 2, we get an upper bound of 0.25:

$$P(|X - 1| \geq 2) = P(|X - \mu| \geq \epsilon) \leq \frac{\text{Var}(X)}{\epsilon^2} = \frac{1}{4} = 0.25$$

The *exact* answer, though, is considerably less:

$$P(|X - 1| \geq 2) = P(X \geq 3) = \int_3^\infty e^{-x}\, dx = e^{-3} = 0.050$$

Table 5.6.1 shows a set of similar computations done on five different distributions. Notice that the Bienaymé-Chebyshev upper bound *is* attained for $P(|X - \mu| \geq \sigma)$ when X is a Bernoulli random variable with $p = \frac{1}{2}$.

Comment

Sometimes the "quality" of the upper bound given by Corollary 2 can be markedly improved by redefining the $g(X)$ of Theorem 5.6.1 and incorporating additional information about the higher moments of the random variable. For example, suppose we let $g(X) = (X - \mu)^4$. Then the event

$$(X - \mu)^4 \geq \epsilon^4$$

is equivalent to

$$|X - \mu| \geq \epsilon$$

and we can write

$$P\{(X - \mu)^4 \geq \epsilon^4\} = P\{|X - \mu| \geq \epsilon\} \leq \frac{E[(X - \mu)^4]}{\epsilon^4} = \frac{\mu_4'}{\epsilon^4}$$

TABLE 5.6.1

Curve type	Exact probabilities	B-C upper bound
$f_X(x) = e^{-x}, \; x > 0$	$P(\lvert X - \mu \rvert \geq \sigma) \;= 0.135$ $P(\lvert X - \mu \rvert \geq 2\sigma) = 0.050$ $P(\lvert X - \mu \rvert \geq 3\sigma) = 0.018$	1 0.250 0.111
$f_X(x) = \dfrac{1}{\sqrt{2\pi}} \cdot e^{-x^2}, \; -\infty < x < \infty$	$P(\lvert X - \mu \rvert \geq \sigma) \;= 0.317$ $P(\lvert X - \mu \rvert \geq 2\sigma) = 0.046$ $P(\lvert X - \mu \rvert \geq 3\sigma) = 0.003$	1 0.250 0.111
$f_X(x) = \lvert x \rvert, \; -1 \leq x \leq +1$	$P(\lvert X - \mu \rvert \geq \sigma) \;= 0.500$ $P(\lvert X - \mu \rvert \geq 2\sigma) = 0$ $P(\lvert X - \mu \rvert \geq 3\sigma) = 0$	1 0.250 0.111
$f_X(x) = \begin{cases} 0 & q = \frac{1}{2} \\ 1 & p = \frac{1}{2} \end{cases}$	$P(\lvert X - \mu \rvert \geq \sigma) \;= 1$ $P(\lvert X - \mu \rvert \geq 2\sigma) = 0$ $P(\lvert X - \mu \rvert \geq 3\sigma) = 0$	1 0.250 0.111
$f_X(x) = \dfrac{e^{-2}2^x}{x!}, \; x = 0, 1, \ldots$	$P(\lvert X - \mu \rvert \geq \sigma) \;= 0.143$ $P(\lvert X - \mu \rvert \geq 2\sigma) = 0.053$ $P(\lvert X - \mu \rvert \geq 3\sigma) = 0.004$	1 0.250 0.111

Of course, from Corollary 2,

$$P\{\lvert X - \mu \rvert \geq \epsilon\} \leq \frac{\sigma^2}{\epsilon^2}$$

Which bound is sharper—μ_4'/ϵ^4 or σ^2/ϵ^2—depends on the size of ϵ. Specifically,

$$\frac{\mu_4'}{\epsilon^4} < \frac{\sigma^2}{\epsilon^2}$$

whenever

$$\epsilon > \sqrt{\frac{\mu_4'}{\sigma^2}}$$

(See Question 5.6.8.)

Comment

Additional information about the shape of $f_X(x)$ can lead to still other Chebyshev-type inequalities. For example, suppose $f_X(x)$ is unimodal with its mode being located at m. If we set

$$\lambda = \frac{\mu - m}{\sigma}$$

it can be shown that for $t > \lvert \lambda \rvert$,

$$P\{\lvert X - \mu \rvert > t\sigma\} \leq \frac{4(1 + \lambda^2)}{9(t - \lvert \lambda \rvert)^2} \qquad (5.6.6)$$

(see (93)). Applying Inequality 5.6.6 to the standard normal distribution ($\lambda = 0$) gives

$$P\{|X - \mu| > \sigma\} \leq \tfrac{4}{9} = 0.44$$
$$P\{|X - \mu| > 2\sigma\} \leq \tfrac{4}{36} = 0.11$$
$$P\{|X - \mu| > 3\sigma\} \leq \tfrac{4}{81} = 0.049$$

Note that these latter three upper bounds are less than half as large as their counterparts in Table 5.6.1.

Example 5.6.2

A student makes 100 check transactions between receiving his January and February bank statements. Rather than subtract the amounts he spends exactly, he rounds each checkbook entry off to the nearest dollar. Use the Bienaymé-Chebyshev inequality to get an upper bound for the probability that the student's accumulated error (either positive or negative) after his 100 transactions is $5.00 or more.

Let Y_i denote the round-off error associated with the ith transaction, $i = 1$, $2, \ldots, 100$. It can be assumed (why?) that the Y_i's are independent random variables and follow a uniform pdf over the interval $(-\$\tfrac{1}{2}, +\$\tfrac{1}{2})$. Therefore,

$$E(Y_i) = 0$$

and

$$\text{Var}(Y) = \tfrac{1}{12}$$

(see Question 5.4.5).

Let $Y = Y_1 + Y_2 + \cdots + Y_{100}$ denote the student's total accumulated error. There is no simple way to find the pdf for Y, but its expected value and variance can be easily derived: By the Corollary to Theorem 5.3.1,

$$E(Y) = \mu = 0$$

and, by Theorem 5.4.3,

$$\text{Var}(Y) = \frac{100}{12} = 8.3$$

Substituting $E(Y)$ and Var(Y) into Corollary 2, we can write

$$P\{|Y| \geq \$5\} = P\{|Y - \mu| \geq \epsilon\} \leq \frac{8.3}{25} = 0.33$$

There is at most a 33% chance that the student's total will be off by as much as $5.00.

Question 5.6.1 A fair die is tossed 100 times. Let X_k denote the outcome on the kth roll. Use Corollary 2 to get a *lower* bound for the probability that $Y = X_1 + X_2 + \cdots + X_{100}$ is between 300 and 400.

Quesiton 5.6.2 Suppose the distribution of scores on an IQ test has mean 100 and standard deviation 16. Show that the probability of a student having an IQ above 148 or below 52 is at most $\frac{1}{9}$.

Question 5.6.3 Use Corollary 2 to determine how many times a fair coin must be tossed in order for the probability to be at least 0.90 that the ratio of the observed number of heads to the total number of tosses be between 0.4 and 0.6.

Question 5.6.4 For the random variables having the following pdf's, compute $P(|X - \mu| \geq \sigma)$, $P(|X - \mu| \geq 2\sigma)$, and $P(|X - \mu| \geq 3\sigma)$, as well as the corresponding Bienaymé-Chebyshev upper bounds.

a. $f_X(x) = \begin{cases} \frac{1}{2} & -2 < x < -1 \\ \frac{1}{2} & 1 < x < 2 \end{cases}$

b. $f_X(x) = (\frac{2}{3})^{x-1} \cdot \frac{1}{3}, \; x = 1, 2, \ldots$

Question 5.6.5 Recall the X-ray inspection problem described in Case Study 5.2. Show that the probability that the unit is out of compliance is no greater than 0.24.

Question 5.6.6 Let $X_1, X_2, \ldots, X_n$ be a random sample of continuous random variables drawn from a population having mean μ and variance σ^2. Show that

$$\lim_{n \to \infty} P(|\overline{X} - \mu| > \epsilon) = 0$$

(This result is known as the *weak law of large numbers*. It is discussed at some length in Chapter 7.)

Question 5.6.7 Fill in the details to show how Corollaries 1, 2, and 3 follow from Theorem 5.6.1.

Question 5.6.8 If X is a uniform random variable defined over the unit interval, for what values of ϵ will μ'_4/ϵ^4 be a sharper bound for $P\{|X - \mu| \geq \epsilon\}$ than σ^2/ϵ^2?

Question 5.6.9 Let X be any random variable with $E(X) = 0$ and $\text{Var}(X) = \sigma^2$. Show that

$$P(X > \epsilon) \leq \frac{\sigma^2}{\sigma^2 + \epsilon^2} \qquad \text{if } \epsilon > 0$$

(*Hint:* Write $g(t) = (t + c)^2$, $c > 0$. Then $g(t) \geq (x + c)^2$ for $t > x > 0$. How does the probability of the event "$X > x$" compare with the probability of the event "$g(X) \geq (x + c)^2$"? See (75).)

5.7 CONDITIONAL EXPECTATION

The notion of a conditional pdf—$f_{Y|x}(y)$—was developed in Section 4.6: Whether X and Y are discrete or continuous, we can write

$$f_{Y|x}(y) = \frac{f_{X,Y}(x, y)}{f_X(x)}$$

If Y is discrete,

$$f_{Y|x}(y) = P(Y = y | X = x) \qquad (5.7.1)$$

If Y is continuous, $f_{Y|x}(y)$ is a function that describes the probabilistic behavior of Y in the sense that for any interval $[a, b]$,

$$P(a \leq Y \leq b \mid X = x) = \int_a^b f_{Y|x}(y)\, dy \tag{5.7.2}$$

Also, for discrete Y,

$$\sum_{\text{all } y} f_{Y|x}(y) = 1 \tag{5.7.3}$$

and for continuous Y,

$$\int_{-\infty}^{\infty} f_{Y|x}(y)\, dy = 1 \tag{5.7.4}$$

Equations 5.7.1 through 5.7.4 suggest that any random variable defined by an $f_{Y|x}(y)$ should have all the properties we have heretofore associated with *unconditional* pdf's. Specifically, it should make sense to talk about the *conditional expectation of Y given x*.

Definition 5.7.1. Let X and Y be random variables. The conditional expectation of Y given x, $E(Y \mid x)$, is given by

1. $\sum_{\text{all } y} y \cdot f_{Y|x}(y)$ if Y is discrete
2. $\int_{-\infty}^{\infty} y \cdot f_{Y|x}(y)\, dy$ if Y is continuous

provided the sum and integral converge absolutely.

Comment

More generally, if Y is continuous and $g(Y)$ is any function of Y, then

$$E[g(Y) \mid x] = \int_{-\infty}^{\infty} g(y) \cdot f_{Y|x}(y)\, dy$$

Thus, if we let $\mu_{Y|x} = E(Y \mid x)$ and define

$$g(Y) = (Y - \mu_{Y|x})^2$$

the obvious expression for the *conditional variance of Y given x*, $\sigma_{Y|x}^2$, would be

$$\sigma_{Y|x}^2 = E[(Y - \mu_{Y|x})^2 \mid x] = \int_{-\infty}^{\infty} (y - \mu_{Y|x})^2 \cdot f_{Y|x}(y)\, dy$$

Example 5.7.1

Suppose X and Y are two random variables jointly distributed over the first quadrant of the xy-plane according to the pdf

$$f_{X,Y}(x, y) = \begin{cases} y^2 e^{-y(x+1)} & x \geq 0,\ y \geq 0 \\ 0 & \text{elsewhere} \end{cases}$$

Find $E(Y \mid x)$.

First, we note that

$$f_X(x) = \int_{-\infty}^{\infty} f_{X,Y}(x, y)\, dy = \int_0^{\infty} y^2 e^{-y(x+1)}\, dy$$

Make the substitution $u = y(x + 1)$. Then $du = (x + 1)\, dy$ and

$$f_X(x) = \frac{1}{x + 1} \int_0^{\infty} \frac{u^2}{(x + 1)^2} \cdot e^{-u}\, du$$

$$= \frac{1}{(x + 1)^3} \int_0^{\infty} u^2 e^{-u}\, du$$

After integrating by parts (twice), we obtain

$$f_X(x) = \frac{1}{(x + 1)^3} [-u^2 e^{-u} - 2u e^{-u} - 2e^{-u}] \Big|_0^{\infty}$$

$$= \frac{1}{(x + 1)^3} \left[2 - \lim_{\mu \to \infty} \left(\frac{u^2}{e^u} + \frac{2u}{e^u} + \frac{2}{e^u} \right) \right]$$

$$= \frac{2}{(x + 1)^3}$$

Having found the marginal pdf for X, we can immediately write down the conditional pdf for Y given x:

$$f_{Y|x}(y) = \frac{y^2 e^{-y(x+1)}}{2/(x + 1)^3} = \frac{1}{2}(x + 1)^3 y^2 e^{-y(x+1)}$$

Figure 5.7.1 shows a sketch of $f_{Y|x}(y)$ versus x for the particular x-values 1, 3, and 5.

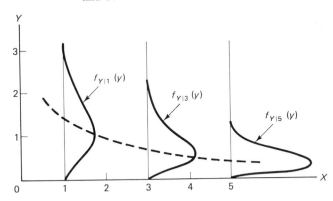

Figure 5.7.1

One more integration gives us the conditional expectation of Y given x. By Definition 5.7.1,

$$E(Y|x) = \int_0^{\infty} y \cdot \frac{1}{2}(x + 1)^3 y^2 e^{-y(x+1)}\, dy$$

$$= \frac{1}{2(x + 1)} \int_0^{\infty} u^3 e^{-u}\, du$$

the latter expression resulting from the substitution $u = y(x + 1)$. But

$$\int_0^\infty u^3 e^{-u} \, du = \Gamma(4) = 3! = 6$$

so the conditional expectation reduces to

$$E(Y|x) = \frac{3}{x + 1}$$

(The equation $Y = E(Y|x) = 3/(x + 1)$ is plotted on the conditional pdf's sketched in Figure 5.7.1.)

Comment

In statistics, a graph of $E(Y|x)$ versus x—such as the one shown in Figure 5.7.1—is called a *regression curve*. An important problem in data analysis is estimating the regression curve of a relationship on the basis of a random sample of size n, (x_1, y_1), (x_2, y_2), . . . , (x_n, y_n), drawn from $f_{X,Y}(x, y)$. Of particular importance is the case where X and Y have the *bivariate normal distribution* (see Figure 5.7.2), meaning that

$$f_{X,Y}(x, y) = \frac{1}{\sqrt{2\pi}\sigma_X\sigma_Y\sqrt{1 - \rho^2}}$$

$$\cdot \exp\left\{-\frac{1}{2}\left(\frac{1}{1 - \rho^2}\right)\left[\frac{(x - \mu_X)^2}{\sigma_X^2} - 2\rho\frac{x - \mu_X}{\sigma_X}\frac{y - \mu_Y}{\sigma_Y} + \frac{(y - \mu_Y)^2}{\sigma_Y^2}\right]\right\}$$

where $-\infty < x < \infty$, $-\infty < y < \infty$, and μ_X, μ_Y, and ρ are parameters such that $-\infty < \mu_X < \infty$, $-\infty < \mu_Y < \infty$, $0 < \sigma_X < \infty$, $0 < \sigma_Y < \infty$, and $-1 \leq \rho \leq 1$. It can be shown for the bivariate normal that $E(Y|x)$ is a straight line (see (48)).

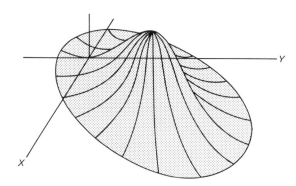

Figure 5.7.2

Given that X and Y are jointly distributed, it follows that $E(Y|x)$ can be viewed as a function of the random variable X—that is,

$$E(Y|X) = g(X)$$

in which case it makes sense to talk about the expected value of $g(X)$. But

$$E[g(X)] = E[E(Y|X)]$$

where

$$E[E(Y|X)] = \sum_{\text{all } x} E(Y|x)P(X = x)$$

if X is discrete, and

$$E[E(Y|X)] = \int_{-\infty}^{\infty} E(Y|x)f_X(x) \, dx$$

if X is continuous. The importance of expressions like $E[E(Y|X)]$ derives from a relationship they share with single expected values.

Theorem 5.7.1. Let the two random variables X and Y be jointly distributed and let the kth moment of Y exist. Then

$$E[E(Y^k|X)] = E(Y^K)$$

Proof. We give an argument for the continuous case only; the derivation for discrete X and Y follows similarly.

Let $f_{X,Y}(x, y)$ be the joint pdf for X and Y and let $f_X(x)$ be the marginal pdf for X. Then

$$f_{Y|x}(y) = \frac{f_{X,Y}(x, y)}{f_X(x)} \qquad \text{if } f_X(x) \neq 0$$

By definition,

$$E(Y^k|x) = \int_{-\infty}^{\infty} y^k \cdot \frac{f_{X,Y}(x, y)}{f_X(x)} \, dy$$

and

$$E[E(Y^k|X)] = \int_{-\infty}^{\infty} \left[\int_{-\infty}^{\infty} y^k \cdot \frac{f_{X,Y}(x, y)}{f_X(x)} \, dy \right] \cdot f_X(x) \, dx$$

Assuming the order of integration can be interchanged, we can write

$$E[E(Y^k|X)] = \int_{-\infty}^{\infty} \int_{-\infty}^{\infty} y^k \cdot f_{X,Y}(x, y) \, dx \, dy$$

$$= \int_{-\infty}^{\infty} y^k \cdot f_Y(y) \, dy \qquad \text{(Why?)}$$

$$= E(Y^k)$$

Corollary. Let $A_1, A_2, \ldots, A_n$ be a set of events partitioning a sample space S, where $P(A_i) > 0$ for all i. Let Y be a random variable defined on S. Then

$$E(Y) = \sum_{i=1}^{n} E(Y|A_i)P(A_i)$$

Example 5.7.2

A biased coin, where $P(\text{heads}) = p$, is to be flipped until two consecutive heads appear. On the average, how many tosses will it take for that to happen?

Let S denote the sample space of all possible toss sequences. Define the three "subsequences,"

$$A_1: \quad H \quad H$$
$$A_2: \quad H \quad T$$
$$A_3: \quad T$$

Clearly, all sequences in S begin with either A_1, A_2, or A_3, meaning the A_i's partition the sample space. For any $s \in S$, let $Y = Y(s)$ denote the number of trials to get the first two consecutive heads. From the corollary to Theorem 5.7.1,

$$E(Y) = E(Y|A_1)P(A_1) + E(Y|A_2)P(A_2) + E(Y|A_3)P(A_3)$$

or, filling in the conditional expected values by inspection,

$$E(Y) = 2 \cdot p^2 + [E(Y) + 2] \cdot pq + [E(Y) + 1] \cdot q$$

Solving for $E(Y)$ gives

$$E(Y) = \frac{p + 1}{p^2}$$

Example 5.7.3

A waitress is responsible for n customers sitting along a lunch counter. If the customers are equally spaced, each 3 ft apart, what is the average distance she walks from customer to customer, assuming that at any given time each customer is equally likely to ask for service?

Imagine the customers to be numbered from 1 to n and assume the waitress is currently taking an order from customer k. Suppose customer i is the next person requesting service. If, in general, the random variable Y denotes the distance the waitress walks in going from her present customer to her next customer, then

$$Y = \begin{cases} 3(k - i) & k \geq i \\ 3(i - k) & k < i \end{cases}$$

if she goes from customer k to customer i.

Now, let X be a random variable denoting the location of the waitress' present customer (so X takes on the values $1, 2, \ldots, n$). Then

$$E(Y|X = k) = \sum_{\text{all } y} y \cdot P(Y = y|X = k)$$

$$= \sum_{i=1}^{k} 3(k - i) \cdot \frac{1}{n} + \sum_{i=k+1}^{n} 3(i - k) \cdot \frac{1}{n}$$

$$= \frac{3}{n} \sum_{j=0}^{k-1} j + \frac{3}{n} \sum_{j=1}^{n-k} j$$

$$= \frac{3}{n} \cdot \frac{(k - 1)(k)}{2} + \frac{3}{n} \cdot \frac{(n - k)(n - k + 1)}{2}$$

$$= \frac{3}{2n}[2k^2 - 2(n + 1)k + n(n + 1)]$$

(Does $P(Y = y|X = k) = 1/n$?) Finally, from Theorem 5.7.1,

$$E(Y) = E[E(Y|X)] = \sum_{k=1}^{n} \frac{3}{2n}[2k^2 - 2(n + 1)k + n(n + 1)] \cdot \frac{1}{n}$$

$$= \frac{3}{n^2} \sum_{k=1}^{n} k^2 - \frac{3(n + 1)}{n^2} \sum_{k=1}^{n} k + \frac{3(n + 1)}{2n} \cdot n$$

$$= \frac{3}{n^2} \left[\frac{n(n + 1)(2n + 1)}{6} \right]$$

$$\qquad - \frac{3(n + 1)}{n^2} \left[\frac{n(n + 1)}{2} \right] + \frac{3(n + 1)}{2}$$

$$= n - \frac{1}{n}$$

(What would $E(Y)$ equal if the n customers sat r feet apart?)

Question 5.7.1 Let X and Y have the joint pdf

$$f_{X,Y}(x, y) = \begin{cases} x + y & 0 \le x \le 1, 0 \le y \le 1 \\ 0 & \text{elsewhere} \end{cases}$$

Find $E(Y|x)$ and graph it as a function of x.

Question 5.7.2 Find $E(Y|X = \frac{1}{2})$ if X and Y are jointly distributed according to the pdf

$$f_{X,Y}(x, y) = \frac{6}{7}\left(x^2 + \frac{xy}{2}\right), \qquad 0 < x < 1, 0 < y < 2$$

Question 5.7.3 Let X, Y, and Z be random variables. Show that

$$E(aX + bY|Z = z) = a \cdot E(X|Z = z) + b \cdot E(Y|Z = z)$$

Question 5.7.4 Prove that $E(X) < \infty$ implies that $E(X|A) < \infty$ if $P(A) > 0$.

Question 5.7.5 An urn contains n chips numbered 1 through n. One chip is picked at random. If the chip selected is number k, a second urn is made up containing only the chips numbered 1 through k, and a chip is drawn from that urn. If Y denotes the number of the chip drawn from the second urn, find $E(Y)$. (*Hint:* Let X denote the number of the chip drawn from the first urn and use the corollary to Theorem 5.7.1.)

Question 5.7.6 Use the corollary to Theorem 5.7.1 to derive the expected value for a geometric random variable having pdf

$$f_X(x) = q^{x-1} \cdot p, \qquad x = 1, 2, \ldots$$

(Recall Example 5.2.4.)

Question 5.7.7 A prisoner is in a cell with three doors—D_1, D_2, and D_3. At random, she opens one of the doors in an attempt to escape. Unfortunately, doors D_1 and D_2 are traps. If she chooses door D_1, she will wander aimlessly in a tunnel for 13 days and return to her cell with a complete loss of memory; if she chooses door D_2, she will be in a tunnel for 8 days and again will return to her cell with a complete loss of memory. Door D_3 leads to immediate freedom. Persistent, she will keep trying to escape until she is lucky enough to choose door D_3. Let Y be a random variable denoting the number of days it takes her to break out. Find $E(Y)$.

Question 5.7.8 Recall Example 5.7.3. Use Theorem 5.7.1 to find the *variance* of the distance walked by the waitress between consecutively served customers.

Question 5.7.9 Let $X_1, X_2, \ldots$ be a sequence of random variables and let Y be a random variable taking on only positive integer values. Define

$$X = X_1 + X_2 + \cdots + X_Y$$

Assume (1) the X's are independent of Y, (2) $E(Y) < \infty$, and (3) $\sum_{i=1}^{\infty} |X_i| P(Y \geq i) < \infty$. Show that

$$E(X) = \sum_{i=1}^{\infty} E(X_i) \cdot P(Y \geq i)$$

Also, show that $E(X) = \mu \cdot E(Y)$ If the X_i's are identically distributed and have $E(X_i) = \mu$, for all i.

6

Special Distributions

A reasonable probability is the only certainty.
Howe

6.1 INTRODUCTION

Certain probability functions occur so often, both in theoretical and applied contexts, that they warrant being studied in greater depth. These functions almost invariably occur in *families,* where they are indexed by one or more unknown parameters. To single out a particular member of such a family, numerical values must be assigned to each of the pdf's parameters. We have already encountered the one-parameter exponential family, $f_X(x) = (1/\lambda)e^{-x/\lambda}$, in precisely this kind of setting: Recall that $f_X(x) = (1/179)e^{-x/179}$ was used in Example 2.5.4 as a model for the life distribution of aircraft radar tubes. Another example—by now, quite familiar—is the binomial distribution, $P(X = k) = \binom{n}{k}p^k(1-p)^{n-k}$, $k = 0, 1, \ldots, n$; this is a family indexed by *two* parameters, n and p.

For the statistician, there are many families of probability functions meriting special attention—far more than can be discussed in a text of this length. Here, only those finding particularly extensive application will be dealt with. For a more complete survey, see (36) or the Johnson and Kotz series (41, 42, 43).

Altogether, five families of probability functions—the Poisson, normal, geometric, negative binomial, and gamma—are profiled in the next several sections. A few others will be introduced but not examined in any great detail. In some cases it will prove enlightening to recount the history of a distribution—how it came to be

305

discovered and what its first applications were. Some of the results presented here date back more than 150 years and are now part of the bedrock of modern probability and statistics.

6.2 THE POISSON DISTRIBUTION

Simeon Denis Poisson (1781–1840) was an eminent French mathematician and physicist, an academic administrator of some note, and, according to an 1826 letter from the mathematician Abel to a friend, a man who knew "how to behave with a great deal of dignity." One of Poisson's many interests was the application of probability to the law, and in 1837 he wrote *Recherches sur la Probabilite de Judgements*. This text contained a good deal of mathematics, including a limit theorem for the binomial distribution.[1] Although initially viewed as little more than a welcome approximation for hard-to-compute binomial probabilities, this particular result was destined for bigger things: It was the analytical seed out of which grew what is now one of the most important of all probability models, the Poisson distribution.

Theorem 6.2.1. Let λ be a fixed number and n be an arbitrary positive integer. For each nonnegative integer x,

$$\lim_{n \to \infty} \binom{n}{x} p^x (1 - p)^{n-x} = \frac{e^{-\lambda} \lambda^x}{x!}$$

where $p = \lambda/n$.

Proof. We begin by rewriting the left-hand side of the equation appearing in Theorem 6.2.1, substituting λ for np:

$$\lim_{n \to \infty} \binom{n}{x} p^x (1 - p)^{n-x} = \lim_{n \to \infty} \binom{n}{x} \left(\frac{\lambda}{n} \right)^x \left(1 - \frac{\lambda}{n} \right)^{n-x}$$

$$= \lim_{n \to \infty} \frac{n!}{x! \, (n - x)!} \lambda^x \left(\frac{1}{n^x} \right) \left(1 - \frac{\lambda}{n} \right)^{-x} \left(1 - \frac{\lambda}{n} \right)^n$$

$$= \frac{\lambda^x}{x!} \lim_{n \to \infty} \frac{n!}{(n - x)! \, (n - \lambda)^x} \left(1 - \frac{\lambda}{n} \right)^n$$

Since $(1 - \lambda/n)^n \to e^{-\lambda}$ as $n \to \infty$, we need only show that $n!/((n - x)! \, (n - \lambda)^x) \to 1$ to prove the theorem. But after dividing $(n - x)!$ into $n!$, we see that

$$\frac{n!}{(n - x)! \, (n - \lambda)^x} = \frac{n(n - 1) \cdots (n - x + 1)}{(n - \lambda)(n - \lambda) \cdots (n - \lambda)}$$

a quantity which, indeed, tends to 1 as $n \to \infty$.

[1]Although credit for this theorem is given to Poisson, there is some evidence that DeMoivre may have discovered it almost a century earlier (16).

Example 6.2.1

Tables 6.2.1 and 6.2.2 give an indication of the *rate* at which

$$\binom{n}{x} p^x (1 - p)^{n-x}$$

converges to

$$\frac{e^{-np}(np)^x}{x!}$$

In both cases, $\lambda = np$ is the same; however, in the former, n is set equal to 5 and in the latter, $n = 100$. We see in Table 6.2.1 ($n = 5$) that for some values of x the agreement between the binomial probability and Poisson's limit is not very good. With n raised to 100, though, agreement is remarkably good for *all* x (Table 6.2.2).

TABLE 6.2.1 BINOMIAL PROBABILITIES AND POISSON LIMITS
$n = 5$ AND $p = \frac{1}{5}$ ($\lambda = 1$)

x	$\binom{5}{x}(0.2)^x(0.8)^{5-x}$	$\dfrac{e^{-1}(1)^x}{x!}$
0	0.328	0.368
1	0.410	0.368
2	0.205	0.184
3	0.051	0.061
4	0.006	0.015
5	0.000	0.003
6+	0	0.001
	1.000	1.000

TABLE 6.2.2 BINOMIAL PROBABILITIES AND POISSON LIMITS
$n = 100$ AND $p = \frac{1}{100}$ ($\lambda = 1$)

x	$\binom{100}{x}(0.01)^x(0.99)^{100-x}$	$\dfrac{e^{-1}(1)^x}{x!}$
0	0.366032	0.367879
1	0.369730	0.367879
2	0.184865	0.183940
3	0.060999	0.061313
4	0.014942	0.015328
5	0.002898	0.003066
6	0.000463	0.000511
7	0.000063	0.000073
8	0.000007	0.000009
9	0.000001	0.000001
10	0.000000	0.000000
	1.000000	0.999999

The next example is typical of how convenient the Poisson limit can be in approximating binomial probabilities. The data concern the incidence of childhood leukemia and whether or not the disease is contagious, a question that has been around for quite a few years but still has medical researchers baffled (recall Case Study 3.2).

CASE STUDY 6.1

Leukemia is a rare but highly virulent form of cancer whose cause and mode of transmission remain largely unknown. While evidence abounds that excessive exposure to radiation can increase a person's risk of contracting the disease, most occurrences are still found in people whose history contains no such overexposure. A related issue, perhaps even more basic than the causality question, concerns the *spread* of the disease. It is safe to say that the prevailing medical opinion holds that most forms of leukemia are not contagious—still, the hypothesis persists that *some* forms of the disease, particular the childhood variety, might be. What continues to fuel such speculation are discoveries of so-called leukemia clusters, aggregations in time and space of unusually large numbers of cases.

To date, one of the most frequently cited leukemia clusters in the medical literature occurred during the late 1950s and early 1960s in Niles, Illinois, a suburb of Chicago (37). In the $5\frac{1}{3}$-year period from 1956 to the first four months of 1961, physicians in Niles reported a total of eight cases of leukemia among children less than 15 years of age. The population at risk (that is, the set of residents in that age range) numbered 7076. To assess the likelihood of eight cases occurring in such a small community, we need to look first at the incidence of leukemia in neighboring towns and use the latter as a benchmark. For all Cook County, *excluding Niles,* there were 1,152,695 children less than 15; among those, there were 286 diagnosed cases of leukemia. Taking the quotient of those two figures gives an average $5\frac{1}{3}$-year leukemia rate of 24.8 cases per 100,000:

$$\frac{286 \text{ cases for } 5\frac{1}{3} \text{ years}}{1,152,695 \text{ children}} \times \frac{100,000}{100,000} = \frac{24.8 \text{ cases per } 100,000 \text{ children}}{5\frac{1}{3} \text{ years}}$$

Now, imagine the 7076 children in Niles to be a series of (independent) Bernoulli trials, each child having the same probability of contracting leukemia. Using already-familiar terminology, we can easily rephrase our initial question in the guise of a probability calculation involving the binomial random variable X: Given $n = 7076$ and $p = 24.8/100,000 = 0.000248$, how likely is it that $X = 8$ "successes" will occur? For statistical reasons, a related event—that eight *or more* cases will occur—is a bit more meaningful. If the probability associated with the latter is very small, it can be argued that leukemia did *not* occur randomly in Niles and that contagion may have been a factor.

Using the binomial distribution, we can express the probability that $X \geq 8$ as

$$P(8 \text{ or more cases}) = \sum_{k=8}^{7076} \binom{7076}{k} (0.000248)^k (0.999752)^{7076-k} \qquad (6.2.1)$$

a computation not entirely pleasant to contemplate. With the aid of Poisson's theorem, though, the problem is trivial. By letting $\lambda = np = (7076)(0.000248) = 1.75$, we can quickly come up with 0.00049 as the Poisson limit:

$$P(8 \text{ or more cases}) \doteq \sum_{k=8}^{\infty} \frac{e^{-1.75}(1.75)^k}{k!}$$

$$= 1 - \sum_{k=0}^{7} \frac{e^{-1.75}(1.75)^k}{k!}$$

$$= 1 - 0.99951$$

$$= 0.00049$$

Have we lost much in precision here by approximating the right-hand side of Equation 6.2.1? No; considering the accuracy of the Poisson limit for an n as small as 100 (recall Table 6.2.2), we should feel very confident with an n of 7076 that 0.00049 is extremely close to the true binomial sum.

Comment

Interpreting this result physically is not as easy as assessing its accuracy mathematically. The fact that the probability is so very small tends to denigrate the hypothesis that leukemia in Niles occurred at random. On the other hand, rare events such as clusters *do* happen. The basic difficulty in putting the probability associated with a given cluster in any meaningful kind of perspective is not knowing in how many similar communities leukemia did *not* exhibit a tendency to cluster. That there is no obvious way to do this is one reason the leukemia controversy is still with us.

Question 6.2.1 A chromosome mutation believed to be linked with colorblindness is known to occur, on the average, once in every 10,000 births. If 20,000 babies are born this year in a certain city, what is the probability that at least one will develop colorblindness? What is the exact probability model that applies here?

Question 6.2.2 Given 400 people, estimate the probability that 3 or more will have a birthday on July 4. Assume there are 365 days in a year and each is equally likely to be the birthday of a randomly chosen person.

Question 6.2.3 Suppose that 1% of all items in a supermarket are unmarked. A customer buys 10 items and proceeds to check out through the express lane. Estimate the probability that the customer will be delayed because one or more of the items require a price check. What assumptions are you making?

Question 6.2.4 Use the Poisson approximation to calculate the probability that at most 1 person in 500 will have a birthday on Christmas. Assume there are 365 days in a year.

Question 6.2.5 Astronomers estimate that as many as 100 billion stars in the Milky Way galaxy may be encircled by planets. Let p denote the probability that any such solar system contains intelligent life. How small can p be and still give a 50-50 chance that there is intelligent life in at least one other solar system in our galaxy?

Poisson's theorem certainly sent no shock waves through the scientific community; on the contrary, it languished in mathematical limbo for over 50 years, attracting little interest and finding no real application. Its resurrection finally came in 1898 when Ladislaus von Bortkiewicz, a German professor born in Russia of Polish ancestry, wrote a monograph entitled *Das Gesetz der Kleinen Zahlen* (*The Law of Small Numbers*). Included in this work was Poisson's theorem (with due credit), possibly the first consideration of the Poisson limit as a probability distribution in its own right, and, most importantly, the use of the Poisson to model real-world phenomena.

Definition 6.2.1. A discrete random variable X is said to be *Poisson* if

$$f_X(x) = \frac{e^{-\lambda}\lambda^x}{x!} \quad \text{for } x = 0, 1, 2, \ldots$$

Definition 6.2.1 implicitly assumes that the expression we recognize as Poisson's limit actually qualifies as a pdf. It does, but we need to establish that fact formally; details are given in the next result.

Theorem 6.2.2. Let $p(x, \lambda)$ denote Poisson's limit,

$$p(x, \lambda) = \frac{e^{-\lambda}\lambda^x}{x!} \quad \text{for } x = 0, 1, \ldots$$

Then $p(x, \lambda)$ is a probability function. Also, if X is a random variable with pdf $p(x, \lambda)$, $E(X) = \lambda$ and $\text{Var}(X) = \lambda$.

Proof. To demonstrate that $p(x, \lambda)$ qualifies as a probability function, we note, first of all, that $p(x, \lambda) \geq 0$ for all nonnegative integers x. To complete the argument, we need to show that its sum over *all* x is 1. But

$$\sum_{x=0}^{\infty} p(x, \lambda) = \sum_{x=0}^{\infty} \frac{e^{-\lambda}\lambda^x}{x!} = e^{-\lambda} \sum_{x=0}^{\infty} \frac{\lambda^x}{x!} = e^{-\lambda} \cdot e^{\lambda} = 1$$

since $\sum_{x=0}^{\infty} \lambda^x/x!$ is the Taylor series expansion of e^{λ}.

The expected value for X is derived in much the same spirit that $E(X)$ was determined for a binomial (recall Example 5.2.3):

$$E(X) = \sum_{x=0}^{\infty} xp(x, \lambda) = \sum_{x=0}^{\infty} x\frac{e^{-\lambda}\lambda^x}{x!}$$

$$= \lambda \sum_{x=1}^{\infty} \frac{e^{-\lambda}\lambda^{x-1}}{(x-1)!}$$

$$= \lambda \sum_{y=0}^{\infty} \frac{e^{-\lambda}\lambda^y}{y!} = \lambda \cdot 1 = \lambda$$

where $y = x - 1$. The proof that $\text{Var}(X) = \lambda$ is similar and will be left as an exercise.

Question 6.2.6 Derive the variance for a Poisson random variable. (*Hint:* Consider the factorial moment, $E(X(X-1))$.)

Question 6.2.7 Show that the moment-generating function for a Poisson random variable is

$$M_X(t) = e^{-\lambda+\lambda e^t}$$

and use $M_X(t)$ to verify the formulas for $E(X)$ and $\text{Var}(X)$.

Question 6.2.8 If the random variable X has a Poisson distribution such that $P(X=1) = P(X=2)$, find $P(X=4)$.

Question 6.2.9 Suppose X and Y are two independent random variables with moment-generating functions

$$M_X(t) = e^{-4+4e^t} \quad \text{and} \quad M_Y(t) = e^{-6+6e^t}$$

respectively. Let $Z = X + Y$. Find $f_Z(z)$. (*Hint:* See Question 6.2.7.)

Question 6.2.10 If X and Y are two independent random variables with moment-generating functions

$$e^{-4+4e^t} \quad \text{and} \quad (\tfrac{1}{3} + \tfrac{2}{3}e^t)^2$$

respectively, find $P(X = 2Y)$. (*Hint:* See Question 6.2.7.)

Among the several phenomena that Bortkiewicz successfully fit with the Poisson model, the best-remembered example involved the Prussian cavalry and the frequency with which soldiers were kicked to death by their horses.

CASE STUDY 6.2

During the latter part of the nineteenth century, Prussian officials gathered information on the hazards that horses posed to cavalry soldiers. A total of 10 cavalry corps were monitored over a period of 20 years (6). Recorded for each year and each corps was x, the number of fatalities due to kicks. Table 6.2.3 shows the empirical distribution of X for these 200 corps-years.

TABLE 6.2.3

x = number of deaths	Observed number of corps-years in which x fatalities occurred
0	109
1	65
2	22
3	3
4	1
	200

Altogether there were 122 fatalities ($109(0) + 65(1) + 22(2) + 3(3) + 1(4)$), meaning the observed fatality *rate* was 122/200, or 0.61 fatalities per corps-year. Bortkiewicz proceeded to set the parameter of the Poisson model equal to 0.61 (Is doing that intuitively reasonable?) and proposed as the pdf for X,

$$P(X = x) = p(x, 0.61) = \frac{e^{-0.61}(0.61)^x}{x!}$$

Multiplying $p(x, 0.61)$ by 200 yields the *expected* number of years in which x fatalities occurred (see Table 6.2.4). Clearly, the agreement between the second and third columns of the table is excellent. We would infer from this that the phenomenon of Prussian soldiers being kicked to death by their horses was modeled very well by the Poisson distribution.

TABLE 6.2.4

x	Observed number of corps-years	Expected number of corps-years
0	109	108.7
1	65	66.3
2	22	20.2
3	3	4.1
4	1	0.6
	200	199.9

With Bortkiewicz's work pointing the way, not many years passed before the Poisson distribution was firmly entrenched in the repertoire of mathematicians, statisticians, and scientists of all types. What began as a simple limiting form of the binomial has since been shown to be an excellent model for the enumeration of such diverse phenomena as radioactive emission, outbreaks of wars, the positioning of stars in space, telephone calls originating at pay telephones, traffic accidents along given stretches of road, mine disasters, and even misprints in books.

We might well ask what causes this distribution to appear so often. Or, turning the question around, what can all these various phenomena, so apparently different, possibly have in common? To answer that, we need to reexamine the implications of Poisson's theorem.

Consider one of the applications just mentioned: calls dialed from a pay telephone. Suppose the calls are placed at an average rate of λ per unit time. We imagine a time interval of length T partitioned into n subintervals, each having length T/n. We suppose that these subintervals are so small that the probability of more than one call originating during any particular subinterval is negligible. Finally, we assume that events defined on these intervals are independent—that is, a call made during the ith subinterval has no effect on the probability of a call made during the jth subinterval, $i \neq j$.

If p_n denotes the probability of a call being placed during any subinterval, the probability of exactly k calls originating during the entire interval of length T is given by the binomial

$$P(X = k) = b(k, n, p_n) = \binom{n}{k} p_n^k (1 - p_n)^{n-k}$$

It follows that the total number of calls placed during an interval of length T will be approximately np_n (the mean for a binomial random variable); on the other hand, since λ is the call rate per unit time, the number made should be approximately λT. We may assume, therefore, that $np_n \to \lambda T$ as $n \to \infty$. Equivalently,

$$b(k, n, p_n) \doteq b\left(k, n, \frac{\lambda T}{n}\right)$$

but, by Poisson's theorem,

$$b\left(k, n, \frac{\lambda T}{n}\right) \longrightarrow p(k, \lambda t) \qquad (6.2.2)$$

Equation 6.2.2, then, is what all these phenomena have in common. They are all basically a series of independent Bernoulli trials where n is large and p is small, conditions allowing the underlying binomial to be well-approximated by the Poisson limit.

Comment

In light of the above discussion, it may be helpful to think of Theorem 6.2.2 in more-general terms. If an event is occurring at a constant rate of λ per unit time and if X denotes the number occurring in a period of length T, then

$$P(X = x) = \frac{e^{-\lambda T}(\lambda T)^x}{x!} \qquad x = 0, 1, 2, \ldots$$

By replacing λ with λT, the periods spanned by the random variable and by the assigned parameter are made to agree, as they must. Of course, if X refers to the number of occurrences in a unit of time, as was the case with the cavalry data, T is set equal to 1.

Question 6.2.11 The Brown's Ferry incident of 1975 focused national attention on the ever-present danger of fires breaking out in nuclear power plants. The Nuclear Regulatory Commission has estimated that with present technology there will be, on the average, one fire for every 10 nuclear-reactor years. Suppose a certain state put two reactors on line in 1985. Assuming the incidence of fires can be described by a Poisson distribution, what is the probability that by 1990 at least two fires will have occurred? Does the Poisson assumption seem reasonable here?

Question 6.2.12 Flaws in a particular kind of metal sheeting occur at an average rate of one per 10 ft^2. What is the probability of two or more flaws in a 5-by-8-foot sheet?

We conclude this section with two examples illustrating the diversity of Poisson applications—the first is from particle physics; the second, from sociology.

CASE STUDY 6.3

Among the early research efforts in radioactivity was a 1910 study of alpha emission done by Rutherford and Geiger (78). Their experimental setup consisted of a polonium source placed a short distance from a small screen. For each of 2608 $\frac{1}{8}$-min intervals, the two physicists recorded the number of α particles impinging on the screen (see the first two columns of Table 6.2.5).

TABLE 6.2.5

Number of α particles recorded, x	Observed frequency	Expected frequency
0	57	54
1	203	211
2	383	407
3	525	526
4	532	508
5	408	394
6	273	254
7	139	140
8	45	68
9	27	29
10	10	11
11+	6	6
	2608	2608

The *average* number of counts per $\frac{1}{8}$ min is 3.87:

$$\frac{57(0) + 203(1) + 383(2) + \cdots}{2608} = 3.87$$

As we did in Case Study 6.2, we use the sample average as a numerical estimate for λ. The last column of Table 6.2.5 then shows the entire set of expected frequencies generated by the hypothesized model,

$$P(X = x) = \frac{e^{-3.87}(3.87)^x}{x!}$$

At first glance, the agreement between the observed and expected frequencies may not seem especially impressive. But it is. The large number of data points (2608) distorts our perception of the extent to which the last two columns disagree. Analyzed statistically (see (48)), these figures lend strong support to our initial presumption that radiation is a Poisson phenomenon.

CASE STUDY 6.4

In the 432 years from 1500 to 1931, war broke out somewhere in the world a total of 299 times. (By definition, a military action was a war if it either was legally declared, involved over 50,000 troops, or resulted in significant boundary realignments. To achieve greater uniformity from war to war, major confrontations were split into smaller "subwars": World War I, for example, was treated as five separate wars.) Table 6.2.6 gives the distribution of the number of years in which x wars broke out (72). The last column gives the expected frequencies based on the Poisson model, $P(X = x) = e^{-0.69}(0.69)^x/x!$.

TABLE 6.2.6

Numbers of wars (x) beginning in a given year	Observed frequency	Expected frequency
0	223	217
1	142	149
2	48	52
3	15	12
4+	4	2
	432	432

Comment

One explanation for the excellent fit evident in Table 6.2.6 is that human society has a certain hostility level—as measured by the number of new wars initiated per year (λ)—and that that hostility level remains constant. A breakdown similar to Table 6.2.6, though, showing the distribution of wars *ending* in a given year also proves to be well described by a Poisson (see (72)). Putting those two pieces of information together, it could be speculated that what is constant in society is not a desire for war or a desire for peace but a desire for change.

Question 6.2.13 A radioactive source is metered for 2 h, during which time the total number of alpha particles counted is 482. What is the probability that exactly three particles will be counted during the next minute? No more than three?

Question 6.2.14 A certain young lady is quite popular with her male classmates. She receives, on the average, four phone calls a night. What is the probability that tomorrow night the number of calls she receives will exceed her average by more than one standard deviation?

Question 6.2.15 Records were kept during World War II of the number of bombs falling in south London and their precise points of impact. The particular part of the city studied was divided into 576 areas, each of area $\frac{1}{4}$ km^2. The numbers of areas experiencing x hits, $x = 0, 1, 2, 3, 4, 5$, are listed (11).

Number of hits, x	Frequency
0	229
1	211
2	93
3	35
4	7
5	1
	576

Compute the expected frequencies to see how well the Poisson model applies. Use

$$P(X = x) = \frac{e^{-0.93}(0.93)^x}{x!}, \qquad x = 0, 1, 2, \ldots$$

Question 6.2.16 In a certain published book of 520 pages, 390 typographical errors occur. What is the probability that one page, selected randomly by the printer as a sample of her work, will be free from errors?

Question 6.2.17 Let X be a Poisson random variable with parameter λ. Show that the probability X is even is $\frac{1}{2}(1 + e^{-2\lambda})$.

Question 6.2.18 If X is Poisson with parameter λ and the conditional pdf of Y given x is binomial with parameters x and p, show that the marginal pdf of Y is Poisson with parameter λp.

Question 6.2.19 Let Y denote the interval length between consecutive occurrences of an event described by a Poisson distribution. If the Poisson events are occurring at a rate of λ per unit time, show that

$$f_Y(y) = \lambda e^{-\lambda y}, \qquad y > 0$$

Question 6.2.20 Among the most famous of all meteor showers are the Perseids, which occur each year in early August. In some areas the frequency of visible Perseids can be as high as 40 per hour. Assume the sighting of these meteors is a Poisson event. Use Question 6.2.19 to find the probability that an observer will have to wait at least 5 min between successive sightings.

Question 6.2.21 Suppose that in a certain country commercial airplane crashes occur at the rate of 2.5 per year. Assuming the frequency of crashes per year to be a Poisson random variable, find the probability that four or more crashes will occur next year. Also, find the probability that the next two crashes will occur within three months of one another. (*Hint:* See Question 6.2.19.)

Question 6.2.22 Let X be a Bernoulli random variable where the probability of a success is equal to p. Suppose Y is such that $f_{Y|0}$ is Poisson with parameter λ and $f_{Y|1}$ is Poisson with parameter μ. Show that

$$f_Y(y) = pe^{-\lambda}\frac{\lambda^y}{y!} + (1 - p)e^{-\mu}\frac{\mu^y}{y!}$$

Question 6.2.23 Suppose that X is Poisson with parameter λ and Y is Poisson with parameter μ. If X and Y are independent, show that $X + Y$ is Poisson with parameter $\lambda + \mu$.

Question 6.2.24 Let X and Y be two independent random variables, each described by a Poisson pdf with $\lambda = 2$. Let $Z = X + Y$. Show that the conditional pdf of X given z is binomial.

6.3 THE NORMAL DISTRIBUTION

The limit proposed by Poisson was not the only—or even the first—approximation to the binomial: DeMoivre had already derived a quite different one in his 1718 tract, *Doctrine of Chances*. Like Poisson's work, DeMoivre's theorem did not initially attract the attention it deserved; it did catch the eye of Laplace, though, who proved a generalization and included the result in his influential *Theorie Analytique des Probabilites,* published in 1812.

Theorem 6.3.1 (DeMoivre-Laplace). Let X be a binomial random variable with parameters n and p. For any numbers c and d,

$$\lim_{n \to \infty} P\left(c < \frac{X - np}{\sqrt{npq}} < d\right) = \frac{1}{\sqrt{2\pi}} \int_c^d e^{-x^2/2} \, dx$$

Proof. We derive a more general theorem in Chapter 7. For a proof of Theorem 6.3.1 using only basic calculus, see (26).

Comment

The integral in Theorem 6.3.1 cannot be expressed in closed form and must be approximated by numerical techniques. As early as 1783, Laplace recognized the pressing need for its tabulation and suggested an appropriate computational formula. The first substantial tables, though, were published in 1799 by the French physicist C. Kramp. A table of

$$\frac{1}{\sqrt{2\pi}} \int_{-\infty}^d e^{-x^2/2} \, dx$$

for values of d ranging from -3.90 to $+3.90$ is given in the appendix.

Question 6.3.1 Use the appendix to evaluate each of the following.

a. $\dfrac{1}{\sqrt{2\pi}} \displaystyle\int_{1.25}^{2.50} e^{-x^2/2} \, dx$

b. $\dfrac{1}{\sqrt{2\pi}} \displaystyle\int_{1.17}^{\infty} e^{-x^2/2} \, dx$

Question 6.3.2 Suppose X is a binomial random variable with $n = 100$ and $p = 0.4$. Use Theorem 6.3.1 to approximate the probability that X is between 38 and 43.

Question 6.3.3 Suppose X is a binomial random variable defined on n independent trials, each trial having a success probability of $p = \frac{1}{2}$. Find

$$P\left(-1 < \frac{X - n\left(\frac{1}{2}\right)}{\sqrt{n\left(\frac{1}{2}\right)\left(\frac{1}{2}\right)}} < +1\right)$$

for $n = 2, 5, 8, 10, 12,$ and 15, and compare the results with the DeMoivre-Laplace limit.

Case Study 6.5 shows how the DeMoivre-Laplace theorem is typically put into practice. As we did in the leukemia clustering problem (Case Study 6.1), we assess the statistical significance of an observed event—this time the number of correct guesses in an ESP experiment—by computing the probability of getting, by chance, the number actually observed, *or more*.

CASE STUDY 6.5

Research in extrasensory perception (ESP) has ranged from the slightly unconventional to the downright bizarre. Toward the latter part of the nineteenth century and even well into the twentieth century, much of what was done involved spiritualists and mediums. But beginning around 1910, experimenters moved out of the seance parlors and into the laboratory, where they began setting up controlled studies that could be analyzed statistically. In 1938, Pratt and Woodruff, working out of Duke University, did an experiment that became a prototype for an entire generation of ESP research (35).

The investigator and a subject sat at opposite ends of a table. Between them was a screen with a large gap at the bottom. Five blank cards, visible to both participants, were placed side by side on the table beneath the screen. On the subject's side of the screen, one of the standard ESP symbols was hung over each of the blank cards.

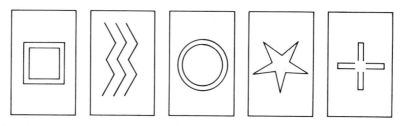

The experimenter shuffled a deck of ESP cards, picked up the top one, and concentrated. The subject tried to read the experimenter's mind: If he or she thought the card being concentrated on was a circle, he or she would point to the blank card on the table that was beneath the circle card hanging on his or her side of the screen. The experimenter then picked another card, and the procedure was repeated. Altogether, a total of 32 subjects, all students, took part in the experiment. They made a total of 60,000 guesses—and were correct 12,489 times.

With five denominations involved, the probability of a subject making a correct identification just by chance is $\frac{1}{5}$. Assuming a binomial model, the expected number of correct guesses would be $60,000 \cdot \frac{1}{5}$, or 12,000. The question is, how "near" to 12,000 is 12,489? Should we write off the observed excess of 489 as nothing more than luck, or can we conclude that ESP has been demonstrated?

To effect a resolution between the conflicting luck and ESP hypotheses, we need to compute the probability of the students' getting 12,489 or more correct answers *under the presumption that $p = \frac{1}{5}$*. Only if that probability is very small can 12,489 be construed as evidence in support of ESP.

Let the random variable X denote the number of correct responses in 60,000 tries. Then

$$P(X \geq 12{,}489) = \sum_{x=12{,}489}^{60{,}000} \binom{60{,}000}{x} \left(\frac{1}{5}\right)^x \left(\frac{4}{5}\right)^{60{,}000-x} \tag{6.3.1}$$

Rather than compute the 47,512 binomial probabilities indicated in Equation 6.3.1 (or the 12,490 making up its complement), we appeal to the theorem of DeMoivre and

Laplace:

$$P(X \geq 12{,}489) = P\left(\frac{X - np}{\sqrt{npq}} \geq \frac{12{,}489 - 60{,}000(\frac{1}{5})}{\sqrt{60{,}000(\frac{1}{5})(\frac{4}{5})}}\right)$$

$$= P\left(\frac{X - np}{\sqrt{npq}} \geq 4.99\right)$$

$$\doteq \frac{1}{\sqrt{2\pi}} \int_{4.99}^{\infty} e^{-x^2/2} \, dx$$

$$= 0.0000003$$

this last value being obtained from a more extensive version of the appendix.

What should we conclude? In the absence of any other information, we conclude that ESP exists. The fact that $P(X \geq 12{,}489)$ is so extremely small makes the hypothesis of luck ($p = \frac{1}{5}$) untenable. It would appear that something other than chance had to be responsible for the occurrence of so many correct guesses.

Comment

The small probability obtained in Case Study 6.5 suggests that ESP exists. Still, not many scientists believe in ESP, and there are many rebuttals in the literature to the evidence presented by Pratt and Woodruff. Criticisms center around the design and execution of the experiment and the probability model chosen. For an example of such a critique, see (25).

Comment

A slightly better approximation to $P(c < (X - np)/\sqrt{npq} < d)$ is obtained by taking the lower limit of the integral in the DeMoivre-Laplace theorem to be $c - 1/2\sqrt{npq}$ and the upper limit, $d + 1/2\sqrt{npq}$. This is known as the *continuity correction* (see (19)).

Question 6.3.4 There is a theory embraced by certain parapsychologists that hypnosis can bring out a person's ESP. In an experiment designed to test that hypothesis (9), 15 students were hypnotized and then asked to guess the identity of an ESP card on which another person (also hypnotized) was concentrating. Each of the students made 100 guesses during the course of the experiment. The total number correct (out of 1500) was 326. Write the binomial expression giving the probability of the event $X \geq 326$ and then use the DeMoivre-Laplace theorem to approximate the probability numerically. What would your conclusion be regarding the efficacy of hypnosis as a technique for improving a person's ESP?

Question 6.3.5 Airlines A and B offer identical service on two flights leaving at the same time (meaning the probability of a passenger choosing either is $\frac{1}{2}$). Suppose both airlines are competing for the same pool of 400 potential passengers. Airline A sells tickets to everyone who requests one, and the capacity of its plane is 230. Approximate the probability that Airline A overbooks.

Question 6.3.6 A random sample of some 747 obituaries published in Salt Lake City newspapers revealed that 46% of the decedents died in the 3-month period following their birthdays (62).

Assess the statistical significance of that finding by estimating the probability that 46% *or more* would die in that particular interval if deaths occurred randomly throughout the year. What would you conclude on the basis of your answer?

Question 6.3.7 A certain type of seed has a probability of 0.8 of germinating. In a package of 100 seeds, what is the probability that at least 75% will germinate?

Question 6.3.8 A political poll of some 200 registered voters shows the Democratic candidate in a gubernatorial race is favored by 110 voters and the Republican candidate by 90, a margin of 10%. But suppose sentiment for the two candidates in the general population is evenly split. Let Y denote the margin percentage in the polls. Use the DeMoivre-Laplace theorem to estimate the probability that Y is as large or larger than 0.10.

Question 6.3.9 A basketball team has a 70% foul-shooting percentage.
    a. Write a formula for the exact probability that out of their next 100 free throws, they will make between 75 and 80, inclusive. Assume the throws are independent.
    b. Using an appropriate approximation, estimate the probability asked for in part (a).

We saw in Theorem 6.2.2 that Poisson's limit to the binomial is a probability distribution. The same is true of the DeMoivre-Laplace limit. To verify the latter, we need to show that $(1/\sqrt{2\pi}) \int_{-\infty}^{\infty} e^{-x^2/2} \, dx = 1$. (Clearly, $(1/\sqrt{2\pi})e^{-x^2/2}$ is greater than or equal to 0 for all x, the other condition a pdf must satisfy.) Proving that the limit integrates to 1 is not obvious and relies on the trick of changing to polar coordinates and showing that the *square* of the integral is 1.

> **Theorem 6.3.2.** $\quad \dfrac{1}{\sqrt{2\pi}} \displaystyle\int_{-\infty}^{\infty} e^{-x^2/2} \, dx = 1$

Proof. The theorem will be established if we can show that

$$\int_{-\infty}^{\infty} e^{-x^2/2} \, dx \cdot \int_{-\infty}^{\infty} e^{-y^2/2} \, dy = 2\pi \qquad (6.3.2)$$

To begin, note that the product of the integrals in Equation 6.3.2 is equal to a double integral:

$$\int_{-\infty}^{\infty} e^{-x^2/2} \, dx \cdot \int_{-\infty}^{\infty} e^{-y^2/2} \, dy = \int_{-\infty}^{\infty} \int_{-\infty}^{\infty} e^{-(x^2+y^2)/2} \, dx \, dy$$

Let $x = r \cos \theta$ and $y = r \sin \theta$, making $dx \, dy = r \, dr \, d\theta$. Then,

$$\int_{-\infty}^{\infty} \int_{-\infty}^{\infty} e^{-(x^2+y^2)/2} \, dx \, dy = \int_{0}^{2\pi} \int_{0}^{\infty} e^{-r^2/2} r \, dr \, d\theta$$

$$= \int_{0}^{\infty} r e^{-r^2/2} \, dr \cdot \int_{0}^{2\pi} d\theta$$

$$= -e^{-r^2/2} \Big|_{0}^{\infty} \cdot \theta \Big|_{0}^{2\pi}$$

$$= 1 \cdot (2\pi)$$

$$= 2\pi$$

The next definition introduces a critical generalization of the density at which we have just looked. Not having any parameters to lend it flexibility, $f_X(x) = (1/\sqrt{2\pi})e^{-x^2/2}$ would be of little practical value as a model for data. Transformed in the manner stated in Definition 6.3.1, though, it suddenly becomes the most useful of *all* probability models.

Definition 6.3.1. A random variable Z is said to have a *standard normal distribution* if its pdf is given by

$$f_Z(z) = \frac{1}{\sqrt{2\pi}}e^{-z^2/2}, \qquad -\infty < z < \infty$$

A random variable X is said to have a *normal distribution with parameters μ and σ* if

$$f_X(x) = \frac{1}{\sqrt{2\pi}\sigma}e^{-(1/2)[(X-\mu)/\sigma]^2}, \qquad -\infty < x < \infty$$

We denote the latter random variable by the symbol $N(\mu, \sigma^2)$. (It follows that a standard normal random variable is denoted by $N(0, 1)$.)

Comment

The two pdf's of Definition 6.3.1 are related through Theorem 4.5.1. If X is an $N(\mu, \sigma^2)$ random variable and we define the transformation

$$Z = \frac{X - \mu}{\sigma}$$

$f_Z(z)$ will be the pdf for a standard normal. Put another way, if $X \sim N(\mu, \sigma^2)$, then $(X - \mu)/\sigma \sim N(0, 1)$. The significance of the relationship between X, $(X - \mu)/\sigma$, and Z is demonstrated in the examples presented later in this section.

Mathematicians and scientists were quick to recognize the usefulness of the two-parameter normal. One of its first applications was due to Gauss, who used it to model observational errors in astronomy. (It appeared in this context in Gauss' celebrated work of 1809, *Theoria Motus Corporum Coelestium in Sectionibus Conicis Solem Ambientium (Theory of the Motion of Heavenly Bodies Moving about the Sun in Conic Sections)*.) By the 1830s, the normal distribution was in quite general use by physicists, as well. It was the social scientist Quetelet, though, who significantly broadened the *scope* of the normal.

Lambert Adolphe Jacques Quetelet (1796–1874) was a Belgian scholar of broad training and interests—mathematics, statistics, astronomy, and poetry, just to name a few. In preparation for establishing the royal observatory at Brussels, he studied probability under Laplace in Paris during the winter of 1823–1824. Quetelet had a passion for collecting data, and he was able to demonstrate that the pdf $f_X(x) = (1/\sqrt{2\pi}\sigma)e^{-(1/2)[(X-\mu)/\sigma]^2}$ described a wide range of sociological and anthropological phenomena. Peripatetic (he belonged to more than a hundred scholarly societies), Quetelet was remarkably effective in spreading his enthusiasm about the utility of the normal distribution throughout all Europe.

Another scholar of the nineteenth century often associated with this same distribution is the noted English biologist Sir Francis Galton (although to label him a biologist does a disservice to the breadth of his intellectual activities). That Galton was influenced by the work of Quetelet is readily apparent from a statement he made in 1889 (29):

> I need hardly remind the reader that the Law of Error upon which these Normal Values are based, was excogitated for the use of astronomers and others who are concerned with extreme accuracy of measurement, and without the slightest idea until the time of Quetelet that they might be applicable to human measures. But Errors, Differences, Deviations, Divergencies, Dispersions, and individual Variations, all spring from the same kind of causes. Objects that bear the same name, or can be described by the same phrase, are thereby acknowledged to have common points of resemblance, and to rank as members of the same species. . . .

Before discussing one of Quetelet's applications of the normal distribution, it would be well to clarify the roles played by μ and σ. As the notation would suggest, one is related to the distribution's location; the other, to its dispersion.

Theorem 6.3.3. Let the random variable X have a normal distribution with parameters μ and σ. Then

$$E(X) = \mu$$

and

$$\text{Var}(X) = \sigma^2$$

Proof. By virtue of Theorems 5.3.1 and 5.4.2, it suffices to show that

$$E\left(\frac{X - \mu}{\sigma}\right) = 0 \quad \text{and} \quad \text{Var}\left(\frac{X - \mu}{\sigma}\right) = 1$$

The statement about μ follows immediately since the integrand for the expected value is an odd function:

$$E\left(\frac{X - \mu}{\sigma}\right) = \int_{-\infty}^{\infty} y \cdot \frac{1}{\sqrt{2\pi}} e^{-y^2/2}\, dy = 0$$

To prove the second part of the theorem, we note that

$$\text{Var}\left(\frac{X - \mu}{\sigma}\right) = E\left(\left(\frac{X - \mu}{\sigma}\right)^2\right) - 0^2$$

$$= \int_{-\infty}^{\infty} y^2 \cdot \frac{1}{\sqrt{2\pi}} e^{-y^2/2} \, dy \qquad (6.3.3)$$

After an integration by parts, the right-hand side of Equation 6.3.3 reduces to

$$\int_{-\infty}^{\infty} y^2 \cdot \frac{1}{\sqrt{2\pi}} e^{-y^2/2} \, dy = \frac{1}{\sqrt{2\pi}} y e^{-y^2/2} \Big|_{-\infty}^{\infty} + \frac{1}{\sqrt{2\pi}} \int_{-\infty}^{\infty} e^{-y^2/2} dy$$

$$= 0 + 1 = 1$$

and the result follows.

Question 6.3.10 Show that the moment-generating function for the normal distribution with parameters μ and σ is

$$M_X(t) = e^{\mu t + \sigma^2 t^2/2}$$

Compute $M_X'(0)$ and $M_X''(0)$ and verify Theorem 6.3.3.

Question 6.3.11 If e^{3t+8t^2} is the moment-generating function for the random variable X, find $P(-1 \le X \le 9)$. (*Hint:* See Question 6.3.10.)

Question 6.3.12 Suppose X is $N(\mu, \sigma^2)$. Use the properties of moment-generating functions to show that

$$Z = \frac{X - \mu}{\sigma}$$

is $N(0, 1)$.

CASE STUDY 6.6

Among Quetelet's many anthropometric applications of the normal pdf was one where he fit that distribution to the chest measurements of 5738 Scottish soldiers. Table 6.3.1 is a facsimile of Quetelet's data as they appeared in an 1846 book of letters to the Duke of Saxe-Cobourg and Gotha (70). Column 1 lists, in inches, the range of possible chest measurements. Column 2 gives the corresponding observed frequencies, which are then divided by 5738 and reappear in Column 3 as *relative* frequencies (the decimal points are omitted).

In the last column are the expected probabilities, assuming a normal distribution. Specifically, the parameter μ was estimated to be 39.8 inches and σ was estimated to be 2.05 inches, so these final entries were based on the model

$$f_X(x) = \frac{1}{\sqrt{2\pi}(2.05)} e^{-(1/2)[(x-39.8)/2.05]^2}$$

The expected probability of a Scottish soldier having a chest measurement x of, for instance, 42 inches is determined by making the Z transformation and using the

TABLE 6.3.1 QUETELET'S CHEST MEASUREMENTS OF SCOTTISH SOLDIERS

Mea-sures de la poitrine	Nom-bre d'hom-mes	Nom-bre propor-tionnel	Prob-abilité d'après l'obser-vation	Rang dans la table	Rang d'après le calcul.	Prob-abilité d'après la table	Nombre d'obser-vations calculé
Pouces.							
33	3	5	0,5000			0,5000	7
34	18	31	0,4995	52	50	0,4993	29
35	81	141	0,4964	42,5	42,5	0,4964	110
36	185	322	0,4823	33,5	34,5	0,4854	323
37	420	732	0,4501	26,0	26,5	0,4531	732
38	749	1305	0,3769	18,0	18,5	0,3799	1333
39	1073	1867	0,2464	10,5	10,5	0,2466	1838
			0,0597	2,5	2,5	0,0628	
40	1079	1882	0,1285	5,5	5,5	0,1359	1987
41	934	1628	0,2913	13	13,5	0,3034	1675
42	658	1148	0,4061	21	21,5	0,4130	1096
43	370	645	0,4706	30	29,5	0,4690	560
44	92	160	0,4866	35	37,5	0,4911	221
45	50	87	0,4953	41	45,5	0,4980	69
46	21	38	0,4991	49,5	53,5	0,4996	16
47	4	7	0,4998	56	61,8	0,4999	3
48	1	2	0,5000			0,5000	1
	5738	1,0000					1,0000

appendix:

$$P(X \text{ is } 42) = P(41.5 < X < 42.5)$$

$$= P\left(\frac{41.5 - 39.8}{2.05} < \frac{X - \mu}{\sigma} < \frac{42.5 - 39.8}{2.05}\right)$$

$$= P\left(0.829 < \frac{X - \mu}{\sigma} < 1.317\right)$$

$$= \frac{1}{\sqrt{2\pi}} \int_{0.829}^{1.317} e^{-y^2/2} \, dy$$

$$= 0.1096$$

this latter figure being the tenth entry appearing in the last column.

Question 6.3.13 The Stanford-Binet IQ test is scaled to give a mean score of 100 with a standard deviation of 16. Suppose children having IQ's of less than 80 or greater than 145 are deemed to need special attention. Given a population of 2000 children, what will be the expected demand for these additional services?

Question 6.3.14 A criminologist has developed a questionnaire for predicting whether a teenager will become a delinquent. Scores on the questionnaire can range from 0 to 100, with higher values reflecting a presumably greater criminal tendency. As a rule of thumb, the criminologist decides to classify a teenager as a potential delinquent if his or her score exceeds 75. The criminologist has already tested the questionnaire on a large sample of teenagers, both delinquent and nondelinquent. Among those considered nondelinquent, scores were normally distributed with a mean of 60 and a standard deviation of 10. Among those considered delinquent, scores were normally distributed with a mean of 80 and a standard deviation of 5.

a. What proportion of the time will the criminologist misclassify a nondelinquent as a delinquent?

b. What proportion of the time will the criminologist misclassify a delinquent as a nondelinquent?

Question 6.3.15 Assume that the number of miles a driver gets on a set of radial tires is normally distributed with a mean of 30,000 mi and a standard deviation of 5000 mi. Would the manufacturer of these tires be justified in claiming that 90% of all drivers will get at least 25,000 mi?

Question 6.3.16 The diameter of the connecting rod in the steering mechanism of a certain foreign sports car must be between 1.480 and 1.500 cm, inclusive, to be usable. The distribution of connecting-rod diameters produced by the manufacturing process is normal with a mean of 1.495 cm and a standard deviation of 0.005 cm. What percentage of rods will have to be scrapped?

To put the normal distribution in its proper perspective, we should leave the nineteenth century—and Quetelet—and look at an example of more-recent vintage. In fields such as education and psychology, a frequent application of the normal—one certainly familiar to students—is the description of test scores: IQ's, ACT's and CEEB's, to name just a few, are all standardized exams whose distributions are known to resemble strongly the bell-shaped curve of Definition 6.3.1. The next example is a variation on that same theme, the data coming this time from an experiment in role analysis.

CASE STUDY 6.7

On the basis of which factors influence them, decision-makers can be assigned positions along a *motivation scale*. At one end of the scale are the *moralists*, individuals who always do what they believe to be right, regardless of the consequences. At the other end sit the *expedients*, persons whose decisions reflect more pragmatic concerns—peer-group pressure, short-range gain, the desire to be liked, and so on. Most decision-makers, of course, would fall somewhere between these extremes, sometimes acting out of righteousness but other times out of expediency. In the experiment to be described (32), a questionnaire was used to quantify the nature of a person's motivation. The purpose was to investigate—and, possibly, to model—the resulting distribution.

A group of 106 subjects was given a test that consisted of 37 different conflict situations. A possible solution was proposed for each one. The subjects had to decide, if it were up to them, whether *they* would resolve the conflict in the manner that was suggested. But instead of answering just yes or no, they were required to indicate the strengths of their convictions in taking (or not taking) the recommended course of action. Their possible responses were:

1. Absolutely must (take the action)
2. Preferably should (take the action)
3. May or may not (take the action)
4. Preferably should not (take the action)
5. Absolutely must not (take the action)

It was felt that the *absolutely must* and *absolutely must not* responses were probably moralistic in origin, and a subject was awarded a point each time he or she gave either one. The total number of points accumulated over the 37 questions was then taken as an index of the extent to which a subject sought moral solutions. The scores are listed in Table 6.3.2.

TABLE 6.3.2

			Motivation scores			
13	12	11	19	24	2	13
17	15	2	17	15	7	15
13	27	4	16	13	9	5
8	19	4	17	12	5	28
7	23	13	13	6	21	20
10	6	10	7	17	18	19
10	2	13	9	27	17	14
21	9	19	12	3	18	11
18	11	25	11	10	12	14
17	5	14	30	7	15	4
19	18	11	19	1	13	8
15	20	4	4	14	13	10
15	24	14	11	22	15	7
23	15	12	18	16	6	23
12	14	23	18	10	25	18
						24

Estimates for μ and σ in this case were taken to be 13.8 and 6.5, respectively. Figure 6.3.1 shows the particular normal curve,

$$f_X(x) = \frac{1}{\sqrt{2\pi}(6.5)} e^{-\frac{1}{2}[(x-13.8)/6.5]^2}$$

superimposed over a histogram of the data from Table 6.3.2. The fit is excellent.

It should be noted here that the ordinate for the normal curve shown in Figure 6.3.1 is simply $f_X(x)$. For the histogram, though, the ordinate is not just relative

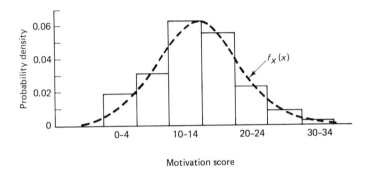

Figure 6.3.1

frequency but rather

$$\frac{\text{Class frequency}}{(\text{Total number of observations}) \times (\text{class width})}$$

The observed frequency, for example, associated with the 20–24 class is 12; therefore, the height of the 20–24 bar is $12/(106 \times 5)$, or 0.0226. Including class width as a factor in the ordinate's denominator scales the area of the histogram to 1, a necessary step if the histogram is to be compared with a probability model (whose area is necessarily 1).

Question 6.3.17 For any distribution, the *interquartile range* is defined to be the difference between its 75th and 25th percentiles. Find the interquartile range for the motivation data of Case Study 6.7. What property of a distribution does the magnitude of its interquartile range reflect?

Question 6.3.18 For the motivation data of Case Study 6.7, compute the expected frequency for the 10–14 class. Assume the appropriate model is a normal curve with a mean of 13.8 and a standard deviation of 6.5.

Applications of probabilistic reasoning are not always confined to laboratory research or field surveys, nor do they always appear in technical journals. The next example is a case in point.

CASE STUDY 6.8

The following letter appeared in a well-known column giving advice to the lovelorn (86):

> Dear Abby: You wrote in your column that a woman is pregnant for 266 days. Who said so? I carried my baby for ten months and five days, and there is no doubt about it because I know the exact date my baby was conceived. My husband is in the Navy and it couldn't

possibly have been conceived any other time because I saw him only once for an hour, and I didn't see him again until the day before the baby was born.

I don't drink or run around, and there is no way this baby isn't his, so please print a retraction about that 266-day carrying time because otherwise I am in a lot of trouble.

San Diego Reader

While the full implications of the plight of the San Diego reader are beyond the scope of this text, it *is* possible to assess quantitatively the statistical likelihood of a pregnancy being 310 days long (10 months and 5 days). By the same reasoning used in Case Study 6.1, this would be done by computing the probability that, by chance alone, a pregnancy will be 310 days long *or longer*. That is, if Y denotes the duration of an arbitrary pregnancy, we want to compute $P(Y \geq 310)$—the smaller that probability is, the less credible San Diego reader becomes. Of course, despite our best-intentioned efforts to come up with a relevant and objective probability, the interpretation of that probability will still be a bit subjective, since the particular value of $P(Y \geq 310)$ that delineates what we choose to accept as the truth from what we reject as an untruth is necessarily arbitrary.

According to well-documented norms, the mean and standard deviation for Y are 266 days and 16 days, respectively. Furthermore, if it can be assumed that the distribution of pregnancy durations is normal, then

$$P(Y \geq 310) = P\left(\frac{Y - 266}{16} \geq \frac{310 - 266}{16}\right) = P(Z \geq 2.75)$$

But, from the appendix,

$$P(Z \geq 2.75) = 1 - 0.9970$$

$$= 0.003$$

Figure 6.3.2 shows the two equivalent areas involved in our analysis, the original one under the $N(266, (16)^2)$ distribution and the transformed one under the $N(0, 1)$. It is left to the reader to decide how the 0.003 should be interpreted in this particular context.

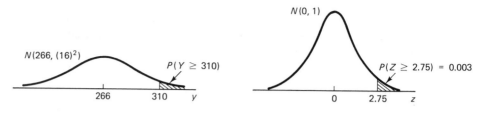

Figure 6.3.2

Comment

Insurance companies sometimes use probability computations of the kind just described in deciding whether or not to reimburse married couples for maternity costs. Certain policies contain a clause to the effect that if Y is too much *less* than

266 (as measured from the couple's date of marriage), the company can refuse to pay.

The final example in this section is typical of an entire class of problems where tables of the standard normal distribution have to be used *backward*. What we need to find is not an area, but a percentile.

Example 6.3.1

Mensa is an international society whose membership is limited to individuals having IQ's above the 98th percentile of the general population. It is well known that the average IQ for the general population is 100, the standard deviation is 16, and the distribution itself is normal (recall Question 6.3.13). What is the *lowest* IQ that will qualify a person to belong to Mensa?

Let Y denote the IQ of a random person. By definition, the lowest qualifying value for Y, which we call Y^*, must satisfy the equation

$$P(Y \geq Y^*) = 0.02$$

Transforming Y to Z gives

$$P(Y \geq Y^*) = P\left(\frac{Y - 100}{16} \geq \frac{Y^* - 100}{16}\right) = P\left(Z \geq \frac{Y^* - 100}{16}\right) = 0.02$$

From the appendix, the 98th percentile of the $N(0, 1)$ distribution is 2.05:

$$P(Z \geq 2.05) = 0.02$$

Therefore, setting the two expressions for the 98th percentile equal gives

$$\frac{Y^* - 100}{16} = 2.05$$

from which it follows that a person must have an IQ of at least 133 to be eligible for membership in Mensa:

$$Y^* = 100 + 2.05(16) = 133$$

Question 6.3.19 Dental structure provides an effective criterion for classifying certain fossils. Not long ago, a baboon skull of unknown origin was discovered in a cave in Angola (59); the length of its third molar was 9.0 mm. Speculation arose that the baboon in question might be a "missing link" and belong to the genus *Papio*. Members of that genus have third molars measuring, on the average, 8.18 mm long with a standard deviation of 0.47 mm. Quantify the significance of the 9.0-mm molar. What would your inference be regarding the baboon's lineage?

Question 6.3.20 A college professor teaches Chemistry 101 each fall to a large class of first-year students. For tests, she uses standardized exams that she knows from past experience produce bell-shaped grade distributions with a mean (μ) of 70 and a standard deviation (σ) of 12. Her philosophy of grading is to impose standards that will yield, in the long run, 14% A's, 20% B's, 32% C's, 20% D's, and 14% F's. Where should the cutoff be between the A's and the B's? Between the B's and the C's?

Question 6.3.21 The systolic blood pressure of 18-year-old women is normally distributed with a mean of 120 mm Hg and a standard deviation of 12 mm Hg. What is the probability that the blood pressure of a randomly selected 18-year-old woman will be greater than 150? Less than 115? Between 110 and 130?

Question 6.3.22 At a certain Ivy League college, the average score of a first-year student on the verbal part of the SAT is 565, with a standard deviation of 75. If the distribution of scores is normal, what proportion of that school's first-year students have verbal SAT's over 650? Under 500?

Question 6.3.23 Find the 60th percentile for the verbal SAT scores described in Question 6.3.22. If there are 1000 first-year students at that particular school, what is the standard deviation of the number of students scoring between 570 and 590, inclusive?

Question 6.3.24 It is estimated that 80% of all 18-year-old women have weights ranging from 103.5 to 144.5 lb. Assuming the weight distribution can be adequately approximated by a normal curve and assuming that 103.5 and 144.5 are equal distances from the average weight μ, calculate σ.

Question 6.3.25 A recent year's traffic-death rates (fatalities per 100 million motor-vehicle miles) are given for each of the 50 states (58).

Ala	6.4	La	7.1	Ohio	4.5
Alaska	8.8	Maine	4.6	Okla	5.0
Ariz	6.2	Mass	3.5	Ore	5.3
Ark	5.6	Md	3.9	Pa	4.1
Cal	4.4	Mich	4.2	RI	3.0
Colo	5.3	Minn	4.6	SC	6.5
Conn	2.8	Miss	5.6	SD	5.4
Del	5.2	Mo	5.6	Tenn	7.1
Fla	5.5	Mont	7.0	Tex	5.2
Ga	6.1	NC	6.2	Utah	5.5
Hawaii	4.7	ND	4.8	Va	4.5
Idaho	7.1	Nebr	4.4	Vt	4.7
Ill	4.3	Nev	8.0	WVa	6.2
Ind	5.1	NH	4.6	Wash	4.3
Iowa	5.9	NJ	3.2	Wisc	4.7
Kans	5.0	NM	8.0	Wy	6.5
Ky	5.6	NY	4.7		

Make a histogram of these observations using as classes 2.0–2.9, 3.0–3.9, and so on. Assuming the distribution of rates can be approximated by a normal distribution, determine the expected frequencies for each of the classes. Let μ and σ have the values 5.3 and 1.3, respectively.

Question 6.3.26 The army is developing a new missile and is concerned about its precision. By observing points of impact, launchers can adjust the missile's initial trajectory, thereby controlling the mean of its impact distribution. If the standard deviation of that distribution is too large, though, the missile will be ineffective. Suppose the Pentagon requires that at least 95% of the missiles will fall within $\frac{1}{8}$ mi of the target when the missiles are aimed properly. What is the maximum allowable standard deviation for the impact distribution? (Assume the latter is normal.)

The normal distribution, as we have seen, provides a singularly important model of broad applicability. But even when a random variable X is *not* normal, there is frequently a transformation $Y = g(X)$ that is. The most useful such transformation is $Y = g(X) = \log X$.

Definition 6.3.2. A random variable X is said to have the *log-normal* distribution if $Y = \log (X - \theta)$ is normally distributed, where θ is a constant such that $X > \theta$. If Y is an $N(\mu, \sigma^2)$ random variable, the pdf for X is given by

$$f_X(x) = \frac{1}{\sqrt{2\pi} \cdot \sigma(x - \theta)} e^{-\frac{1}{2}[(\log (x-\theta)-\mu)/\sigma]^2}, \qquad x > \theta$$

The derivation of the lognormal pdf in Definition 6.3.2 provides an opportunity to recall one of the techniques introduced in Section 4.5: finding a pdf by differentiating its cdf. Here,

$$
\begin{aligned}
F_X(x) = P(X \le x) &= P(X - \theta \le x - \theta) \\
&= P(\log (X - \theta) \le \log (x - \theta)) \\
&= P(Y \le \log (x - \theta)) \\
&= F_Y(\log (x - \theta))
\end{aligned}
$$

Setting $f_X(x) = F_X'(x)$, we can write

$$f_X(x) = f_Y(\log (x - \theta)) \cdot \frac{1}{x - \theta}$$

Since f_Y is a normal density with mean μ and variance σ^2, the expression for $f_X(x)$ given in Definition 6.3.2 follows readily.

Comment

We are deferring for the moment the interesting question of *why* the normal and lognormal pdf's arise so frequently as data models. We come back to that point in Section 7.4.

Question 6.3.27 Suppose X is lognormal with $\theta = 0$. Show that $X = e^{\sigma Z + \mu}$, where Z is a standard normal.

Question 6.3.28 Let X be lognormal with $\theta = 0$. Show that $E(X) = e^{\mu + \sigma^2/2}$ and $\text{Var}(X) = e^{2\mu + 2\sigma^2} - e^{2\mu + \sigma^2}$. (*Hint:* Use Question 6.3.27 to express both $E(X)$ and $E(X^2)$ in terms of the moment-generating function of Z.)

6.4 THE GEOMETRIC DISTRIBUTION

Given a series of independent Bernoulli trials, we are accustomed to thinking of n and p as fixed, and X, the number of successes, as the (binomial) random variable. Suppose the problem is turned around, though, and the question is asked, how many trials are required in order to achieve the first success? Put this way, n is the random variable and X is fixed. Clearly, the first success will occur on the very first trial with probability p, on the second trial with probability $(1 - p)p$, on the third trial with probability $(1 - p)^2 p$, and so on. In general, the pdf for N, the trial number of the first success, is

$$f_N(n) = p(1 - p)^{n-1} = pq^{n-1}, \qquad n = 1, 2, \ldots$$

This is called the *geometric distribution* (with parameter p).

Intuitively, the average number of trials required for the first success to appear—that is, $E(N)$—should be inversely proportional to p. The next theorem shows that $E(N)$ is, in fact, *exactly* equal to $1/p$.

Theorem 6.4.1. Let N be a geometric random variable with parameter p (where $p = P(\text{success at any trial})$). Then

$$E(N) = \frac{1}{p}$$

and

$$\text{Var}(N) = \frac{q}{p^2}$$

Proof. By definition,

$$E(N) = \sum_{n=1}^{\infty} npq^{n-1} = p \sum_{n=1}^{\infty} nq^{n-1} \tag{6.4.1}$$

Evaluating the right-hand side of Equation 6.4.1 is not a trivial problem. Probably the quickest solution, although one that may seem a bit roundabout, is to take the derivative of the sum's integral. Let

$$h(q) = \sum_{n=1}^{\infty} nq^{n-1}$$

Then

$$\int h(q)\, dq = \sum_{n=1}^{\infty} q^n = \sum_{n=0}^{\infty} q^n - q^0$$

$$= \frac{1}{1 - q} - 1$$

$$= \frac{q}{1 - q}$$

By using the rule for the derivative of a quotient, we get that $h(q) = 1/p^2$:

$$h(q) = \frac{d}{dq} \int h(q) \, dq = \frac{(1-q) - (-q)}{(1-q)^2} = \frac{1}{(1-q)^2} = \frac{1}{p^2}$$

Substituting $h(q)$ into Equation 6.4.1 gives the first part of the theorem:

$$E(N) = p\left(\frac{1}{p^2}\right) = \frac{1}{p}$$

The proof that $\mathrm{Var}(N) = q/p^2$ is left as an exercise.

Question 6.4.1 Show that the moment-generating function for the geometric distribution is

$$M_N(t) = \frac{pe^t}{1 - qe^t}$$

and use $M_N(t)$ to find $E(N)$ and $\mathrm{Var}(N)$.

Although not nearly so ecumenical as the Poisson or the normal, the geometric distribution does find a variety of applications. An unusual hybrid geometric is developed in Case Study 6.9. The data are meteorological—the variables being observed are the lengths (in days) of weather cycles in Tel Aviv.

CASE STUDY 6.9

A *wet spell of x days* is defined to be a succession of x days, on each of which occurs some measurable precipitation. Under the assumption that the weather tomorrow depends only on the weather today, there is a probability p_0 that a wet day will be followed by a dry day. The probability of an x-day-long wet spell is $p_0(1 - p_0)^{x-1} = p_0 q_0^{x-1}$, a pdf we recognize as the geometric. In a similar way, there is a random variable Y associated with the lengths of *dry* spells; furthermore, $f_Y(y) = p_1 q_1^{y-1}$, where p_1 is the probability that a dry day is followed by a wet one.

A weather *cycle* is defined to be a wet spell followed by a dry spell. If Z denotes the length of such a cycle, then $Z = X + Y$. We leave it as an exercise to verify that

$$f_{X+Y}(z) = p_0 p_1 \frac{q_0^{z-1} - q_1^{z-1}}{q_0 - q_1}, \qquad z = 2, 3, \ldots$$

(*Hint:* Write $P(X + Y = z) = P(X = 1)P(Y = z - 1) + \cdots + P(X = z - 1)P(Y = 1)$ and simplify.)

The plausibility of its derivation notwithstanding, $f_{X+Y}(z)$ is a useful model only if it can accurately predict weather-cycle distributions. A set of data suitable for testing $f_{X+Y}(z)$ comes from Israel: the lengths of 351 weather cycles in Tel Aviv were recorded during the months of December, January, and February from the winter of 1923–1924 through the winter of 1949–1950 (28). The model's two parameters, p_0 and p_1, were estimated to be 0.338 and 0.250, respectively.

Table 6.4.1 lists the observed and expected distributions for Z, the latter based on

$$f_{X+Y}(z) = (0.338)(0.250)\frac{(0.662)^{z-1} - (0.750)^{z-1}}{0.662 - 0.750}$$

Considering the simplicity of some of our assumptions, the agreement between the last two columns is quite good.

TABLE 6.4.1

Cycle length	Observed frequency	Expected frequency
2	33	30
3	38	42
4	33	44
5	50	42
6	39	37
7	26	32
8	30	26
9	24	21
10	16	17
11	14	14
12	15	11
13	9	8
14	3	6
15+	21	21

Question 6.4.2 A professional football team has three quarterbacks on its traveling squad. Suppose that each quarterback has an 80% chance of completing a game without getting injured. If the team plays a schedule of 12 games, what is the probability the third-string quarterback makes his first starting appearance in the eleventh game? What is the probability the team has to find a *fourth* quarterback before the season is over? (Assume the probability of two or more injuries in any one game is zero.)

Question 6.4.3 A young couple plans to continue having children until they have their first girl. Suppose the probability that a child is a girl is $\frac{1}{2}$ and the outcome of each birth is an independent event. What is their expected family size?

Question 6.4.4 Show that a geometric random variable, X, is *memoryless*. That is, prove that

$$P(X = n - 1 + k \mid X > n - 1) = P(X = k)$$

Question 6.4.5 A somewhat uncoordinated man is attempting to get a driver's license. Having successfully completed the written exam, he needs only to pass the road test to achieve his objective. His abilities in that particular area, though, are less than exceptional: An unbiased observer would give him no more than a 10% chance of passing. What is the probability he will have to take the test at least seven times in order to get his license? Assume his driving skills remain at the same level, regardless of how many times he fails the test. What is the expected number of times he will have to take the test?

Question 6.4.6 A fair die is tossed. What is the probability the first 5 occurs on the fourth roll?

6.5 THE NEGATIVE BINOMIAL DISTRIBUTION

In this section we consider a natural generalization of the geometric distribution. Rather than focus on the number of Bernoulli trials required for the *first* success to occur, we now look at X, the number required for the occurrence of the *r*th success. It should be clear that the only possible scenario for X being equal to x is for a total of $r - 1$ successes (and $x - 1 - (r - 1)$ failures) to occur somewhere during the first $x - 1$ trials and for one additional success (the *r*th) to occur on the very next (that is, the *x*th) trial. Assuming all trials to be independent, we can easily write a formula for $f_X(x)$:

$$f_X(x) = \binom{x - 1}{r - 1} p^{r-1} q^{x-1-(r-1)} p$$

$$= \binom{x - 1}{r - 1} p^r q^{x-r}, \qquad x = r, r + 1, \ldots$$

In practice, the random variable X is often replaced by Y, where Y is the number of trials *in excess of r* required to produce the *r*th success. That is, $Y = X - r$. Let n be the argument for Y. Then

$$f_Y(n) = \binom{n + r - 1}{r - 1} p^r q^n = \binom{n + r - 1}{n} p^r q^n, \quad n = 0, 1, \ldots$$

Since we derived $f_Y(n)$ as a pdf, we fully expect its values to sum to one. It will be instructive, though, to prove that property independently. In so doing, we show that $f_Y(n)$ is a pdf even when r is not an integer. This last statement makes no sense unless $\binom{n+r-1}{n}$ is defined for all real r. Up to this point, the symbol $\binom{k}{n}$ has been used as shorthand for the quantity $k(k - 1) \cdots (k - n + 1)/n!$ only when k was a non-negative integer. But $k(k - 1) \cdots (k - n + 1)/n!$ is well defined for *any* real k, so there is no problem in extending the definition of $\binom{k}{n}$.

Theorem 6.5.1. $\displaystyle\sum_{n=0}^{\infty} f_Y(n) = \sum_{n=0}^{\infty} \binom{n + r - 1}{n} p^r q^n = 1$

Proof. We begin by establishing a combinatorial identity involving $\binom{n+r-1}{n}$. Since

$$\binom{-r}{n} = \frac{-r(-r - 1) \cdots (-r - n + 1)}{n!}$$

$$= (-1)^n \frac{r(r + 1) \cdots (n + r - 1)}{n!}$$

$$= (-1)^n \binom{n + r - 1}{n}$$

we can write

$$\binom{n + r - 1}{n} = (-1)^n \binom{-r}{n} \tag{6.5.1}$$

With Equation 6.5.1 we have our key for evaluating $\sum_{n=0}^{\infty} f_Y(n)$. Note, first of all, that

$$\sum_{n=0}^{\infty} \binom{n + r - 1}{n} p^r q^n = p^r \sum_{n=0}^{\infty} \binom{n + r - 1}{n} q^n$$

$$= p^r \sum_{n=0}^{\infty} \binom{-r}{n} (-1)^n q^n$$

$$= p^r \sum_{n=0}^{\infty} \binom{-r}{n} (-q)^n \tag{6.5.2}$$

Now, recall Newton's binomial formula: For any real k,

$$(1 + t)^k = \sum_{n=0}^{\infty} \binom{k}{n} t^n \tag{6.5.3}$$

Applying Equation 6.5.3 to Equation 6.5.2 gives the result:

$$\sum_{n=0}^{\infty} \binom{n + r - 1}{n} p^r q^n = p^r (1 - q)^{-r}$$

$$= p^r p^{-r} = 1$$

What we have shown to be true about Y, the number of Bernoulli trials in excess of r required to produce the rth success, is stated formally in Definition 6.5.1. The mean and variance of Y are given in Theorem 6.5.2.

Definition 6.5.1. The random variable Y is said to have the negative binomial distribution with parameters r and p if

$$f_Y(n) = \binom{n + r - 1}{n} p^r q^n, \qquad n = 0, 1, \ldots$$

Theorem 6.5.2. If the random variable Y has the negative binomial distribution with parameters r and p, then

$$E(Y) = \frac{rq}{p}$$

and

$$\text{Var}(Y) = \frac{rq}{p^2}$$

Proof. The theorem is true for arbitrary $r > 0$, but we consider only the case where r is a positive integer. Let $X_1, X_2, \ldots, X_r$ be independent geometric random variables, each with the same parameter p. Since each X_i can be thought of as the number of trials required for the first success, it seems intuitively reasonable that their sum, $X = X_1 + X_2 + \cdots + X_r$, should have the same distribution as the number of trials required for the rth success—that is, $Y = X - r$ should be negative binomial with parameters r and p (see Question 6.5.4). It follows from Theorem 6.4.1 and the properties of sums of independent random variables that

$$E(X) = rE(X_1) = r\left(\frac{1}{p}\right) = \frac{r}{p}$$

and

$$\text{Var}(X) = r\text{Var}(X_1) = r\left(\frac{q}{p^2}\right) = \frac{rq}{p^2}$$

Therefore,

$$E(Y) = E(X - r) = E(X) - r = \left(\frac{r}{p}\right) - r = \frac{rq}{p}$$

and

$$\text{Var}(Y) = \text{Var}(X - r) = \text{Var}(X) = \frac{rq}{p^2}$$

Question 6.5.1 Suppose an underground military installation is fortified to the extent that it can withstand up to four direct hits from air-to-surface missiles and still function. Enemy aircraft can score direct hits with these particular missiles 7 times out of 10. What is the probability that a plane will require fewer than 8 shots to destroy the installation? Assume all firings are independent.

Question 6.5.2 Let Y be a negative binomial random variable. Prove directly that the expected value of $X = Y + r$ is r/p by evaluating the sum

$$\sum_{x=r}^{\infty} x\binom{x-1}{r-1}p^r q^{x-r}$$

(Hint: Reduce the sum to one involving negative binomial probabilities with parameters $r + 1$ and p.)

When r is a positive integer, its interpretation in the negative binomial model is clear. When r is positive but not an integer, we lose that physical interpretation, but the model itself may still be very useful. The next set of data is a case in point.

CASE STUDY 6.10

An early application of the negative binomial is attributed to the famous British statistician Sir Ronald A. Fisher. Pursuing a question of somewhat less gravity than was his custom, Fisher was able to show that Y, the number of ticks found on sheep, could be described remarkably well by a function of the form,

$$f_Y(n) = \binom{n + r - 1}{n} p^r q^n$$

His data consisted of 60 sheep and their uninvited entourage of some 200 ticks. Table 6.5.1 shows the observed and expected tick distributions (27).

TABLE 6.5.1

Number of ticks	Observed frequency	Expected frequency
0	7	6
1	9	10
2	8	11
3	13	10
4	8	8
5	5	5
6	4	4
7	3	2
8	0	1
9	1	1
10+	2	2
	60	60

The parameters r and p were estimated to be 3.75 and 0.536, respectively, so the entries in the last column were found by multiplying 60 times $\binom{n+3.75-1}{n}(0.536)^{3.75} \cdot (0.464)^n$.

Question 6.5.3 Suppose Y_1 and Y_2 are two independent negative binomial random variables, each with parameters r and p. Show that $Y_1 + Y_2$ is negative binomial with parameters $2r$ and p.

Question 6.5.4 Let $X_1, X_2, \ldots, X_r$ be r independent random variables, each having the geometric pdf with parameter p. Show that $X = X_1 + X_2 + \cdots + X_r$ has pdf

$$f_X(x) = \binom{x - 1}{r - 1} p^x q^{x-r}, \qquad x = r, r + 1, \ldots$$

and hence that $Y = X - r$ has the negative binomial pdf with parameters r and p.

Question 6.5.5 A door-to-door encyclopedia salesperson is required to document five in-home visits each day. Suppose she has a 30% chance of being invited into any given home, with each home representing an independent trial. If she selects, ahead of time, the addresses of 10 households upon which to call, what is the probability her fifth success occurs on the tenth trial? What is the probability she requires fewer than 8 addresses to record her fifth success? Suppose she goes to only 7 homes: What is the probability she will meet or exceed her quota?

Question 6.5.6 The moment-generating function for the negative binomial pdf as given in Definition 6.5.1 is not the same as the moment-generating function for

$$f_X(x) = \binom{x-1}{r-1} p^r q^{x-r}, \qquad x = r, r+1, \ldots$$

yet both pdf's relate to the occurrence of the rth success in a series of independent Bernoulli trials. Why are the two mgf's different?

Question 6.5.7 When a machine is improperly adjusted, it has probability 0.15 of producing a defective item. Each day the machine is run until three defective items are produced. Then it is stopped and checked for adjustment. What is the probability that an improperly adjusted machine will produce five or more items before being stopped? What is the average number of items the machine will produce before being stopped?

6.6 THE GAMMA DISTRIBUTION

In Section 6.2, the useful Poisson distribution was introduced as the limiting form of the binomial. The Poisson distribution concerns the *number* of occurrences of an event during a fixed time period. We use the ideas of Section 6.2 to derive the distribution of the *time* required for a fixed number of events to occur. This important distribution, the *gamma*, is the subject of this section.

Suppose a certain event occurs homogeneously over a time interval of length x at an average rate of λ per unit time. We wish to examine the continuous variable X measuring the length of time required for r events to occur. Assume, as in Section 6.2, that the interval $[0, x]$ can be subdivided into small, independent subintervals so that the probability of more than one event in a given subinterval is negligible. Let W be the random variable counting the number of occurrences of the event in the interval $[0, x]$. Then W is a Poisson random variable with parameter λx. The desired cdf of X can be obtained using W as follows:

$$
\begin{aligned}
F_X(x) &= P(X \leq x) \\
&= 1 - P(X > x) \\
&= 1 - P(\text{less than } r \text{ events occur in the interval } [0, x]) \\
&= 1 - F_w(r-1) \\
&= 1 - \sum_{k=0}^{r-1} e^{-\lambda x} \frac{(\lambda x)^k}{k!}
\end{aligned}
$$

Therefore,

$$f_X(x) = -\frac{d}{dx} \sum_{k=0}^{r-1} e^{-\lambda x} \frac{(\lambda x)^k}{k!}$$

$$= -\left[-\lambda e^{-\lambda x} + \sum_{k=1}^{r-1} (-\lambda) e^{-\lambda x} \frac{(\lambda x)^k}{k!} + e^{-\lambda x}(\lambda) \frac{(\lambda x)^{k-1}}{(k-1)!} \right]$$

$$= \sum_{k=0}^{r-1} \lambda e^{-\lambda x} \frac{(\lambda x)^k}{k!} - \sum_{k=1}^{r-1} \lambda e^{-\lambda x} \frac{(\lambda x)^{k-1}}{(k-1)!}$$

$$= \sum_{k=0}^{r-1} \lambda e^{-\lambda x} \frac{(\lambda x)^k}{k!} - \sum_{k=0}^{r-2} \lambda e^{-\lambda x} \frac{(\lambda x)^k}{k!}$$

$$= \lambda e^{-\lambda x} \frac{(\lambda x)^{r-1}}{(r-1)!}$$

$$= \frac{\lambda^r}{(r-1)!} x^{r-1} e^{-\lambda x} \tag{6.6.1}$$

What we have just derived for $f_X(x)$ is a special case of a *gamma* density, a family of probability distributions that is formally introduced in Definition 6.6.2. Before examining the gamma in any detail, though, we generalize Equation 6.6.1 to include those cases where r is positive but not necessarily an integer. To do this, it is necessary to replace $(r-1)!$ with a continuous function of (nonnegative) r, $\Gamma(r)$, the latter reducing to $(r-1)!$ when r is a positive integer. Such a function was first defined in 1731 by Euler, who reexpressed it 50 years later in the form it takes in Definition 6.6.1. Legendre is credited with giving it the name *gamma function*.

Definition 6.6.1. For any real number $r > 0$, the gamma function (of r) is given by

$$\Gamma(r) = \int_0^\infty x^{r-1} e^{-x} \, dx$$

Six properties of the gamma function are stated in Theorem 6.6.1. The first five are left as homework exercises; the sixth is proved in Appendix 6.A.

Theorem 6.6.1. Let $\Gamma(r) = \int_0^\infty x^{r-1} e^{-x} \, dx$. Then:

1. $\Gamma(1) = 1$
2. $\Gamma\left(\frac{1}{2}\right) = \sqrt{\pi}$
3. $\Gamma(r+1) = r\Gamma(r)$
4. $\Gamma(r+1) = r!$ if r is an integer
5. $\dbinom{n+r-1}{n} = \dfrac{\Gamma(n+r)}{\Gamma(n+1)\Gamma(r)}$
6. $\dfrac{\Gamma(r)\Gamma(s)}{\Gamma(r+s)} = \displaystyle\int_0^1 u^{r-1}(1-u)^{s-1} \, du$

Question 6.6.1 Show that $\Gamma(\frac{7}{2}) = (15\sqrt{\pi})/8$.

Having found the appropriate generalization of $(r - 1)!$ for arbitrary r, we can now define the gamma family of probability functions.

Definition 6.6.2. Let X be a random variable such that

$$f_X(x) = \frac{\lambda^r}{\Gamma(r)} x^{r-1} e^{-\lambda x}, \qquad x > 0$$

Then X is said to have a gamma distribution with parameters r and λ. Both r and λ must be greater than 0.

Question 6.6.2 Show that $f_X(x)$ as stated in Definition 6.6.2 is a true probability density; verify that

$$\int_0^\infty \frac{\lambda^r}{\Gamma(r)} x^{r-1} e^{-\lambda x}\, dx = 1$$

CASE STUDY 6.11

Records of daily rainfall in Sydney, Australia, for the period from October 17 to November 7 were examined for the years 1859 through 1952 (15). It was hypothesized that these data might have a gamma distribution, the reason being that precipitation occurs only if water particles can coalesce around dust of sufficient mass, but the accumulation of such dust is similar to the waiting-time aspect implicit in the gamma model.

The first two columns of Table 6.6.1 summarize the observed precipitation distribution. Estimates for r and λ were found to be 0.105 and 0.013, respectively.

To compute the expected probability of daily rainfall measuring between 0 and 5 mm, for instance, we compute

$$p = \frac{(0.013)^{0.105}}{\Gamma(0.105)} \int_0^{5.5} x^{0.105-1} e^{-0.013x}\, dx$$

or, with the substitution $u = 0.013x$,

$$p = \frac{1}{\Gamma(0.105)} \int_0^{0.0715} u^{0.105-1} e^{-u}\, du \tag{6.6.2}$$

The integral in Equation 6.6.2 can be evaluated using tables of the incomplete gamma function (see (85)). In this case, $p = 0.793$, so the expected number of days with

TABLE 6.6.1

Rainfall (mm)	Observed frequency	Expected frequency
0–5	1631	1639
6–10	115	106
11–15	67	62
16–20	42	44
21–25	27	32
26–30	26	26
31–35	19	21
36–40	14	17
41–45	12	14
46–50	18	12
51–60	18	20
61–70	13	15
71–80	13	12
81–90	8	9
91–100	8	7
101–125	16	12
125–150	7	7
150–425	14	13

rainfall between 0 and 5 mm is equal to 2068×0.793, or 1639. Column 3 of Table 6.6.1 shows the entire set of expected frequencies. The gamma pdf obviously models these data extremely well.

Theorem 6.6.2. Let X be a random variable having a gamma distribution with parameters r and λ. Then

$$E(X) = \frac{r}{\lambda}$$

and

$$\mathrm{Var}(X) = \frac{r}{\lambda^2}$$

Proof. The proof is a straightforward exercise in substitution. By definition,

$$E(X) = \int_0^\infty x \frac{\lambda^r}{\Gamma(r)} x^{r-1} e^{-\lambda x} \, dx$$

$$= \frac{\lambda^r}{\Gamma(r)} \int_0^\infty x^r e^{-\lambda x} \, dx$$

Note that the integrand is the variable part of a gamma density with parameters $r + 1$ and λ. Therefore, the integral equals $\Gamma(r + 1)/\lambda^{r+1}$ and

$$E(X) = \frac{\lambda^r}{\Gamma(r)} \cdot \frac{\Gamma(r+1)}{\lambda^{r+1}} = \frac{r}{\lambda}$$

The derivation of $\text{Var}(X)$ is similar; we omit the details.

Question 6.6.3 Let X be the gamma random variable of Theorem 6.6.2. Show that $E(X^2) = (r+1)r/\lambda^2$. Then $\text{Var}(X) = E(X^2) - [E(X)]^2 = r/\lambda^2$.

Question 6.6.4 Let X be a gamma random variable. Derive a formula for $E(X^m)$, where m is any positive integer. (*Hint:* Follow the technique used in the proof of Theorem 6.6.2.)

Question 6.6.5 For a gamma random variable X, show that $M_X(t) = [\lambda/(\lambda - t)]^r$. Use the moment-generating function to verify Theorem 6.6.2.

Question 6.6.6 Suppose X is a gamma random variable with $E(X) = 2$ and $\text{Var}(X) = 7$. Find r and λ.

We conclude this section with an application taken from psychology that shows how the Poisson, the gamma, and the negative binomial are all related to one another.

CASE STUDY 6.12

Psychologists use a two-handed coordination test to study performance speed and error proneness. Experience has shown that individuals given such a test exhibit character-istic error liabilities, as measured by the number of errors they make per unit time. A person with an error rate of λ will make x errors, x being the realization of a Poisson random variable:

$$P(\text{subject with error rate } \lambda \text{ makes } x \text{ errors}) = f_{X|\lambda}(x) = \frac{e^{-\lambda}\lambda^x}{x!}$$

Since the parameter λ varies from person to person, it can be considered the value of a random variable, Λ. There are many situations where a reasonable assumption is that Λ is gamma with parameters β and r:

$$f_\Lambda(\lambda) = \frac{\beta^r}{\Gamma(r)}\lambda^{r-1}e^{-\beta\lambda}$$

What we examine here is the error distribution produced by a sample of 504 subjects, each given the same coordination test (81). Our objective is to see whether the resulting data can be adequately fit with a model that incorporates the assumptions implicit in $f_{X|\lambda}(x)$ and $f_\Lambda(\lambda)$.

As a first step, we need to find the *unconditional* pdf for X. From Section 4.6, we can write

$$f_{X,\Lambda}(x, \lambda) = f_{X|\lambda}(x) \cdot f_\Lambda(\lambda)$$

from which it follows that the marginal probability function for X is given by

$$f_X(x) = \int_0^\infty f_{X,\Lambda}(x, \lambda)\, d\lambda = \int_0^\infty f_{X|\lambda}(x) \cdot f_\Lambda(\lambda)\, d\lambda \qquad (6.6.3)$$

Here, Equation 6.6.3 reduces to

$$
\begin{aligned}
f_X(x) &= \int_0^\infty \frac{\lambda^x}{x!} e^{-\lambda} \frac{\beta^r}{\Gamma(r)} \lambda^{r-1} e^{-\beta\lambda}\, d\lambda \\[2mm]
&= \frac{1}{x!} \frac{\beta^r}{\Gamma(r)} \int_0^\infty \lambda^{x+r-1} e^{-(\beta+1)\lambda} \lambda\, d\lambda \\[2mm]
&= \frac{1}{x!} \frac{\beta^r}{\Gamma(r)} \frac{\Gamma(n+r)}{(\beta+1)^{x+r}} \\[2mm]
&= \left(\frac{\beta}{\beta+1}\right)^r \left(\frac{1}{\beta+1}\right)^x \frac{\Gamma(x+r)}{x!\,\Gamma(r)} \\[2mm]
&= \left(\frac{\beta}{\beta+1}\right)^r \left(\frac{1}{\beta+1}\right)^x \binom{x+r-1}{x} \qquad (6.6.4)
\end{aligned}
$$

Setting $p = \beta/(\beta+1)$ and $q = 1/(\beta+1)$, we can rewrite Equation 6.6.4 in a more familiar form:

$$f_X(x) = \binom{x+r-1}{x} p^r q^x$$

Thus, if the conditional distribution of X is Poisson and if the parameter (λ) of that distribution is gamma, then the unconditional distribution of x is negative binomial.

Table 6.6.2 shows the observed and expected error distributions for the 504 subjects. Values in the third column were determined from a negative binomial distribution with estimated parameters $r = 0.7751$ and $p = 0.0962$. Quite clearly, the model fits very well.

Question 6.6.7 Suppose X_1 and X_2 are independent gamma random variables, X_1 with parameters r and λ and X_2 with parameters s and λ. Show that $X_1 + X_2$ is a gamma random variable with parameters $r + s$ and λ.

Question 6.6.8 Use Question 6.6.7 to show that the sum of r independent exponential random variables each with parameter λ is a gamma random variable with parameters r and λ.

Question 6.6.9 Suppose an Antarctic weather station has three electronic wind gauges, an original and two backup gauges. The lifetime of each gauge is exponentially distributed with a mean of 1000 h. What is the pdf for Y, the random variable measuring the time until the last gauge wears out?

Question 6.6.10 Let Z be a standard normal random variable. Show that Z^2 has the gamma pdf with parameters $r = \frac{1}{2}$ and $\lambda = \frac{1}{2}$.

TABLE 6.6.2

Number of errors	Observed frequency	Expected frequency
0	74	82
1	58	57
2	51	46
3	49	39
4	42	33
5	23	28
6	30	25
7	20	22
8	17	19
9	18	17
10	16	15
11	11	13
12	7	12
13	10	10
14	7	9
15	6	8
16	5	7
17	7	6
18	2	6
19	4	5
20	5	5
21	3	4
22	5	4
23	1	3
24	3	3
25	1	3
26	2	2
27	2	2
28	3	2
29	3	2
30	3	2
31–73	16	13
Totals	504	504

APPENDIX 6.A A PROPERTY OF THE GAMMA FUNCTION

Theorem 6.A.1. $\dfrac{\Gamma(r)\Gamma(s)}{\Gamma(r + s)} = \displaystyle\int_0^1 u^{r-1}(1 - u)^{s-1}\, du$

Proof. By definition,

$$\Gamma(r)\Gamma(s) = \int_0^\infty x^{r-1} e^{-x}\, dx \int_0^\infty y^{s-1} e^{-y}\, dy$$

$$= \int_0^\infty \int_0^\infty x^{r-1} y^{s-1} e^{-(x+y)}\, dx\, dy$$

Let $u = x/(x + y)$, so that $x = uy/(1 - u)$. Then $dx = (y\,du)/(1 - u)^2$ and

$$\Gamma(r)\Gamma(s) = \int_0^\infty \int_0^1 \left(\frac{uy}{1-u}\right)^{r-1} y^{s-1} e^{-(y/(1-u))} \frac{y}{(1-u)^2} \, du \, dy$$

Now, make the substitution $v = y/(1 - u)$, or, equivalently, $y = (1 - u)v$. This reduces the double integral to

$$\Gamma(r)\Gamma(s) = \int_0^\infty \int_0^1 (uv)^{r-1}(1-u)^{s-1} v^{s-1} e^{-v} v \, du \, dv$$

$$= \int_0^\infty \int_0^1 u^{r-1}(1-u)^{s-1} v^{r+s-1} e^{-v} \, dv \, du$$

$$= \left(\int_0^\infty v^{r+s-1} e^{-v} \, dv\right)\left(\int_0^1 u^{r-1}(1-u)^{s-1} \, du\right)$$

$$= \Gamma(r+s) \int_0^1 u^{r-1}(1-u)^{s-1} \, du$$

and the theorem follows.

7

Limit Theorems

Hold Infinity in the palm of your hand.
Blake

7.1 INTRODUCTION

Suppose an experiment is performed under presumably identical conditions n times, and the event A occurs on m of those n repetitions. What can we say about the relationship between the ratio m/n and the probability of the event A? We can say two things: Recalling the intuitive definition of probability introduced in Chapter 2, we expect, first of all, that m/n will be "close" to $P(A)$. Second, as n tends to infinity, we expect the ratio to converge, in some sense, to the true probability. In this chapter we want to formulate those notions rigorously by developing what are referred to as *limit theorems*.

In *Ars Conjectandi,* published in 1713, James Bernoulli derived probability's first limit theorem when he described the asymptotic behavior of the sequence of random variables now named in his honor.

Theorem 7.1.1. Let B_n denote the number of successes in n Bernoulli trials, and let p be the probability of success at any given trial. For any $\epsilon > 0$,

$$\lim_{n \to \infty} P\left(\left| \frac{B_n - np}{n} \right| < \epsilon \right) = 1$$

Other results were quick to follow. Section 7.2 deals with *convergence in probability,* the type of asymptotic behavior implied in Theorem 7.1.1. Two generalizations of Bernoulli's theorem are singled out, one due to Chebyshev; the other, to Khinchin.

In Section 7.3 we introduce a stronger type of asymptotic behavior known as *almost-sure convergence.* Limit properties based on almost-sure convergence are generically called *strong laws of large numbers* (as opposed to the *weak* laws of large numbers coming out of Section 7.2).

Still a third type of asymptotic behavior—and, in many ways, the most important—gives rise to the *central limit theorems,* the prototype of which was credited to DeMoivre in 1718 and appeared in the last chapter as Theorem 6.3.1. Central limit theorems state conditions under which cdf's of sequences of random variables converge to the cdf of a normal random variable. The evolution, derivation, and application of central limit theorems are discussed in Sections 7.4 and 7.5.

Because of the nature of the subject matter, much of the material in this chapter is considerably more difficult than anything we have seen thus far. The statements of the theorems and the discussions and examples that connect them are essentially at the same level as the earlier chapters in the book; the proofs, though, presuppose a familiarity with real analysis and complex variables.

7.2 CONVERGENCE IN PROBABILITY

Poisson called Bernoulli's result (Theorem 7.1.1) a *law of large numbers* in a paper generalizing the theorem to the case of repeated independent trials with varying probability of success. In 1866, the Russian mathematician Chebyshev, using what is now called the Bienaymé-Chebyshev inequality (Theorem 5.6.1), derived an even more general law of large numbers. Before stating and proving Chebyshev's theorem, we need to examine in more detail the mode of convergence that all these results have in common.

Definition 7.2.1. Suppose W_1, W_2, . . . is an infinite sequence of random variables, all defined on the sample space S. The sequence *converges in probability* to the random variable W if for each $\epsilon > 0$,

$$\lim_{n \to \infty} P(|W_n - W| < \epsilon) = 1$$

(For convenience, we write "W_n converges in pr.")

It is easier to work with Definition 7.2.1 if we introduce an abbreviation. Let $C_{n, \epsilon} = \{s \in S \mid W_n(s) - W(s)| < \epsilon\}$. Convergence in probability can then be defined by the condition

$$\lim_{n \to \infty} P(C_{n, \epsilon}) = 1, \qquad \text{all } \epsilon > 0$$

or, equivalently,

$$\lim_{n\to\infty} P(C_{n,\epsilon}^c) = 0, \qquad \text{all } \epsilon > 0$$

Note that $C_{n,\epsilon} \subset C_{n,\epsilon_1}$ if $\epsilon < \epsilon_1$.

The next several examples should help clarify the notion of *convergence in pr.*

Example 7.2.1

Suppose $S = [0, \infty)$ and let $f(x)$ be any continuous pdf on S. Define $W_n = 1$ if $s \in (n, \infty)$ and $W_n = 0$, otherwise. Show that W_n converges in probability to the zero random variable. (W is the zero random variable if $W(s) = 0$ for all $s \in S$.)

For any ϵ, $0 < \epsilon < 1$,

$$C_{n,\epsilon} = \{s \mid |W_n(s) - 0| < \epsilon\} = [0, n]$$

and $P(C_{n,\epsilon}) = \int_0^n f(x)\, dx$. But

$$\lim_{n\to\infty} P(C_{n,\epsilon}) = \lim_{n\to\infty} \int_0^n f(x)\, dx = \int_0^\infty f(x)\, dx = 1$$

Thus, by definition, W_n converges in pr. to W.

Example 7.2.2

Suppose $S = [0, 1]$ and $f(x)$ is the uniform pdf. Take $W_n(s) = s^n$ for each $s \in [0, 1]$. Show that W_n converges in pr. to the zero random variable.

Here,

$$C_{n,\epsilon} = \{s \mid |s^n| < \epsilon\} = \{s \mid s < \epsilon^{1/n}\}$$
$$= [0, \epsilon^{1/n}]$$

Since $f(x)$ is uniform,

$$\lim_{n\to\infty} P(C_{n,\epsilon}) = \lim_{n\to\infty} \epsilon^{1/n} = \lim_{n\to\infty} e^{(1/n)\log \epsilon} = e^0 = 1$$

and the convergence is established.

Example 7.2.3

Let $S = [0, 1]$ and take $f(x)$ to be the uniform pdf. Define $W_n(s) = n$ for all $s \in [0, 1]$. Show that W_n has no limit.

Assume the W_i's converge to W. Choose $\epsilon = 1$. Then

$$C_{n,1} = \{s \mid |W_n(s) - W(s)| < 1\}$$
$$= \{s \mid |n - W(s)| < 1\}$$
$$= \{s \mid W(s) \in (n - 1, n + 1)\}$$
$$\subset D_n = \{s \mid W(s) \in (n - 1, \infty)\}$$

Since $D_1 \supset D_2 \supset D_3 \supset \cdots$,

$$\lim_{n\to\infty} P(C_{n,1}) \le \lim_{n\to\infty} P(D_n) = P\left(\bigcap_{n=1}^\infty D_n\right) = P(\emptyset) = 0$$

But $\lim_{n \to \infty} P(C_{n,1})$ must be 1 if W_n is to converge to W. It follows by contradiction that no limit random variable W can exist.

Question 7.2.1 Suppose $\{W_n\}$ and $\{V_n\}$ are sequences of random variables such that $W_n \to W$ in pr. and $V_n \to V$ in pr. Show that $W_n + V_n \to W + V$ in pr.

Question 7.2.2 Unlike limits of sequences of points, limits of functions that converge in probability are not unique. Suppose, for example, the sequence $\{W_n\}$ converges in pr. to W. Let V be a random variable with $V = W$ except on a nonempty set of probability zero. Show that $\{W_n\}$ also converges in pr. to V.

Question 7.2.3 On any probability space, define the random variables X and Y ($X \neq Y$) over some set of positive probability. Let $W_n = X$ if n is odd, $W_n = Y$ if n is even. Show that the sequence of W_i's does not converge in pr.

Next we state and prove Chebyshev's weak law of large numbers. Included as special cases of this result are Theorem 7.1.1 and Poisson's generalization of Theorem 7.1.1, which was cited at the beginning of this section.

> **Theorem 7.2.1 (Chebyshev).** Let Y_1, Y_2, . . . be an infinite sequence of independent random variables. Suppose there is a constant d such that $\mathrm{Var}(Y_k) \leq d$, $k = 1, 2, \ldots$. Let
>
> $$\bar{Y}_n = \frac{1}{n}(Y_1 + \cdots + Y_n)$$
>
> For each $\epsilon > 0$, $\lim_{n \to \infty} P(|\bar{Y}_n - E(\bar{Y}_n)| < \epsilon) = 1$.

Proof. By Corollary 2 of the Bienaymé-Chebyshev inequality (see Section 5.6),

$$P(|\bar{Y}_n - E(\bar{Y}_n)| < \epsilon) > 1 - \frac{\mathrm{Var}(\bar{Y}_n)}{\epsilon^2}$$

$$= 1 - \frac{\sum_{k=1}^{n} \mathrm{Var}(Y_k)}{n^2 \epsilon^2}$$

$$\geq 1 - \frac{nd}{n^2 \epsilon^2}$$

$$= 1 - \frac{d}{n \epsilon^2}$$

Taking limits as $n \to \infty$ gives the result.

Question 7.2.4 Let Y_1, Y_2, . . . be an infinite sequence of random variables such that

$$\lim_{n \to \infty} \frac{\mathrm{Var}(Y_1 + \cdots + Y_n)}{n^2} = 0$$

This is known as *Markov's condition*. Show that Markov's condition implies that $\bar{Y}_n - E(\bar{Y}_n)$ converges in pr. to 0.

Case Study 7.1 shows an unusual application of Chebyshev's weak law of large numbers to a famous problem in analysis. What is interesting about the use of Chebyshev's theorem here is that the result being proved has nothing to do with probability.

CASE STUDY 7.1

The idea of approximating complicated functions by simple ones has a long history. Probably the two most familiar methods are the series expansions credited to Taylor and Fourier. Unfortunately, not every continuous function admits such an expansion. Another approach makes use of interpolation—finding a polynomial, or other desirable function, that agrees with the target function at a set of specified points. The well-known Lagrange interpolation formula falls into this latter category. However, for even so simple a function as $f(x) = |x|$ on $[-1, 1]$, we can find a sequence of nth-degree polynomials that agree with $|x|$ at $n + 1$ points, yet do not converge to $f(x)$ anywhere except at -1, 0, and $+1$.

All this suggests that the approximation problem is not a trivial one: Finding general results is not easy. That being the case, the theorem to be discussed in this particular case study—that every continuous function on a closed interval can be uniformly approximated by a polynomial—is all the more amazing. Not surprisingly, the man who discovered the result was, himself, a bit out of the ordinary.

Karl Weierstrass (1815–1897) was a gifted mathematics teacher in several provincial German gymnasia by day and a formidable research mathematician by night. (Gymnasia are academies roughly equivalent to American prep schools.) His first publication appeared in 1854 in the prestigious *Crelle's Journal*. Weierstrass' work astonished the mathematical world, coming as it did from an "unknown schoolmaster in an obscure village" (4). But mathematical genius requires no pedigree, and the German mathematical community responded swiftly. The University of Konigsberg conferred a doctoral degree, *honoris causa,* on Weierstrass, and by 1856 he had a professorial post at the University of Berlin.

Some 30 years later, Weierstrass posed and solved the problem of using polynomials to uniformly approximate continuous functions. His result is now known as the Weierstrass approximation theorem (see (77)).

Theorem 7.2.2. Suppose $f(x)$ is continuous on the closed interval $[a, b]$. For any $\epsilon > 0$, there exists a polynomial $Q(x)$ such that

$$|f(t) - Q(t)| < \epsilon$$

for all t in $[a, b]$.

Proof. First, note that it suffices to prove the theorem for functions on $[0, 1]$. Given a continuous $g(x)$ over the interval $[a, b]$, we can always define $f(x) = g((b - a)x + a)$. If $Q(x)$ satisfies the statement of Theorem 7.2.2 for $[0, 1]$, $Q((x - a)/(b - a))$ will be within ϵ of $g(x)$ on $[a, b]$.

Recall that a continuous function on a closed interval is bounded and uniformly continuous. Thus there is a constant M such that $|f(t)| < M$ for all t in $[0, 1]$. Also, for each $\epsilon > 0$, there is a number $\delta > 0$ such that $|t_1 - t_2| < \delta$ implies $|f(t_1) - f(t_2)| < \epsilon/2$ for any t_1, t_2 in $[0, 1]$.

The approximating polynomials we use are defined by

$$Q_n(t) = \sum_{k=0}^{n} f\left(\frac{k}{n}\right)\binom{n}{k}t^k(1 - t)^{n-k}$$

These are known as *Bernstein polynomials,* named for the Russian mathematician S. N. Bernstein (1880–1968). For any t in $[0, 1]$,

$$|f(t) - Q_n(t)| = \left| \sum_{k=0}^{n} \left[f(t) - f\left(\frac{k}{n}\right)\right]\binom{n}{k}t^k(1 - t)^{n-k} \right| \qquad \text{(Why?)}$$

$$\leq \sum_{|k/n-t|\leq\delta} \left| f(t) - f\left(\frac{k}{n}\right)\right| \binom{n}{k}t^k(1 - t)^{n-k}$$

$$+ \sum_{|k/n-t|>\delta} \left| f(t) - f\left(\frac{k}{n}\right)\right| \binom{n}{k}t^k(1 - t)^{n-k}$$

$$\leq \sum_{|k/n-t|\leq\delta} \frac{\epsilon}{2}\binom{n}{k}t^k(1 - t)^{n-k} + \sum_{|k/n-t|>\delta} 2M\binom{n}{k}t^k(1 - t)^{n-k}$$

$$\leq \frac{\epsilon}{2} + 2M \cdot P\left(\left|\frac{B_n}{n} - t\right| > \delta\right) \qquad (7.2.1)$$

where B_n is the number of successes in n Bernoulli trials. Write $B_n = Y_1 + \cdots + Y_n$, letting $Y_k = 1$ or 0 depending on whether the kth trial is a success or a failure. By inspection, the probability of a success on any given trial is t, so $\text{Var}(Y_k) = t(1 - t) \leq \frac{1}{4}$. The bound d ($d = \frac{1}{4}$) of Theorem 7.2.1 therefore exists and is *independent of t.* From that same theorem, there exists an $n(\epsilon, \delta)$ such that $n > n(\epsilon, \delta)$ implies that

$$P\left(\left|\frac{B_n}{n} - t\right| > \delta\right) < \frac{\epsilon}{4M}$$

for all t. Applying this $\epsilon/4M$ bound to Inequality 7.2.1 gives

$$|f(t) - Q_n(t)| < \frac{\epsilon}{2} + 2M \cdot \frac{\epsilon}{4M} = \epsilon$$

and the theorem is proved. (What was gained here by showing at the outset that it was sufficient to establish the result for functions defined over the unit interval?)

Comment

The Weierstrass approximation theorem has spawned a number of remarkable generalizations to abstract spaces. The first of these appeared in a 1937 paper of M. H. Stone. A full account of Stone's work and references to further extensions of the theorem are given in (8).

Question 7.2.5 Suppose $f(x)$ is continuous on $[a, b]$ and $\int_a^b f(x)x^n \, dx = 0$ for $n = 0, 1, 2, \ldots$. Show that $f(x) = 0$ on $[a, b]$. (*Hint:* First prove that $\int_a^b f(x)Q(x) \, dx = 0$ for every polynomial $Q(x)$. Then use the Weierstrass approximation theorem to show that $\int_a^b (f(x))^2 \, dx = 0$.)

In 1909 Emile Borel produced the next important generalization of Theorem 7.1.1. What concerned Borel was not the type of random variable being considered (unlike Chebyshev, he restricted himself to Bernoulli random variables) but, rather, the *way* in which the sequence converged to its limit. However, we defer Borel's contribution to the next section and instead consider one final weak law. This one, due to Khinchin, eliminates—for a certain class of random variables—Chebyshev's restriction that the variance of the Y_i's be bounded. All that Khinchin needs to establish convergence in probability is for $E(Y_n)$ to exist. The random variables, though, must be independent *and identically distributed*. (The proof we give here is due to Markov and uses his technique of truncating variables.)

Theorem 7.2.3. Suppose $Y_1, Y_2, \ldots$ is an infinite sequence of independent, identically distributed random variables with $E(|Y_n|)$ finite. Let $\mu = E(Y_n)$. For each $\epsilon > 0$,

$$\lim_{n \to \infty} P(|\bar{Y}_n - \mu| < \epsilon) = 1$$

Proof. Assume the underlying distribution $f_Y(y)$ is continuous. Reworking the proof to accommodate discrete pdf's is straightforward.

Given $\epsilon > 0$, we can prove the theorem by showing that n can be chosen large enough to make

$$P(|\bar{Y}_n - \mu| > \epsilon) < \delta$$

for any $\delta > 0$. Let m be a positive integer such that

$$m > \frac{1}{\delta}\left(\frac{4E(|Y|)}{\epsilon^2} + 1\right)$$

(This particular choice of m is not relevant until the end of the proof.) For any positive integer n, define the random variables $U_1, U_2, \ldots, U_n$ by

$$U_k = \begin{cases} Y_k & |Y_k| \leq \dfrac{n}{m} \\ 0 & \text{otherwise} \end{cases}$$

$k = 1, 2, \ldots, n$. Define $X_k = Y_k - U_k$ so that $Y_k = U_k + X_k$, $k = 1, \ldots, n$. The event of interest, $(|\bar{Y}_n - \mu| > \epsilon)$, is a subset of the union

$$(|\bar{U}_n - \mu| > \epsilon) \cup \bigcup_{k=1}^{n} (X_k \neq 0) \qquad \text{(Why?)}$$

Thus

$$P(|\bar{Y}_n - \mu| > \epsilon) \leq P(|\bar{U}_n - \mu| > \epsilon) + \sum_{k=1}^{n} P(X_k \neq 0) \qquad (7.2.2)$$

The probability that X_k is nonzero can be expressed as an integral:

$$P(X_k \neq 0) = P\left(|Y_k| > \frac{n}{m}\right)$$

$$= \int_{n/m}^{\infty} f_{|Y|}(y)\, dy$$

$$\leq \int_{n/m}^{\infty} \frac{m}{n} \cdot y \cdot f_{|Y|}(y)\, dy$$

$$= \frac{m}{n} \int_{n/m}^{\infty} y \cdot f_{|Y|}(y)\, dy$$

Now, by the main hypothesis of the theorem, $E(|Y|) < \infty$. It follows that $\int_{n/m}^{\infty} y \cdot f_{|Y|}(y)\, dy \to 0$ as $n \to \infty$, so we can choose n large enough to make $\int_{n/m}^{\infty} f_{|Y|}(y)\, dy < 1/m^2$. The second term on the right-hand side of Inequality 7.2.2, then, is bounded by $1/m$:

$$\sum_{k=1}^{n} P(X_k \neq 0) \leq n \cdot \frac{m}{n} \cdot \frac{1}{m^2} = \frac{1}{m} \qquad (7.2.3)$$

Next, observe that for any k, $1 \leq k \leq n$,

$$E(U_k) = \int_{-n/m}^{n/m} y f_Y(y)\, dy \longrightarrow \mu \quad \text{as} \quad n \longrightarrow \infty$$

Therefore, we can always find an n large enough to make

$$|E(\bar{U}_n) - \mu| < \frac{\epsilon}{2} \qquad (7.2.4)$$

Also, for each k,

$$\text{Var}(U_k) \leq E(U_k^2)$$

$$= \int_{-n/m}^{n/m} y^2 f_Y(y) \, dy$$

$$\leq \frac{n}{m} \int_{-n/m}^{n/m} |y| f_Y(y) \, dy$$

$$\leq \frac{n}{m} E(|Y|)$$

It follows that

$$\text{Var}(\overline{U}_n) \leq \frac{1}{n^2} \sum_{k=1}^{n} \text{Var}(U_k) \leq \frac{1}{n^2} \cdot n \cdot \frac{n}{m} E(|Y|)$$

$$= \frac{1}{m} E(|Y|)$$

By the Bienaymé-Chebyshev inequality,

$$P\left(|\overline{U}_n - E(\overline{U}_n)| > \frac{\epsilon}{2}\right) \leq \frac{\text{Var}(\overline{U}_n)}{\epsilon^2/4} \leq \frac{4E(|Y|)}{m\epsilon^2}$$

Using Inequality 7.2.4, we can write

$$P(|\overline{U}_n - \mu| > \epsilon) = P(|(\overline{U}_n - E(\overline{U}_n)) + (E(\overline{U}_n) - \mu)| > \epsilon)$$

$$\leq P\left(|\overline{U}_n - E(\overline{U}_n)| > \frac{\epsilon}{2}\right)$$

Thus, for n sufficiently large,

$$P(|\overline{U}_n - \mu| > \epsilon) \leq \frac{4E(|Y|)}{m\epsilon^2} \tag{7.2.5}$$

Inequalities 7.2.3 and 7.2.5 applied to Inequality 7.2.2 complete the proof: For $n > n(\epsilon, \delta)$,

$$P(|\overline{Y}_n - \mu| > \epsilon) \leq \frac{4E(|Y|)}{m\epsilon^2} + \frac{1}{m} < \delta$$

Question 7.2.6 Markov's condition (see Question 7.2.4) is not necessary for the weak law of large numbers. Consider the following sequence $\{Y_n\}$ of independent random variables:

$$Y_n = \begin{cases} \dfrac{1}{n} & \text{with probability } \dfrac{1}{2}(1 - 2^{-n}) \\[2mm] \dfrac{-1}{n} & \text{with probability } \dfrac{1}{2}(1 - 2^{-n}) \\[2mm] 2^n & \text{with probability } 2^{-n-1} \\[2mm] -2^n & \text{with probability } 2^{-n-1} \end{cases}$$

Show that the conclusion of Theorem 7.2.1 holds for $\{Y_n\}$ despite the fact that Markov's condition is not satisfied.

7.3 ALMOST-SURE CONVERGENCE

By the early part of the twentieth century, mathematicians had begun to recognize that there were a number of valuable definitions of convergence for a sequence of functions. Adapting the notion of "almost everywhere convergence" to a probabilistic setting, Borel gave a stronger version of Bernoulli's theorem. We begin by describing this second type of convergence.

Definition 7.3.1. For a sequence $W_1, W_2, \ldots$ of random variables defined on a sample space S, set $L = \{s \mid \lim_{n \to \infty} W_n(s) = W(s)\}$. *Almost-sure convergence* is the condition that $P(L) = 1$. For convenience, we abbreviate both *almost-sure convergence* and *convergence almost surely* by *convergence a.s.*

Borel's *strong* law of large numbers that appeared in 1909 is the statement that B_n/n converges a.s. to p. We shall be able to prove an even better theorem, but first we need to clarify Definition 7.3.1 with a few examples and comments.

Example 7.3.1

Recall the sequence $W_n(s) = s^n$ for $s \in [0, 1]$ that we discussed in Example 7.2.2. Show that W_n converges a.s. to the zero random variable.

Here, L is the semiopen interval $[0, 1)$:

$$L = \left\{ s \;\middle|\; \lim_{n \to \infty} W_n(s) = 0 \right\} = \left\{ s \;\middle|\; \lim_{n \to \infty} s^n = 0 \right\} = [0, 1)$$

Since $P(L) = P([0, 1)) = 1$, we have the desired convergence. Note that the sequence does not converge to the zero function at *every* point: It converges to 1 at the sample point $s = 1$.

It is also valid to say that the sequence W_n converges a.s. to the limit function V, where

$$V(s) = \begin{cases} 0 & s \text{ irrational} \\ 1 & s \text{ rational} \end{cases}$$

Relative to V, L is the set of irrational points in $[0, 1]$, so $P(L) = 1$. Notice how we can say that W_n converges to V even though the set on which $W_n(s)$ does *not* converge to $V(s)$ is infinite: All Definition 7.3.1 requires is for the set of nonconvergence to have probability zero.

Example 7.3.2

Let $S = (-\infty, \infty)$ and suppose $f(x)$ is any continuous pdf on S. Define the sequence W_n by

$$W_n(s) = \sin(n\pi s)$$

Does W_n converge a.s.?

No. The W_n's converge only when s is an integer, a set of probability zero. At all other points, the sequence oscillates.

Question 7.3.1 Does the trigonometric sequence in Example 7.3.2 converge in pr.?

The next theorem and the example that follows demonstrate an important relationship between convergence in probability and convergence a.s. Implicit in the statement of Theorem 7.3.1 is the justification for distinguishing the *weak* laws of large numbers described in Section 7.2 from the *strong* laws developed a little later in this section.

> **Theorem 7.3.1.** Convergence a.s. implies convergence in pr.

Proof. Assume that $W_n(s)$ converges almost surely to $W(s)$. We need to show that $\lim_{n\to\infty} P(C_{n,\epsilon}) = 1$ for each $\epsilon > 0$, where $C_{n,\epsilon} = \{s \mid |W_n(s) - W(s)| < \epsilon\}$. Define $D_{n,\epsilon} = \cap_{k=n}^{\infty} C_{k,\epsilon}$. Note that $D_{1,\epsilon} \subset D_{2,\epsilon} \subset \cdots$ and $D_{n,\epsilon} \subset C_{n,\epsilon}$. Also, $s \in L = \{s \mid \lim_{n\to\infty} W_n(s) = W(s)\}$ if and only if for each $\epsilon > 0$, there is an n such that $s \in D_{n,\epsilon}$. Thus, for each $\epsilon > 0$, $L \subset \cup_{n=1}^{\infty} D_{n,\epsilon}$. It follows, then, that

$$1 = P(L) \leq P\left(\bigcup_{n=1}^{\infty} D_{n,\epsilon}\right) = \lim_{n\to\infty} P(D_{n,\epsilon})$$

$$\leq \lim_{n\to\infty} P(C_{n,\epsilon}) \leq 1$$

the inequalities implying that $\{W_n\}$ converges in pr.

Example 7.3.3

The converse of Theorem 7.3.1 is false. To prove that, we construct a sequence $\{W_n\}$ that converges in probability to the zero random variable but does not converge a.s.

Let $f(x)$ be the uniform pdf defined over the unit interval $S = [0, 1]$. Write n in the form $n = 2^j + k$, where $0 \leq k < 2^j$—that is, 2^j is the largest power of 2 less than n. Define on S the following sequence of random variables:

$$W_n(s) = \begin{cases} 1 & \dfrac{k}{2^j} \leq s \leq \dfrac{k+1}{2^j} \\ 0 & \text{otherwise} \end{cases}$$

By inspection, the set $C_{n,\epsilon}$ implicit in Definition 7.2.1 is the complement of the interval $[k/2^j, (k+1)/2^j]$. Note that as $n \to \infty$, $2^j \to \infty$, in which case $\lim_{n\to\infty} P(C_{n,\epsilon}) = 1$, and the latter implies that $\{W_n\}$ converges in probability. But convergence a.s. fails badly: The sequence $\{W_n(s)\}$ converges nowhere in $[0, 1]$.

Question 7.3.2 What is $D_{n,\epsilon}$ for the sequence described in Example 7.3.3?

The proof of Theorem 7.3.1 suggests the possibility of characterizing L in terms of $C_{n,\epsilon}$. Establishing such a relationship will be facilitated if we first introduce a new set operation.

Definition 7.3.2. Suppose $A_1, A_2, \ldots$ is an infinite sequence of subsets defined on a given set S. The *limit inferior* of $\{A_n\}$, written A_*, is a union of intersections:

$$A_* = \bigcup_{n=1}^{\infty} \bigcap_{k=n}^{\infty} A_k$$

The *limit superior* of $\{A_n\}$, written A^*, is an intersection of unions:

$$A^* = \bigcap_{n=1}^{\infty} \bigcup_{k=n}^{\infty} A_k$$

Question 7.3.3 Use DeMorgan's law to establish that the limit superior of $\{A_n^c\}$ is A_*^c.

Question 7.3.4 Let $\{A_n\}$ be a sequence of sets. Let $B = \{s \in S \mid \text{there exists an integer } N(s) \text{ so that } s \in A_k \text{ if } k \geq N(s)\}$.
  a. Show that $B = A_*$.
  b. Characterize A^*.

Question 7.3.5 Define $A_n = [0, 1 - 1/n]$ for n even and $A_n = [-1, 1/n]$ for n odd. Find A_* and A^*.

Theorem 7.3.2. Let $A_{*,m}$ denote the limit inferior for $\{C_{n,1/m}\}$. Then $L = \{s \in S \mid \lim_{n\to\infty} W_n(s) = W(s)\} = \bigcap_{m=1}^{\infty} A_{*,m}$.

Proof. Suppose $s \in L$. Then $\lim_{k\to\infty} W_k(s) = W(s)$. That is, for $\epsilon = 1/m$, there is some N such that $|W_k(s) - W(s)| < 1/m$ for $k \geq N$ or, equivalently, $s \in C_{k,1/m}$ for $k \geq N$. But this latter statement can be rewritten as $s \in \bigcap_{k=N}^{\infty} C_{k,1/m} \subset \bigcup_{n=1}^{\infty} \bigcap_{k=n}^{\infty} C_{k,1/m} = A_{*,m}$, and since the inclusion holds for *every* m, $s \in \bigcap_{m=1}^{\infty} A_{*,m}$.

Conversely, assume $s \in \bigcap_{m=1}^{\infty} A_{*,m}$. Let $\epsilon > 0$ be given. Select some integer M so that $1/M < \epsilon$. Then $s \in A_{*,M} = \bigcup_{n=1}^{\infty} \bigcap_{k=n}^{\infty} C_{k,1/M}$, so $s \in \bigcap_{k=N}^{\infty} C_{k,1/M}$ for some integer N. Therefore, if $k \geq N$,

$$|W_k(s) - W(s)| < 1/M < \epsilon$$

Since ϵ is arbitrary, $\lim_{k \to \infty} W_k(s) = W(s)$, implying that $s \in L$.

Corollary. The sequence $\{W_n\}$ converges a.s. to W if and only if

$$\lim_{n \to \infty} P\left(\bigcup_{k=1}^{\infty} C_{n+k, \epsilon}^C \right) = 0$$

for every $\epsilon > 0$.

Proof. We leave the proof as an exercise.

Theorem 7.3.3 (The First Borel-Cantelli Lemma). Suppose $A_1, A_2, \ldots$ is a sequence of events such that $\Sigma_{k=1}^{\infty} P(A_k) < \infty$. Then $P(A^*) = 0$.

Proof. From the definition of A^*, we have that $A^* \subset \bigcup_{k=n}^{\infty} A_k$ for each positive integer n. Therefore, $P(A^*) \leq P(\bigcup_{k=n}^{\infty} A_k) \leq \Sigma_{k=n}^{\infty} P(A_k)$. But the term on the right—the upper bound for $P(A^*)$—is the tail of a convergent series, meaning it goes to 0 as $n \to \infty$.

Question 7.3.6 Prove that $\Sigma_{n=1}^{\infty} P(C_{n, \epsilon}^c) < \infty$ for each ϵ implies that W_n converges a.s. to W.

Question 7.3.7 Prove that $W_n \to W$ a.s. if $\Sigma_{k=1}^{\infty} E(|W_k - W|^2) < \infty$.

Question 7.3.8 Suppose $W_1, W_2, \ldots$ is a sequence of random variables such that for $s \in S$,

$$W_1(s) \geq W_2(s) \geq \cdots \geq 0$$

Show that $W_n \to 0$ in pr. implies $W_n \to 0$ a.s.

Borel (1909), Hausdorff (1913), Hardy-Littlewood (1914), and Cantelli (1917) all proved theorems involving the convergence a.s. of Bernoulli random variables. Hausdorff's theorem is particularly illuminating in that it shows the critical role of dividing by the appropriate power of n. As before, let B_n denote the number of successes in n Bernoulli trials, and let p be the probability of success for any given trial. Hausdorff proved that $(B_n - np)/n^{\frac{1}{2} + \alpha}$ converges a.s. to 0 *for any* $\alpha > 0$. When $\alpha = 0$, the theorem cannot be true: By the DeMoivre-Laplace theorem, $(B_n - np)/\sqrt{n}$ converges in probability to a normal random variable. (Since the sequence does not converge in pr. to 0, it cannot converge a.s. to 0.)

By the 1920s much was known about limit theorems for broad classes of random variables. As mentioned previously, Khinchin proved the weak law of large numbers for identically distributed, independent random variables, assuming only the existence of the mean. Then, in *Grundbegriffe* (46), Kolmogorov proved the *strong* law of large numbers given that same set of assumptions, thus generalizing substantially what his countryman had derived. The last part of this section is devoted to proving Kolmogorov's result (Theorem 7.3.7).

In terms of its basic structure, the derivation of Kolmogorov's theorem parallels the proof of Khinchin's theorem. Recall how the argument establishing Theorem 7.2.3 truncated the random variables to obtain a sequence having finite variances of constrained growth. Chebyshev's inequality was then used to determine the limiting behavior of the truncated sequence. Ultimately, that truncated portion was shown to be negligible. The proof of Kolmogorov's theorem takes a similar approach. For the strong law, the truncated variables are shown to satisfy the hypotheses of another result due to Kolmogorov (Theorem 7.3.6). This latter theorem derives from a strengthened form of Chebyshev's inequality, also due to Kolmogorov (Theorem 7.3.4). As was done in the proof of Khinchin's theorem, the truncated portion is shown not to affect the desired convergence.

Theorem 7.3.4 (Kolmogorov's Inequality).   Let $Y_1, Y_2, \ldots, Y_n$ be mutually independent random variables, each with $E(Y_k) = 0$ and $\mathrm{Var}(Y_k) = E(Y_k^2) = \sigma_k^2$. Let $S_k = Y_1 + \cdots + Y_k$, $1 \le k \le n$. For any $\epsilon > 0$,

$$P\left(\max_{1 \le k \le n} |S_k| > \epsilon \right) \le \frac{\mathrm{Var}(S_n)}{\epsilon^2}$$

Equivalently, the probability of the simultaneous occurrence of the n inequalities $|S_1| \le \epsilon, \ldots, |S_n| \le \epsilon$ is greater than or equal to $1 - \mathrm{Var}(S_n)/\epsilon^2$.

Proof. For each k, $1 \le k \le n$, define the set D_k by

$$D_k = \{s \mid |S_k(s)| \ge \epsilon, |S_1(s)| < \epsilon, \ldots, |S_{k-1}(s)| < \epsilon \}$$

The event on which we want to focus is $D = \{s \mid \max_{1 \le k \le n} |S_k(s)| > \epsilon\}$, or $D = \cup_{k=1}^n D_k$, where the D_k's are pairwise disjoint.

The derivation is simplified if we introduce the *indicator variables* $X_1, \ldots, X_n$ associated with $D_1, \ldots, D_n$. The variable X_k is defined by

$$X_k(s) = \begin{cases} 1 & s \in D_k \\ 0 & s \notin D_k \end{cases}$$

Let X_0 be the indicator variable of D^c. Note that $X_0 + X_1 + \cdots + X_n = 1$ and that exactly one of the variables is nonzero at each sample point. Therefore,

$$P(D) = P(X_1 + \cdots + X_n = 1) = E(X_1 + \cdots + X_n)$$

Since the X_k's sum to 1, we can write

$$E(S_n^2) = E(S_n^2(X_0 + \cdots + X_n)) = \sum_{k=0}^n E(S_n^2 X_k)$$

and the hypothesis that $E(Y_k) = 0$ makes

$$\mathrm{Var}(S_n) = E(S_n^2) \ge \sum_{k=1}^n E(S_n^2 X_k) \tag{7.3.1}$$

The remainder of the derivation hinges on our finding an appropriate lower bound for each $E(S_n^2 X_k)$.

Define $U_k = S_n - S_k = \sum_{j=k+1}^n Y_j$. Note that U_k is independent of $Y_1, \ldots, Y_k$ and, hence, of X_k and S_k. Also, $E(U_k) = \sum_{j=k+1}^n E(Y_j) = 0$. Simple algebra gives

$$E(S_n^2 X_k) = E((S_k + U_k)^2 X_k)$$
$$= E(S_k^2 X_k) + 2E(S_k U_k X_k) + E(U_k^2 X_k)$$

where, by the independence of U_k and $S_k X_k$,

$$E(S_k U_k X_k) = E(S_k X_k) \cdot E(U_k) = E(S_k X_k) \cdot 0 = 0$$

Thus $E(S_n^2 X_k) = E(S_k^2 X_k) + E(U_k^2 X_k) \geq E(S_k^2 X_k)$. Recall that $X_k \neq 0$ implies $|S_k| \geq \epsilon$, so $S_k^2 X_k \geq \epsilon^2 X_k$. A bound for the expected value of $S_n^2 X_k$ follows immediately:

$$E(S_n^2 X_k) \geq \epsilon^2 E(X_k), \qquad k = 1, \ldots, n$$

Combining the above statements with Inequality 7.3.1, we can write

$$E(S_n^2) \geq \sum_{k=1}^n \epsilon^2 E(X_k) = \epsilon^2 E(X_1 + \cdots + X_n) = \epsilon^2 P(D)$$

Equivalently,

$$P(D) \leq \frac{E(S_n^2)}{\epsilon^2} = \frac{\mathrm{Var}(S_n^2)}{\epsilon^2}$$

and the proof is complete.

Before we can proceed with the proof of Kolmogorov's theorem, we need the following property concerning the convergence of sequences. This is a special case of a very useful result known as the Toeplitz lemma (52).

Theorem 7.3.5. Suppose $\{t_n\}$ is a sequence of numbers such that $\lim_{n \to \infty} t_n = t$. Then $\lim_{n \to \infty} (1/n) \sum_{k=1}^n t_n = t$.

Proof. Let $\epsilon > 0$. Since the limit of $\{t_n\}$ is t, there exists an integer N such that $|t_k - t| < \epsilon$ for $k \geq N$. Note that

$$\left| \frac{1}{n} \sum_{k=1}^n t_k - t \right| = \left| \frac{1}{n} \sum_{k=1}^n (t_k - t) \right|$$

$$= \left| \frac{1}{n} \sum_{k=1}^N (t_k - t) + \frac{1}{n} \sum_{k=N+1}^n (t_k - t) \right|$$

$$\leq \left| \frac{1}{n} \sum_{k=1}^N (t_k - t) \right| + \frac{1}{n} \sum_{k=N+1}^n |t_k - t|$$

making

$$\left| \frac{1}{n} \sum_{k=1}^{n} t_k - t \right| \leq \frac{1}{n} \cdot \left| \sum_{k=1}^{N} (t_k - t) \right| + \frac{n-N}{n} \epsilon$$

In the latter inequality, let n tend to ∞ while keeping N and ϵ fixed. It follows that

$$\lim_{n \to \infty} \left| \frac{1}{n} \sum_{k=1}^{n} t_k - t \right| \leq \epsilon$$

Since the inequality here holds for every $\epsilon > 0$, the theorem is proved.

Question 7.3.9 Suppose $\{x_n\}$ is a sequence of numbers with $x_n \to x$. Let $\{a_{ij} \mid i = 1, 2, \ldots; j = 1, 2, \ldots, i\}$ be a set of nonnegative real numbers such that $\sum_{j=1}^{i} a_{ij} = 1$ and $\lim_{i \to \infty} a_{ij} = 0$ for each j. Prove that $\lim_{i \to \infty} \sum_{j=1}^{i} a_{ij} x_j = x$. (This is known as *Toeplitz's theorem;* among its special cases is Theorem 7.3.5.)

Theorem 7.3.6 (The Kolmogorov Condition). Let $Y_1, Y_2, \ldots$ be a sequence of independent random variables such that

$$\sum_{k=1}^{\infty} \frac{\text{Var}(Y_k)}{k^2} < \infty$$

Then $\bar{Y}_n - E(\bar{Y}_n)$ converges a.s. to zero.

Proof. For each k, define the random variable $X_k = (1/k)(Y_k - E(Y_k))$. Then $E(X_k) = 0$ and $\text{Var}(X_k) = \text{Var}(Y_k)/k^2$. Let $S_n = X_1 + \cdots + X_n$. The first step in the proof is to show that S_n converges a.s. to some limit T.

For any $s \in S$, the following statements are equivalent:

1. $\lim_{n \to \infty} S_n(s)$ exists.
2. $\lim_{n \to \infty} \sum_{k=1}^{n} X_k(s)$ exists.
3. $\sum_{k=1}^{\infty} X_k(s)$ converges.
4. $\lim_{m \to \infty} \sum_{k=m}^{\infty} X_k(s) = 0$.
5. $\lim_{m \to \infty} (\max_{m \leq n} |\sum_{k=m+1}^{n} X_k(s)|) = 0$.
6. $\lim_{m \to \infty} (\max_{m \leq n} |S_n(s) - S_m(s)|) = 0$.

For each $\epsilon > 0$, define

$$C_{m,n,\epsilon} = \left\{ s \,\middle|\, \max_{m \leq k \leq n} |S_k(s) - S_m(s)| < \epsilon \right\}$$

Property 4 holds if and only if

$$s \in \bigcap_{\epsilon > 0} \bigcup_{m=1}^{\infty} \bigcap_{n=m}^{\infty} C_{m,n,\epsilon} = C$$

In terms of C, our first objective is to show that $P(C) = 1$.

A sufficient condition for $P(C)$ being 1 is that

$$\lim_{m\to\infty} \lim_{n\to\infty} P(C_{m,n,\epsilon}) = 1$$

for each $\epsilon > 0$. The latter can be established by applying Kolmogorov's inequality to the sequence $X_{m+1}, X_{m+2}, \ldots, X_n$:

$$P\left(\bigcap_{m=k}^{n} C_{m,k,\epsilon}\right) \geq 1 - \frac{\sum\limits_{k=m+1}^{n} \mathrm{Var}(X_k)}{\epsilon^2}$$

which makes

$$\lim_{m\to\infty} \lim_{n\to\infty} P(C_{m,n,\epsilon}) \geq 1 - \lim_{m\to\infty} \frac{\sum\limits_{k=m+1}^{\infty} \mathrm{Var}(X_k)}{\epsilon^2}$$

By hypothesis, $\sum_{k=1}^{\infty} \mathrm{Var}(X_k)$ converges, so the numerator of the last inequality vanishes. It follows, therefore, that the sequence $\{S_n\}$ converges a.s. to some limit variable T.

Note that $Y_k - E(Y_k) = k(S_k - S_{k-1})$, $k = 2, 3, \ldots, n$, so

$$n\overline{Y}_n - nE(\overline{Y}_n) = \sum_{k=1}^{n} (Y_k - E(Y_k)) = S_1 + \sum_{k=2}^{n} k(S_k - S_{k-1}) = nS_n - \sum_{k=1}^{n-1} S_k$$

If we divide the expression for $n\overline{Y}_n - nE(\overline{Y}_n)$ by n, we get

$$\overline{Y}_n - E(\overline{Y}_n) = S_n - \frac{1}{n}\sum_{k=1}^{n-1} S_k = \left(1 + \frac{1}{n}\right)S_n - \frac{1}{n}\sum_{k=1}^{n} S_k$$

Taking limits on n gives, with probability 1,

$$\lim_{n\to\infty} (\overline{Y}_n - E(\overline{Y}_n)) = 1 \cdot T - T = 0$$

and the theorem is proved. (Why does $\lim_{n\to\infty} (1/n) \sum_{k=1}^{n} S_k = T$?)

We have the background now to prove Kolmogorov's statement of the strong law of large numbers. Notice, again, how the conditions here are exactly the same as those assumed earlier by Khinchin when he proved the weak law.

Theorem 7.3.7. Suppose $Y_1, Y_2, \ldots$ is a sequence of independent, identically distributed random variables with $E(|Y_k|) < \infty$. Let μ denote the common mean of the Y_k's. Then $\overline{Y}_n$ converges a.s. to μ.

Proof. For each positive integer k, define T_k by

$$T_k = \begin{cases} Y_k & |Y_k| < k \\ 0 & \text{otherwise} \end{cases}$$

Truncating the Y_k's in this fashion allows us to use Kolmogorov's condition (Theorem 7.3.6) on the T_k's.

Assume the Y_k's are continuous with common pdf $f_Y(y)$. Then

$$\text{Var}(T_k) \le E(T_k^2) = \int_{-k}^{k} y^2 f_Y(y)\, dy$$

By partitioning the original range of integration, we can write $E(T_k^2)$ as a sum:

$$\int_{-k}^{k} y^2 f_Y(y)\, dy = \int_0^k y^2 f_{|Y|}(y)\, dy$$

$$= \sum_{j=1}^{k} \int_{j-1}^{j} y^2 f_{|Y|}(y)\, dy$$

Taking a second sum—this one over k—we get an upper bound for the term that appears in Kolmogorov's condition:

$$\sum_{k=1}^{\infty} \frac{\text{Var}(T_k)}{k^2} \le \sum_{k=1}^{\infty} \frac{1}{k^2} \sum_{j=1}^{k} \int_{j-1}^{j} y^2 f_{|Y|}(y)\, dy$$

$$= \sum_{j=1}^{\infty} \int_{j-1}^{j} y^2 f_{|Y|}(y)\, dy \sum_{k=j}^{\infty} \frac{1}{k^2}$$

But $\sum_{k=j}^{\infty} (1/k^2) \le 2/j$, $j = 1, 2, \ldots$ (To see that for $j \ge 2$, observe that

$$\sum_{k=j}^{\infty} \frac{1}{k^2} \le \sum_{k=j}^{\infty} \frac{1}{k(k-1)} = \sum_{k=j}^{\infty} \left(\frac{1}{k-1} - \frac{1}{k} \right)$$

$$= \frac{1}{j-1} \le \frac{2}{j}$$

For $j = 1$, $\sum_{k=1}^{\infty} (1/k^2) = 1 + \sum_{k=2}^{\infty} (1/k^2) \le 1 + \frac{2}{3} \le \frac{2}{1}$.) The finiteness of $\sum_{k=1}^{\infty} (\text{Var}(T_k)/k^2)$ follows readily:

$$\sum_{k=1}^{\infty} \frac{\text{Var}(T_k)}{k^2} \le \sum_{j=1}^{\infty} \frac{2}{j} \int_{j-1}^{j} y^2 f_{|Y|}(y)\, dy$$

$$\le \sum_{j=1}^{\infty} 2 \int_{j-1}^{j} y f_{|Y|}(y)\, dy$$

$$= 2E(|Y|) < \infty$$

With an upper bound having been established for $\sum_{k=1}^{\infty} (\text{Var}(T_k)/k^2)$, we can use Theorem 7.3.6 to argue that $\bar{T}_n - E(\bar{T}_n)$ converges a.s. to 0. The definition of T_k implies that $E(T_k) \to \mu$. From Theorem 7.3.5, $E(\bar{T}_n) = (1/n) \sum_{k=1}^{n} E(T_k) \to \mu$, so $\bar{T}_n$ converges a.s. to μ.

To complete the proof, we must show that $\{T_k\}$ and $\{Y_k\}$ have the same convergence properties. Let $A_k = \{s \mid T_k(s) \ne Y_k(s)\} = \{s \mid |Y_k(s)| \ge k\}$. If

$$\sum_{k=1}^{\infty} P(A_k) < \infty \tag{7.3.2}$$

it would follow by the Borel-Cantelli lemma that $P(A^*) = 0$, or $P(\lim \inf A_n^C) = 1$. For any s in $\lim \inf A_k^C$, there would be an n such that $s \in A_k^C$ for $k \ge n$ or, equivalently, $T_k(s) = Y_k(s)$ for $k \ge n$. Thus, on a set of probability 1, $\lim_{k \to \infty} T_k =$

$\lim_{k\to\infty} Y_k$. By Theorem 7.3.5, $\lim_{n\to\infty} \bar{T}_n = \lim_{n\to\infty} \bar{Y}_n$, so $\bar{Y}_n$ would converge a.s. to μ.

All that is left to be done is to verify Inequality 7.3.2. The argument is straightforward. Notice how it makes use of the assumed existence of the first moment:

$$\sum_{k=1}^{\infty} P(A_k) = \sum_{k=1}^{\infty} P(|Y_k| \geq k) = \sum_{k=1}^{\infty} \sum_{j=k}^{\infty} \int_j^{j+1} f_{|Y|}(y) \, dy$$

$$\leq \sum_{k=1}^{\infty} \sum_{j=k}^{\infty} \frac{1}{j} \int_j^{j+1} y f_{|Y|}(y) \, dy$$

$$= \sum_{j=1}^{\infty} \frac{1}{j} \int_j^{j+1} y f_{|Y|}(y) \, dy \cdot \sum_{k=1}^{j} 1$$

$$= \sum_{j=1}^{\infty} \int_j^{j+1} y f_{|Y|}(y) \, dy \leq E(|Y|) < \infty$$

Question 7.3.10 Show that the following three statements are equivalent:
  a. $E(|Y|) < \infty$
  b. $\sum_{n=1}^{\infty} P(|Y|) \geq n) < \infty$
  c. $\sum_{n=1}^{\infty} (1/n^2) \int_{-n}^{n} y^2 f_Y(y) \, dy < \infty$

CASE STUDY 7.2

The study of arithmetic passes from the prosaic to the poetic, or perhaps the mysterious, upon the discovery that there are irrational numbers. The word *irrational* itself, which merely means not a ratio (of two integers), is now synonymous with *illogical* or *unreasonable*. Students are often uncomfortable with the lack of any pattern in the decimal expansion of common irrationals like $\sqrt{2}$ or π. But this unpredicability raises an interesting question: How random are the digits in the decimal expansion of a "typical" irrational number? Borel saw that this question could be answered by appealing to the strong law of large numbers.

Suppose a number s is chosen at random from $S = [0, 1)$, that is, s is a value of a uniform random variable. For each $m = 0, 1, 2, \ldots, 9$, define an infinite sequence of random variables by $Y_k^{(m)}(s) = 1$ if the kth digit of the decimal expansion of s is m and $Y_k^{(m)}(s) = 0$ otherwise. Borel declared that a number had a *normal* decimal expansion if, for each m,

$$\lim_{n\to\infty} \frac{Y_1^{(m)} + \cdots + Y_n^{(m)}}{n} = \frac{1}{10} \tag{7.3.3}$$

Since the $Y_k^{(m)}$ are independent and each has mean $\frac{1}{10}$, the strong law of large numbers applies. We conclude that $P(A_m) = 1$ for the set A_m of numbers satisfying Equation

7.3.3. Furthermore, if A denotes the set of normal decimal numbers, $A = \cap_{m=0}^{9} A_m$, and $P(A) = 1$.

Of course, even though we have proved that a *typical* number in $[0, 1)$ is normal, it is difficult to show that any *given* number enjoys that property. Should that surprise us? No, not really. To the never-ending consternation of the practical person, mathematicians are often able to demonstrate that a set with an infinite number of elements has a certain property, yet remain unable to exhibit even a single such element.

7.4 CENTRAL LIMIT THEOREMS

The theorem of Hausdorff mentioned in Section 7.3 points out the crucial role played in limit theorems by the power of n. For example, given any number $b > 0$, we know from Hausdorff's theorem that

$$\lim_{n \to \infty} P\left(-b < \frac{B_n - np}{n^{\frac{1}{2}+\alpha}\sqrt{pq}} < b\right) \longrightarrow 1$$

for the binomial variable B_n and any number $\alpha > 0$. However, if α *equals* 0, the limit is *not* 1: By the DeMoivre-Laplace theorem (Section 6.3),

$$\lim_{n \to \infty} P\left(-b < \frac{B_n - np}{\sqrt{npq}} < b\right) = \frac{1}{\sqrt{2\pi}} \int_{-b}^{b} e^{-x^2/2} \, dx$$

Just as the classical weak law of large numbers has been modified and strengthened and extended to variables other than Bernoulli's, so, too, has the DeMoivre-Laplace-type of limit theorem, in which the random variable is scaled by $\sqrt{n}$ instead of n, evolved. The first major generalization was given in 1887 by Chebyshev, although there were gaps in his proof. One of Chebyshev's students, Markov, simplified the derivation, while making it completely rigorous. The main limitation of the Chebyshev-Markov approach was the requirement that the random variables possess finite moments of all integral orders. In 1901, another of Chebyshev's students, A. M. Lyapunov, was able to weaken those existence criteria (55, 56). Before stating Lyapunov's theorem, we give a definition that will facilitate our discussion.

Definition 7.4.1. Let $\{W_n\}$ be a sequence of random variables. If there exist sequences of numbers $\{m_n\}$ and $\{d_n\}$ such that

$$\lim_{n \to \infty} P\left(a < \frac{W_n - m_n}{d_n} < b\right) = \frac{1}{\sqrt{2\pi}} \int_{a}^{b} e^{-z^2/2} \, dz$$

for any numbers $a < b$, we say the sequence $\{W_n\}$ is *asymptotically normal* (m_n, d_n).

In the terminology of Definition 7.4.1, the DeMoivre-Laplace theorem says that $\{B_n\}$ is asymptotically normal $(np, \sqrt{npq})$. Theorems establishing asymptotic normality are often called *central limit theorems;* the name is due to Polya (67).

Theorem 7.4.1 (Lyapunov).   Suppose $Y_1, Y_2, \ldots$ is a sequence of independent random variables. If there exists an $\alpha > 0$ such that

$$\lim_{n \to \infty} \frac{\sum\limits_{k=1}^{n} E(|Y_k - E(Y_k)|^{2+\alpha})}{\left(\mathrm{Var}\left(\sum\limits_{k=1}^{n} Y_k\right)\right)^{1+\alpha/2}} = 0$$

then the sequence $\{\sum_{k=1}^{n} Y_k\}$ is asymptotically normal $(E(\sum_{k=1}^{n} Y_k), \sqrt{\mathrm{Var}(\sum_{k=1}^{n} Y_k)})$.

Proof. See (52).

Theorem 7.4.2, credited to Lindeberg (50) and Lévy (49), is the statement most often referred to in statistics texts as *the* central limit theorem. Only first and second moments need exist, but the theorem is restricted to identically distributed sequences.

Theorem 7.4.2 (Lindeberg-Lévy).   Let $Y_1, Y_2, \ldots$ be a sequence of independent, identically distributed random variables with mean μ and finite variance σ^2. Then the sequence $\{\sum_{k=1}^{n} Y_k\}$ is asymptotically normal $(n\mu, \sqrt{n}\sigma)$.

Historically, both Theorems 7.4.1 and 7.4.2 were first proved using what are known as *characteristic functions*. Essentially Fourier transforms, characteristic functions owe their genesis to the work of Lagrange, Laplace, and Fourier. In practice, they are used in much the same fashion as moment-generating functions were in Chapter 5.

When Lyapunov used characteristic functions to remove Markov's hypothesis of finite moments of all orders, Markov remarked that Lyapunov had played "a great dirty trick." Markov then rehabilitated his method of moments by inventing the technique of truncation of variables (recall the proofs of Theorems 7.2.3 and 7.3.7), and by 1913 he was able to derive Lyapunov's theorem without using characteristic functions. In the next section we give a proof of Theorem 7.4.2, taking an approach much in the spirit of Lyapunov.

Question 7.4.1 Suppose the variables in Theorem 7.4.2 have $E(|Y_k - \mu|^{2+\alpha}) < \infty$ for some $\alpha > 0$. Show that, in that case, Theorem 7.4.1 implies Theorem 7.4.2.

Question 7.4.2 A taxpayer is allowed to round amounts on the federal income tax return to the nearest dollar. For the kth item on the form, let Y_k be the actual amount minus the rounded value. Assume that Y_k is equally likely to be any of the differences $\{-0.50, -0.49, \ldots, 0.49, 0.50\}$ and that the Y_k's are independent. Suppose 100 items on the form are to be summed. The difference between the actual and the rounded amount is $S_{100} = Y_1 + \cdots + Y_{100}$. Use Theorem 7.4.2 to approximate the probability that $|S_{100}| > 2$.

Question 7.4.3 Compare the approximation calculated in Question 7.4.2 with the one given by the Bienaymé-Chebyshev inequality.

A significant advance in the development of central limit theorems occurred in 1922 when Lindeberg (50) gave a new condition implying the asymptotic normality of $\{\sum_{k=1}^{n} Y_k\}$. As we have seen, there were already a number of sufficient conditions known for the convergence of $\{\sum_{k=1}^{n} Y_k\}$. This, however, was the first such condition to be also necessary, a fact that Feller proved (24) in 1935.

Theorem 7.4.3 (Lindeberg-Feller). Suppose Y_1, Y_2, ... are independent random variables. Let $S_n = Y_1 + \cdots + Y_n$. Let $\sigma_k^2 = \mathrm{Var}(Y_k)$ and $s_n^2 = \mathrm{Var}(S_n) = \sum_{k=1}^{n} \sigma_k^2$. Then $\{S_n\}$ is asymptotically normal and $\lim_{n \to \infty} \max_{k \le n} (\sigma_k^2 / s_n^2) = 0$ if and only if, for each $\epsilon > 0$,

$$\lim_{n \to \infty} \frac{1}{s_n^2} \sum_{k=1}^{n} E[W_k(Y_k - E(Y_k))^2] = 0$$

where $W_k = 1$ if $|Y_k - E(Y_k)| > \epsilon s_n$ and $W_k = 0$ otherwise.

Proof. The derivation of Theorem 7.4.3 is beyond the scope of this text. See (52).

Question 7.4.4 Let Y_1, Y_2, ... be a sequence of independent random variables and suppose $f_{Y_k}(2^k) = \frac{1}{2}$ and $f_{Y_k}(-2^k) = \frac{1}{2}$. Does $\lim_{n \to \infty} (1/s_n^2) \sum_{k=1}^{n} E\{W_k[Y_k - E(Y_k)]^2\} = 0$?

Question 7.4.5 Let Y_1, Y_2, ... be a sequence of independent random variables and suppose $f_{Y_k}(b^k) = f_{Y_k}(-b^k) = 1/b^{2k}$ and $f_{Y_k}(0) = 1 - (2/b^{2k})$, where b is a constant greater than 1. Does $\lim_{n \to \infty} (1/s_n^2) \sum_{k=1}^{n} E\{W_k[Y_k - E(Y_k)]^2\} = 0$?

The volume of published research on asymptotic normality is enormous, and many new results continue to appear. But rather than pursue any further theoretical developments here, we turn instead, to the *use* of central limit theorems in describing nature. As we shall see, these theorems help explain why so many phenomena produce data that are normally distributed.

A case in point is inheritance. If a physical trait is determined by a large number of genes, we can think of the realization of that trait as being the sum S_n of a large number of random variables Y_1, Y_2, ... , each making its own small contribution. The statements of the various central limit theorems would suggest that the distribution of the trait in the general population is likely to be normal. (It can be argued that the Y_i's being postulated here are not independent, but there are central limit theorems that also hold for large classes of dependent variables.)

Another example arises in the study of Brownian motion. In 1827, the Scottish botanist Robert Brown observed under a microscope that pollen grains suspended in water exhibited a "rapid oscillatory behavior." Brown first thought the motion was a characteristic of pollen, but he soon observed that a variety of other particles behaved similarly. In one of his three landmark papers of 1905, Albert Einstein established quantitatively that the motion Brown had discovered was the result of the pollen being bombarded by molecules of the liquid. Einstein's explanation provided major support for the existence of molecules—an existence that more than a few prominent scientists of the time still denied.

The statistical intricacies of Brownian motion lay beyond our intentions. We can benefit, though, by examining one small part of the theory. Consider a three-dimensional space endowed with a rectangular coordinate system, and imagine a suspended particle undergoing Brownian motion placed at the origin. We observe its position at $h, 2h, 3h, \ldots$ seconds, thus obtaining a sequence of three-dimensional vectors that give the position of the particle after each specified time. Let the kth vector be (X_k', Y_k', Z_k'), where X_k', for example, is the particle's net displacement in the x-direction. Associated with the sequence of displacements is a sequence of *increments*. If we define the kth increment to be X_k, then $X_k = X_k' - X_{k-1}'$, $k = 1$, $2, \ldots$, and

$$X_n' = \sum_{k=1}^{n} X_k$$

Assuming the liquid in which the particle is suspended is homogeneous, the X_k's are independent and identically distributed. If such is the case, we might reasonably expect X_n' to be approximately normal.

Question 7.4.6 Empirical studies have shown that the sum of k independent uniform random variables, when suitably standardized, can be used as an approximation to an $N(0, 1)$ random variable. Furthermore, the agreement is quite good even for k as small as 12. Write a formula for generating approximate standard normal deviates.

Another application of central limit theorems derives from the *law of proportionate effect*, which states that *changes* in certain kinds of measurements made on an organism—such as weight—are randomly proportional to the organism's *current* measurement. Conceptually, the law of proportionate effect is a probabilistic version of the more familiar law of exponential growth.

Consider a sequence $Y_0, Y_1, Y_2, \ldots$ of random variables, where Y_k is the dimension of the organism at time k. A second sequence, $X_1, X_2, \ldots$, can be defined to describe the organism's growth *rate:*

$$X_k = \frac{Y_k - Y_{k-1}}{Y_{k-1}} \tag{7.4.1}$$

We assume the X's are independent, identically distributed, and independent of the Y's. What can we infer, then, about the asymptotic distribution of Y_k?

To begin, note that Equation 7.4.1 can be rewritten as

$$Y_k = (1 + X_k)Y_{k-1}$$

Applying that formula recursively gives

$$Y_k = \prod_{j=1}^{k} (1 + X_j) Y_0$$

or, equivalently,

$$\log Y_k = \log Y_0 + \sum_{j=1}^{k} \log (1 + X_j)$$

Suppose k is large and $\log (1 + X_j)$ has finite variance. We would expect $\log Y_k$ to be approximately normal because of its structure or, equivalently, Y_k to be approximately lognormal.

CASE STUDY 7.3

Knowing the distribution of personal incomes is important to a variety of people and institutions for a number of different reasons. Manufacturers of consumer goods, for example, like to have some idea of the purchasing power of their potential clientele; governments need to know the incomes of their constituents when formulating tax policies and social programs. As the data presented here suggest, a good probability model to keep in mind when working with income data is the lognormal.

The Italian economist Vilfredo Pareto (1848–1923) gave the first mathematical model for income distribution. It was his discovery that income data from England, Prussia, Saxony, and Italy were well approximated by the curve

$$\log F = \log A - \alpha \log x \qquad (7.4.2)$$

where F is the cumulative frequency of incomes, x is income, and A and α are parameters. Thinking along purely descriptive lines, we might interpret Equation 7.4.2 to mean that X is a random variable with pdf $f_X(x) = A\alpha/x^{\alpha+1}$, $x > A$ (why?). To Pareto, though, the data suggested something entirely different. He wrote in 1897 (65): "These results are remarkable. It is absolutely impossible to accept that they are due to chance. There is certainly a *cause* which produces the tendency for incomes to lie on a certain curve."

To Pareto, being due to chance meant that the data followed a normal distribution. Since he observed this not to be the case for incomes, he denied a probabilistic explanation. We have no such qualms. We show below that the lognormal density provides not only a better statistical description for incomes than $f_X(x)$, but also explains *why* incomes are distributed the way they are.

In an economic context, the law of proportionate effect holds that a worker's increase in earnings for the next pay period is a random proportion of his or her current earnings. If that postulate is valid, the distribution of personal income should be approximately lognormal. Columns 1 and 2 of Table 7.4.1 (adapted from (87)) give the weekly salary in pounds of 4,443,000 English male manual workers. Let the random variable X represent a worker's weekly earnings. Entries in Column 3 were obtained by assuming $\log X \sim N(2.65, 0.0845)$, the 2.65 and 0.0845 being statistically derived estimates for μ and σ, respectively (see (87)). Except for the lowest-income group, the lognormal model fits the data quite well. (The lognormal's failure to agree with the data for a salary of 9 pounds or less is not entirely unexpected—economists traditionally have had trouble modeling the extreme portions of income distributions, no matter what pdf they try to fit.)

TABLE 7.4.1

Salary (pounds per week)	Observed frequency (in thousands)	Expected frequency (in thousands)
≤9	160.39	263.91
9–12	1079.20	999.23
12–15	1337.35	1310.69
15–18	959.24	953.47
18–21	519.83	522.50
21–24	234.15	236.81
24–27	94.19	97.75
27–30	33.77	36.88
30–35	17.33	17.33
≥35	7.55	4.43

7.5 CHARACTERISTIC FUNCTIONS

The moment-generating function introduced in Section 5.5, $M_Y(t) = E(e^{tY})$, is a useful tool for proving central limit theorems. The random variables to which such proofs extend, though, include only sequences having mgf's that exist. A more powerful central limit theorem can be derived by using *characteristic functions*. The latter have the distinct advantage that they always exist, no matter what the random variable might be. Our goal in this section is to prove the Lindeberg-Lévy result (Theorem 7.4.2) using characteristic functions.

> **Definition 7.5.1.** The *characteristic function* of a random variable Y is denoted by $\phi_Y(t)$, where
> $$\phi_Y(t) = E(e^{itY})$$

Comment

The existence of $\phi_Y(t)$ for any Y is easy to verify. Since e^{ity} is a complex number with absolute value 1, the absolute value of the characteristic function has an obvious finite upper bound:
$$|E(e^{itY})| \leq E(|e^{itY}|) \leq 1$$

Example 7.5.1

Let Y be a binomial random variable with pdf $f_Y(y) = \binom{n}{y}p^y(1 - p)^{n-y}$, $y = 0$, $1, \ldots , n$. Find $\phi_Y(t)$.

From Definition 7.5.1,

$$\phi_Y(t) = \sum_{y=0}^{n} e^{ity} \binom{n}{y} p^y (1-p)^{n-y}$$

$$= \sum_{y=0}^{n} \binom{n}{y} (pe^{it})^y (1-p)^{n-y}$$

$$= [pe^{it} + (1-p)]^n$$

Example 7.5.2

Let Y be a Poisson random variable with pdf $f_Y(y) = e^{-\lambda}\lambda^y/y!$, $y = 0, 1,$
2, ... Use the complex power series expansion for e^z to find $\phi_Y(t)$.
Summing $e^{ity} \cdot f_Y(y)$ over y gives

$$\phi_Y(t) = \sum_{y=0}^{\infty} e^{ity} \cdot \frac{e^{-\lambda}\lambda^y}{y!} = e^{-\lambda} \sum_{y=0}^{\infty} \frac{(\lambda e^{it})^y}{y!}$$

Recall that

$$e^z = \sum_{k=0}^{\infty} \frac{z^k}{k!}$$

By inspection,

$$\phi_Y(t) = e^{-\lambda} \cdot e^{\lambda e^{it}}$$

$$= e^{-\lambda + \lambda e^{it}}$$

(How do the characteristic functions for the binomial and Poisson random
variables compare to those distributions' moment-generating functions?)

The mathematical tools necessary for a rigorous development of characteristic
functions are beyond the level of this text, so we cannot derive all the results support-
ing the Lindeberg-Lévy theorem. Nevertheless, the number of unproved assertions
can be held to a minimum if we make use of a certain well-known result in the theory
of integration.

Theorem 7.5.1 (Lebesgue Dominated Convergence Theorem). Suppose
$\{g_n(y)\}$ is a sequence of real- or complex-valued functions defined on some
(Borel) set A of real numbers. Suppose for each $y \in A$ (a set of measure zero
can be excluded), $\lim_{n \to \infty} g_n(y) = g(y)$. If there is a function $M(y)$ whose
integral over A exists and which is finite and if $|g_n(y)| < M(y)$ for all $y \in A$,
then

1. $\int_A g(y)\, dy < \infty$
2. $\lim_{n \to \infty} \int_A g_n(y)\, dy = \int_A g(y)\, dy$

Proof. See (77).

Theorem 7.5.2. If $E(|Y|^n) < \infty$, then:

1. $\phi_Y^{(k)}(t)$ exists and is continuous, $k = 1, \ldots, n$.
2. $|\phi_Y^{(k)}(t)| \leq E(|Y|^k)$, $k = 1, \ldots, n$.
3. $\phi_Y^{(n)}(0) = i^n E(Y^n)$.

Proof. We consider the case $n = 1$. Repeated applications of the argument gives the result for higher-order derivatives.

Assume $E(|Y|) < \infty$. Choose a sequence $\{h_n\}$ of real numbers with $h_n \to 0$. For a fixed value of t, define the functions $g_n(y)$ by

$$g_n(y) = \frac{1}{h_n}(e^{i(t+h_n)y} - e^{ity})f_Y(y)$$

Then

$$\lim_{n \to \infty} g_n(y) = \frac{\partial}{\partial t}e^{ity}f_Y(y) = iye^{ity}f_Y(y)$$

By the mean value theorem, there is some point k_n, $0 \leq |k_n| \leq |h_n|$, such that

$$g_n(y) = \frac{1}{h_n} \cdot [h_n iye^{i(t+k_n)y}(e^{ih_n y} - 1)f_Y(y)]$$

Thus, $|g_n(y)| \leq 2|y|f_Y(y)$. Since $\int_{-\infty}^{\infty} 2|y|f_Y(y)\,dy = 2E(|Y|) < \infty$, the Lebesgue dominated convergence theorem applies, and $\int_{-\infty}^{\infty} g_n(y)\,dy \to \int_{-\infty}^{\infty} iye^{ity}f_Y(y)\,dy$ for every choice of $\{h_n\}$. But $\int_{-\infty}^{\infty} g_n(y)\,dy = (1/h_n)[\phi_Y(t + h_n) - \phi_Y(t)]$, so $\phi_Y'(t) = \lim_{n\to\infty} \int_{-\infty}^{\infty} g_n(y)\,dy = \int_{-\infty}^{\infty} iye^{ity}f_Y(y)\,dy$. It follows that

$$|\phi_Y'(t)| \leq E(|Y|) \quad \text{and} \quad \phi_Y'(0) = iE(Y)$$

To complete the proof, we must show that $\phi_Y'(t)$ is continuous. Choose any sequence $\{t_n\}$ with $t_n \to t$. Then

$$iye^{it_n y}f_Y(y) \longrightarrow iye^{ity}f_Y(y)$$

and $|iye^{it_n y}f_Y(y)| \leq |y|f_Y(y)$. Again, Legesgue dominated convergence applies and

$$\lim_{n \to \infty} \phi_Y'(t_n) = \lim_{n \to \infty} \int_{-\infty}^{\infty} iye^{it_n y}f_Y(y)\,dy$$

$$= \int_{-\infty}^{\infty} iye^{ity}f_Y(y)\,dy$$

$$= \phi_Y'(t)$$

Question 7.5.1 Let $\psi_Y(t) = \log \phi_Y(t)$. Show that $E(Y) = (1/i)\psi'(0)$ and $\text{Var}(Y) = -\psi''(0)$.

Using the results of Theorem 7.5.2, we can express $\phi_Y(t)$ in terms of Taylor polynomials.

Theorem 7.5.3. Suppose $E(|Y|^n) < \infty$. Then

$$\phi_Y(t) = 1 + \sum_{k=1}^{n} i^k \frac{E(Y^k)}{k!} t^k + R_n(t) t^n$$

where $\lim_{t \to \infty} R_n(t) = 0$.

Proof. Recall that any complex-valued function $\phi(t)$ defined for t in some interval of real numbers can be written $\phi(t) = \phi_1(t) + i\phi_2(t)$, where ϕ_1 and ϕ_2 are real-valued functions. Also, the nth derivative of ϕ, written $\phi^{(n)}(t)$, is equal to $\phi_1^{(n)}(t) + i\phi_2^{(n)}(t)$. Here, we write $\phi_Y(t) = \phi_1(t) + i\phi_2(t)$. Taylor's theorem gives

$$\phi_1(t) = \phi_1(0) + \sum_{k=1}^{n-1} \frac{\phi_1^{(k)}(0)}{k!} t^k + \frac{\phi_1^{(n)}(t_1)}{n!} t^n$$

where t_1 is some point between 0 and t. Equivalently, we can write

$$\phi_1(t) = \phi_1(0) + \sum_{k=1}^{n} \frac{\phi_1^{(k)}(0)}{k!} t^k + \frac{\phi_1^{(n)}(t_1) - \phi_1^{(n)}(0)}{n!} t^n$$

A similar expansion holds for ϕ_2, where t_1 is replaced by a point t_2, the latter also being between 0 and t. Rewritten, then,

$$\phi_Y(t) = \phi_1(t) + i\phi_2(t) = \phi_Y(0) + \sum_{k=1}^{n} \frac{\phi_Y^{(k)}(0)}{k!} t^k + R_n(t) t^n$$

where

$$R_n(t) = \frac{\phi_1^{(n)}(t_1) - \phi_1^{(n)}(0) + \phi_2^{(n)}(t_2) - \phi_2^{(n)}(0)}{n!}$$

Since we hypothesized the existence of $E(|Y|^n)$, Theorem 7.5.2 guarantees the existence and continuity of $\phi_Y^{(n)}(t)$. Therefore, $\phi_1^{(n)}(t)$ and $\phi_2^{(n)}(t)$ are, themselves, continuous. And since $t_1 \to 0$ and $t_2 \to 0$ as $t \to 0$, $\lim_{t \to 0} R_n(t) = 0$. Finally, Theorem 7.5.2 equates $\phi^{(k)}(0)$ and $i^k E(Y^k)$, so the proof is complete.

Theorem 7.5.3 raises the question of which $\phi_Y(t)$'s admit a Taylor *series* expansion. Surprisingly, one sufficient condition involves the existence of the moment-generating function of the random variable.

Theorem 7.5.4. Suppose $M_Y(u)$ and $M_Y(-u)$ exist for some $u > 0$. Then every moment of Y exists and:

1. $M_Y(t) = \sum_{k=0}^{\infty} \frac{E(Y^k)}{k!} t^k$, $\qquad -u \le t \le u$

---

2. $\phi_Y(t) = \sum\limits_{k=0}^{\infty} \dfrac{E(Y^k)}{k!}(it)^k, \qquad -u \le t \le u$

Proof. Let t be a fixed value with $|t| \le u$. For each positive integer n, let

$$g_n(y) = \sum_{k=0}^{n} \frac{(ity)^k}{k!} f_Y(y)$$

Then $|g_n(y)| \le \sum_{k=0}^{n} (|ty|^k/k!) f_Y(y) \le e^{|ty|} f_Y(y) \le e^{u|y|} f_Y(y)$. To be able to apply the Lebesgue dominated convergence theorem here, we must show that $\int_{-\infty}^{\infty} e^{u|y|} f_Y(y)\, dy$ is finite. But

$$\int_{-\infty}^{\infty} e^{u|y|} f_Y(y)\, dy = \int_{-\infty}^{0} e^{u(-y)} f_Y(y)\, dy + \int_{0}^{\infty} e^{uy} f_Y(y)\, dy$$

$$\le \int_{-\infty}^{\infty} e^{-uy} f_Y(y)\, dy + \int_{-\infty}^{\infty} e^{uy} f_Y(y)\, dy$$

$$= M_Y(-u) + M_Y(u) < \infty$$

Also, for each y, $\lim_{n\to\infty} g_n(y) = e^{ity} f_Y(y)$. From Theorem 7.5.1,

$$\phi_Y(y) = \int_{-\infty}^{\infty} e^{ity} f_Y(y)\, dy = \lim_{n\to\infty} \int_{-\infty}^{\infty} g_n(y)\, dy$$

And since

$$\lim_{n\to\infty} \int_{-\infty}^{\infty} g_n(y) = \lim_{n\to\infty} \sum_{k=0}^{n} \frac{1}{k!} \int_{-\infty}^{\infty} y^k f_Y(y)\, dy \cdot (it)^k$$

$$= \sum_{k=0}^{\infty} \frac{E(Y^k)}{k!}(it)^k$$

the proof for part 2 is complete. The derivation for $M_Y(t)$ is identical, except that *it* is replaced by t.

Example 7.5.3

Suppose Y is an exponential random variable with pdf $f_Y(y) = \lambda e^{-\lambda y}$, $y > 0$. Use Theorem 7.5.4 to find $E(Y^k)$ and $\phi_Y(t)$.

The first step is to get the moment-generating function. A simple integration gives $M_Y(t) = 1/[1 - (t/\lambda)]$:

$$M_Y(t) = \int_{0}^{\infty} e^{ty} \lambda e^{-\lambda y}\, dy = \lambda \int_{0}^{\infty} e^{-(\lambda - t)y}\, dy$$

$$= \frac{\lambda}{\lambda - t} = \frac{1}{1 - (t/\lambda)}$$

We recognize the latter expression to be the sum of a geometric series, so we can also write

$$M_Y(t) = \sum_{k=0}^{\infty} \left(\frac{t}{\lambda}\right)^k = \sum_{k=0}^{\infty} \frac{(k! \, \lambda^{-k})}{k!} t^k$$

It follows from Theorem 7.5.4 that

$$E(Y^k) = k! \, \lambda^{-k}, \qquad k = 0, 1, 2, \ldots$$

A second application of Theorem 7.5.4 yields the characteristic function

$$\phi_Y(t) = \sum_{k=0}^{\infty} \frac{k! \, \lambda^{-k}}{k!} (it)^k$$

$$= \sum_{k=0}^{\infty} \left(\frac{it}{\lambda}\right)^k = \frac{\lambda}{\lambda - it}$$

Another way to find $M_Y(t)$ and $\phi_Y(t)$ is by first deriving a recursion formula for $E(Y^k)$. Here, integration by parts gives

$$E(Y^k) = \int_0^{\infty} y^k \lambda e^{-\lambda y} \, dy$$

$$= \left. -y^k e^{-\lambda y} \right|_0^{\infty} + \int_0^{\infty} k y^{k-1} e^{-\lambda y} \, dy$$

or $E(Y^k) = k\lambda^{-1} E(Y^{k-1})$, $k \geq 1$. Knowing that $E(Y^0) = 1$, we can apply the expression for $E(Y^k)$ recursively and find that $E(Y^k) = k! \, \lambda^{-k}$. Suppose it is also known that $M_Y(t)$ exists on some interval about 0; then Theorem 7.5.4 applies, and the moment-generating function can be easily deduced:

$$M_Y(t) = \sum_{k=0}^{\infty} \frac{k! \, \lambda^{-k}}{k!} \cdot t^k = \sum_{k=0}^{\infty} \left(\frac{t}{\lambda}\right)^k = \frac{1}{1 - (t/\lambda)}$$

Comment

Prior knowledge of the existence of the moment-generating function is essential to the argument we have just used. Distributions can be found that have finite moments of every integral order but whose moment-generating functions still do not exist on any open interval containing 0 (see Question 7.5.2).

Question 7.5.2 Let $f_Y(y) = \frac{1}{2} e^{-\sqrt{y}}$, $y > 0$.
    a. Show that $E(Y^k) = (2k + 1)!$.
    b. Verify that $M_Y(t)$ is not finite for $t > 0$.
    c. Show that $\sum_{k=0}^{\infty} (E(Y^k)/k!) \, t^k$ diverges for any $t > 0$.

Given the characteristic function for a random variable Y, we can easily write the characteristic function of X, where $X = aY + b$. Not surprisingly, the statement of

Theorem 7.5.5 parallels the effect a linear transformation has on a moment-generating function (recall Theorem 5.5.4).

Theorem 7.5.5. Suppose the random variable Y has characteristic function $\phi_Y(t)$. Let $X = aY + b$. Then

$$\phi_X(t) = \phi_{aY+b}(t) = e^{ibt}\phi_Y(at)$$

for any constants a and b.

Proof. Let Y be a random variable with characteristic function $\phi_Y(t)$. Two applications of Definition 7.5.1 give the result:

$$\phi_{aY+b}(t) = E(e^{it(aY+b)}) = E(e^{itaY} \cdot e^{itb})$$
$$= e^{ibt}E(e^{i(at)Y}) = e^{ibt}\phi_Y(at)$$

In manipulating moment-generating functions to identify unknown distributions, we frequently use the fact that $M_Y(t) = \prod_{i=1}^{n} M_{Y_i}(t)$, where $Y = \sum_{i=1}^{n} Y_i$ and the Y_i's are independent. The same result holds for characteristic functions.

Theorem 7.5.6. Suppose $Y_1, Y_2, \ldots, Y_n$ are independent random variables with characteristic functions $\phi_{Y_1}(t), \phi_{Y_2}(t), \ldots, \phi_{Y_n}(t)$. Let $Y = Y_1 + Y_2 + \cdots + Y_n$. Then

$$\phi_Y(t) = \phi_{Y_1 + \cdots + Y_n}(t) = \phi_{Y_1}(t) \cdot \phi_{Y_2}(t) \cdots \phi_{Y_n}(t)$$

Proof. Let $Y_1, \ldots, Y_n$ be a set of independent random variables with characteristic functions $\phi_{Y_1}(t), \ldots, \phi_{Y_n}(t)$. Define Y to be the sum of the Y_i's. Then

$$\phi_Y(t) = \phi_{Y_1 + \cdots + Y_n}(t) = E(e^{it(Y_1 + \cdots + Y_n)}) = E(e^{itY_1} \cdot e^{itY_2} \cdots e^{itY_n})$$
$$= E(e^{itY_1}) \cdot E(e^{itY_2}) \cdots E(e^{itY_n}) \qquad \text{(Why?)}$$
$$= \phi_{Y_1}(t) \cdot \phi_{Y_2}(t) \cdots \phi_{Y_n}(t)$$

The next two theorems are crucial if we want to use characteristic functions to prove central limit theorems. The derivations are omitted, but they are quite standard in any theoretical coverage of Fourier integrals. For details, see (14), (31), or (54).

Theorem 7.5.7. Suppose F_Y is continuous at the point x. Then

$$F_Y(x) = \frac{1}{2\pi} \lim_{y \to -\infty} \lim_{c \to \infty} \int_{-c}^{c} \frac{e^{-ity} - e^{-itx}}{it} \phi_Y(t) \, dt$$

In short, ϕ_Y determines F_Y.

Theorems 7.5.6 and 7.5.7 can often simplify the problem of finding the sum of two (or more than two) random variables. Notice in the next example how the much more difficult convolution approach of Chapter 4 is avoided.

Example 7.5.4

Suppose X is a binomial random variable with parameters m and p and Y is binomial with parameters n and p. Assume X and Y are independent. Find the distribution of their sum, $X + Y$.

Using Theorem 7.5.6 and Example 7.5.1, we can easily write the characteristic function for $X + Y$:

$$\phi_{X+Y}(t) = \phi_X(t) \cdot \phi_Y(t)$$
$$= (pe^{it} + 1 - p)^m \cdot (pe^{it} + 1 - p)^n$$
$$= (pe^{it} + 1 - p)^{m+n}$$

What we get looks familiar: We recognize $(pe^{it} + 1 - p)^{m+n}$ to be the characteristic function of a binomial random variable with parameters $m + n$ and p. By Theorem 7.5.7, $f_{X+Y}(z) = \binom{m+n}{z} p^z (1 - p)^{z-m-n}$, $z = 0, 1, \ldots, m + n$.

Question 7.5.3 Suppose X and Y are independent Poisson variables with parameters λ and μ, respectively. Use Theorems 7.5.6 and 7.5.7 to find the distribution of $X + Y$.

Theorem 7.5.8. Suppose $Y_1, Y_2, \ldots$ are random variables whose characteristic functions converge pointwise to some continuous limit function—that is, there is a continuous function $\phi(t)$ such that for each t, $\lim_{n \to \infty} \phi_{Y_n}(t) = \phi(t)$. Then there exists a random variable X such that $\lim_{n \to \infty} F_{Y_n}(x) = F_X(x)$ for every point x at which F_X is continuous. Also, the limit function for the ϕ_{Y_n}'s is the characteristic function for X: $\phi(t) = \phi_X(t)$.

One final result needs to be set into place before we can use characteristic functions to prove the Lindeberg-Lévy central limit theorem.

Theorem 7.5.9. Let Y be a standard normal random variable with pdf

$$f_Y(y) = \frac{1}{\sqrt{2\pi}} e^{-y^2/2}, \qquad -\infty < y < \infty$$

Then $\phi_Y(t) = e^{-t^2/2}$

Proof. We will establish the theorem by using the first technique described in Example 7.5.3. We begin by calculating the moment-generating function for Y:

$$M_Y(t) = \frac{1}{\sqrt{2\pi}} \int_{-\infty}^{\infty} e^{ty} e^{-y^2/2} \, dy$$

$$= \frac{1}{\sqrt{2\pi}} \int_{-\infty}^{\infty} e^{-(y^2 - 2ty)/2} \, dy$$

$$= \frac{1}{\sqrt{2\pi}} e^{t^2/2} \int_{-\infty}^{\infty} e^{-(y^2 - 2ty + t^2)/2} \, dy$$

$$= e^{t^2/2} \cdot \int_{-\infty}^{\infty} \frac{1}{\sqrt{2\pi}} e^{-(y-t)^2/2} \, dy$$

$$= e^{t^2/2} \cdot 1 \qquad\qquad\qquad \text{(Why?)}$$

$$= e^{t^2/2}$$

Written as a Taylor series,

$$M_Y(t) = \sum_{k=0}^{\infty} \frac{1}{k!} \left(\frac{t^2}{2}\right)^k$$

$$= \sum_{k=0}^{\infty} \frac{(2k)!}{k!} \frac{1}{2^k} \cdot \frac{1}{(2k)!} t^{2k}$$

$$= \sum_{k=0}^{\infty} \frac{E(Y^{2k})}{(2k)!} \cdot t^{2k}$$

(recall Example 5.5.1). Appealing to Theorem 7.5.4, we can argue that $\phi_Y(t)$ is the same series as $M_Y(t)$, except that t is replaced by it. Once that substitution is made, the statement of the theorem follows immediately:

$$\phi_Y(t) = \sum_{k=0}^{\infty} \frac{1}{k!} \left(\frac{(it)^2}{2}\right)^k$$

$$= e^{(it)^2/2}$$

$$= e^{-t^2/2}$$

Corollary. Let X be a normal random variable with mean μ and variance σ^2. Then

$$\phi_X(t) = e^{i\mu t - \frac{1}{2}\sigma^2 t^2}$$

Proof. If X is $N(\mu, \sigma^2)$, then $X = \sigma Y + \mu$, where Y is $N(0, 1)$. The corollary follows from Theorems 7.5.5 and 7.5.9.

Question 7.5.4 Suppose X and Y are two independent normal random variables, X with mean μ_1 and variance σ_1^2 and Y with mean μ_2 and variance σ_2^2. Let $Z = X + Y$. Show that Z is $N(\mu_1 + \mu_2, \sigma_1^2 + \sigma_2^2)$.

We now have all the background needed to prove the Lindeberg-Lévy central limit theorem. For convenience, we restate it here as Theorem 7.5.10.

> **Theorem 7.5.10 (Lindeberg-Lévy).** Let $Y_1, Y_2, \ldots$ be a sequence of independent, identically distributed random variables with mean μ and finite variance σ^2. Then the sequence $\{\sum_{k=1}^{n} Y_k\}$ is asymptotically normal $(n\mu, \sqrt{n}\sigma)$.

Proof. It suffices to prove that for each real number x,

$$\lim_{n \to \infty} F_{W_n}(x) = F_Z(x)$$

where

$$W_n = \frac{\sum\limits_{k=1}^{n} Y_k - n\mu}{\sqrt{n}\sigma} = \sum_{k=1}^{n} \frac{1}{\sqrt{n}}\left(\frac{Y_k - \mu}{\sigma}\right)$$

and $Z \sim N(0, 1)$. By Theorem 7.5.8 that convergence holds if

$$\lim_{n \to \infty} \phi_{W_n}(t) = \phi_Z(t) = e^{-t^2/2}$$

Write $X_k = (Y_k - \mu)/\sigma$. Then $W_n = \sum_{k=1}^{n} (1/\sqrt{n})X_k$ and

$$\phi_{W_n}(t) = \sum_{k=1}^{n} \phi_{(1/\sqrt{n})X_k}(t)$$

$$= \prod_{k=1}^{n} \phi_{X_k}\left(\frac{t}{\sqrt{n}}\right)$$

$$= \phi^n\left(\frac{t}{\sqrt{n}}\right)$$

where $\phi(t)$ represents the common characteristic function of the X_k's.

By Theorem 7.5.3, we can express $\phi(t)$ in terms of Taylor polynomials:

$$\phi(t) = 1 + iE(X)t + i^2 \frac{E(X^2)}{2}t^2 + R_2(t) \cdot t^2$$

Since $E(X) = 0$ and $E(X^2) = \text{Var}(X) = 1$,

$$\phi(t) = 1 - \tfrac{1}{2}t^2 + R_2(t)t^2$$

and

$$\phi\left(\frac{t}{\sqrt{n}}\right) = 1 + \frac{-\tfrac{1}{2}t^2}{n} + R_2\left(\frac{t}{\sqrt{n}}\right)\frac{t^2}{n}$$

so

$$\phi^n\left(\frac{t}{\sqrt{n}}\right) = \left(1 + \frac{-\tfrac{1}{2}t^2}{n} + R_2\left(\frac{t}{\sqrt{n}}\right)\frac{t^2}{n}\right)^n \tag{7.5.1}$$

All that remains is to use Equation 7.5.1 to show that $\lim_{n \to \infty} \phi^n(t/\sqrt{n}) = e^{-t^2/2}$. If the term $R_2(t/\sqrt{n}) \cdot t^2/n$ were not present, the limit of the right-hand side of Equation 7.5.1 would be the definition of $e^{-t^2/2}$. To complete the proof, then, it suffices to show that

$$\lim_{n\to\infty} \left| \left(1 + \frac{-\frac{1}{2}t^2}{n} + R_2\left(\frac{t}{\sqrt{n}}\right)\frac{t^2}{n}\right)^n - \left(1 + \frac{-\frac{1}{2}t^2}{n}\right)^n \right| = 0 \qquad (7.5.2)$$

Lemma.   For any real number c and (possibly complex) sequence $\{b_n\}$, where $\lim_{n\to\infty} b_n = 0$,

$$\lim_{n\to\infty} D_n = \lim_{n\to\infty} \left| \left(1 + \frac{c}{n} + \frac{b_n}{n}\right)^n - \left(1 + \frac{c}{n}\right)^n \right| = 0$$

Proof of lemma. Using the formula for a binomial expansion, we can write

$$D_n = \left| \sum_{k=0}^n \binom{n}{k}\left(1 + \frac{c}{n}\right)^k \left(\frac{b_n}{n}\right)^{n-k} - \left(1 + \frac{c}{n}\right)^n \right|$$

$$= \left| \sum_{k=0}^{n-1} \binom{n}{k}\left(1 + \frac{c}{n}\right)^k \left(\frac{b_n}{n}\right)^{n-k} \right|$$

$$\le \sum_{k=0}^{n-1} \binom{n}{k}\left|1 + \frac{c}{n}\right|^k \left|\frac{b_n}{n}\right|^{n-k}$$

Let $0 < \epsilon < 1$. Since $\lim_{n\to\infty} b_n = 0$, we can find an n sufficiently large that $|b_n| \le \epsilon < 1$. Therefore, $|b_n|^{n-k} \le \epsilon$, $k = 0, 1, \ldots, n-1$, and

$$D_n \le \sum_{k=0}^{n-1} \binom{n}{k}\left|1 + \frac{c}{n}\right|^k \epsilon \cdot \left(\frac{1}{n}\right)^{n-k}$$

$$\le \epsilon \cdot \sum_{k=0}^n \binom{n}{k}\left|1 + \frac{c}{n}\right|^k \left(\frac{1}{n}\right)^{n-k}$$

$$\le \epsilon\left(1 + \frac{c+1}{n}\right)^n$$

Taking limits on both sides of the last inequality gives

$$\lim_{n\to\infty} D_n \le \epsilon \cdot e^{c+1}$$

Since ϵ was arbitrary, $\lim_{n\to\infty} D_n = 0$, and the lemma is proved.

The final step in the proof of Theorem 7.5.10—verifying that Equation 7.5.2 is true—now follows immediately. Recall that $\lim_{t\to\infty} R_2(t) = 0$, which means that $\lim_{n\to\infty} R_2(t/\sqrt{n})t^2$ is also 0. By the lemma,

$$\lim_{n\to\infty} \left| \left(1 + \frac{-\frac{1}{2}t^2}{n} + R_2\left(\frac{t}{\sqrt{n}}\right)\frac{t^2}{n}\right)^n - \left(1 + \frac{-\frac{1}{2}t^2}{n}\right)^n \right| = 0$$

so

$$\lim_{n\to\infty} \phi_{W_n}(t) = e^{-t^2/2}$$

implying that $\{\sum_{k=1}^n Y_k\}$ is asymptotically normal $(n\mu, \sqrt{n}\sigma)$.

Bibliography

1. Allen, Raymond L., and Lawrence W. Doubleday. "The Deterrence of Unauthorized Intrusion in Residences, a Systems Analysis." In *First International Electronic Crime Countermeasures Conference* (Edinburgh, Scotland, 1973). Lexington, Ky.: ORES Publications, 32.

2. Ash, Robert B. *Real Analysis and Probability.* New York: Academic Press, 1972.

3. Bartle, Robert G. *The Elements of Real Analysis.* New York: Wiley, 1976.

4. Bell, E. T. *Men of Mathematics.* New York: Simon and Schuster, 1937, 420.

5. Blum, Julius R., and Judah I. Rosenblatt. *Probability and Statistics.* Philadelphia: W. B. Saunders, 1972, 144.

6. Bortkiewicz, L. *Das Gesetz der Kleinen Zahlen.* Leipzig: Teubner, 1898.

7. Bradley, James V. *Probability; Decision; Statistics.* Englewood Cliffs, N.J.: Prentice-Hall, 1976.

8. Buck, R. C., ed. *Studies in Modern Analysis.* Vol. 1. Englewood Cliffs, N.J.: The Mathematical Association of America–Prentice-Hall, 1962.

9. Casler, Lawrence. "The Effects of Hypnosis on GESP." *Journal of Parapsychology* 28 (1964): 126–134.

10. Chattergee, S. "Estimating Wildlife Populations." In *Statistics by Example—Finding Models,* edited by Frederick Mosteller et al. Reading, Mass.: Addison-Wesley, 1973, 25–33.

11. Clarke, R. D. "An Application of the Poisson Distribution." *Journal of the Institute of Actuaries* 22 (1946): 48.

12. Conover, W. J. *Practical Nonparametric Statistics*. New York: John Wiley, 1971, 349–356.

13. Craf, John R. *Economic Development of the U.S.* New York: McGraw-Hill, 1952, 368–371.

14. Cramér, H. *Mathematical Methods of Statistics*. Princeton, N.J.: Princeton University Press, 1946.

15. Das, S. C. "Fitting Truncated Type III Curves to Rainfall Data." *Australian Journal of Physics* 8 (1955): 298–304.

16. David, F. N. *Games, Gods, and Gambling*. New York: Hafner, 1962, 168.

17. Davis, D. J. "An Analysis of Some Failure Data." *Journal of the American Statistical Association* 47 (1952): 113–150.

18. Davis, Harold T. *The Summation of Series*. San Antonio: Principia Press, 1962, 108.

19. Devore, Jay L. *Probability and Statistics for Engineering and the Sciences*. Monterey, Calif.: Brooks/Cole, 1982.

20. Dewey, G. *Relative Frequency of English Spellings*. Columbia University, N.Y.: Teacher's College Press, 1970, 27.

21. Ederer, F., M. H. Myers, and N. A. Mantel. "A Statistical Problem in Space and Time: Do Leukemia Cases Come in Clusters?" *Biometrics* 20 (1964): 626–638.

22. Fairley, William B. "Evaluating the 'Small' Probability of a Catastrophic Accident from the Marine Transportation of Liquefied Natural Gas." In *Statistics and Public Policy,* edited by William B. Fairley and Frederick Mosteller. Reading, Mass.: Addison-Wesley, 1977, 331–353. Copyright © 1977 Addison-Wesley. Reprinted with permission.

23. Fairley, William B., and Frederick Mosteller. "A Conversation about Collins." In *Statistics and Public Policy,* edited by William B. Fairley and Frederick Mosteller. Reading, Mass.: Addison-Wesley, 1977, 369–379.

24. Feller, W. "Über der zentral Grenzwertsatz der Wahrscheinlichkeitsrechnung." *Mathematische Zeitschrift* 40 (1935): 521–559.

25. Feller, W. "Statistical Aspects of ESP." *Journal of Parapsychology* 4 (1940): 271–298.

26. Feller, W. *An Introduction to Probability Theory and Its Applications*. Vol. 1, 2nd ed. New York: John Wiley, 1957.

27. Fisher, R. A. "The Negative Binomial Distribution." *Annals of Eugenics* 11 (1941): 182–187.

28. Gabriel, K. R., and J. Neumann. "On a Distribution of Weather Cycles by Length." *Quarterly Journal of the Royal Meteorological Society* 83 (1957): 375–380.

29. Galton, F. *Natural Inheritance*. London: Macmillan, 1908.

30. Gardner, Geoffrey Y. "Computer Identification of Bullets." In *Carnahan Conference on Crime Countermeasures*. Lexington, Ky.: ORES Publications, 1977, 149–166.

31. Gnedenko, B. V. *The Theory of Probability*. 4th ed. New York: Chelsea, 1967.

32. Gross, Neal, Ward S. Mason, and Alexander W. McEachern. *Explorations in Role Analysis*. New York: John Wiley, 1958, 297.

33. Grosswald, Emil. *Topics from the Theory of Numbers*. New York: Macmillan, 1966, 113–114.

34. Hald, A. *Statistical Theory with Engineering Applications*. New York: John Wiley, 1952, 155–157. Copyright © 1952 by John Wiley & Sons, Inc. Reprinted by permission of John Wiley & Sons, Inc.

35. Hansel, C. E. M. *ESP: A Scientific Evaluation*. New York: Scribner, 1966, 86–89.

36. Hastings, N. A. J., and J. B. Peacock. *Statistical Distributions*. London: Butterworth, 1975.

37. Heath, Clark W., and Robert J. Hasterlik. "Leukemia among Children in a Suburban Community." *The American Journal of Medicine* 34 (1963): 796–812.

38. Herstein, I. N., and I. Kaplansky. *Matters Mathematical*. New York: Harper & Row Pub., 1974, 121–128.

39. Howell, John M. "A Strange Game." *Mathematics Magazine* 47 (1974): 292–294.

40. James, Andrew, and Robert Moncada. "Many Set Color TV Lounges Show Highest Radiation." *Journal of Environmental Health* 31 (1969): 359–360.

41. Johnson, Norman L., and Samuel Kotz. *Discrete Distributions*. Boston: Houghton Mifflin Company, 1969.

42. Johnson, Norman L., and Samuel Kotz. *Continuous Univariate Distributions*. Vol. 1. Boston: Houghton Mifflin Company, 1970.

43. Johnson, Norman L., and Samuel Kotz. *Continuous Univariate Distributions*. Vol. 2. Boston: Houghton Mifflin Company, 1970.

44. Kahn, David. *The Codebreakers*. New York: Macmillan, 1967.

45. Kendall, Maurice G., and Alan Stuart. *The Advanced Theory of Statistics*. Vol. 1. New York: Hafner, 1958, 109–111.

46. Kolmogorov, A. *Foundations of the Theory of Probability*. 2nd ed. New York: Chelsea, 1956.

47. Larsen, Diane K. Personal communication, 1979.

48. Larsen, Richard J., and Morris L. Marx. *An Introduction to Mathematical Statistics and Its Applications*. Englewood Cliffs, N.J.: Prentice-Hall, 1981, 282–287.

49. Lévy, P. *Calcul des probabilités*. Paris, 1925.

50. Lindeberg, J. W. "Eine neue Herleitung des Exponentialgesetzes in der Wahrschein-lichkeitsrechnung." *Mathematische Zeitschrift* 15 (1922): 211–225.

51. Liu, C. L. *Introduction to Combinatorial Mathematics*. New York: McGraw-Hill, 1968.

52. Loève, M. *Probability Theory*. 3rd ed. New York: D. Van Nostrand, 1963.

53. Lomax, K. S. "Business Failures: Another Example of the Analysis of Failure Data." *Journal of the American Statistical Association* 49 (1954): 847–852.

54. Lukacs, E. *Characteristic Functions*. London: Charles Griffin, 1960.

55. Lyapunov, A. M. "Sur une proposition de la théorie des probabilités." *Izv. Akad. Nauk. Ser. 5*, Vol. 13 (1900): 359–386.

56. Lyapunov, A. M. "Nouvelle forme du théorème sur la limite de probabilité." *Mém. Acad. Sc. St-Pétersbourg* 12 (1901): 1–24.

57. Maguire, B. A., E. S. Pearson, and A. H. A. Wynn. "The Time Intervals between Industrial Accidents." *Biometrika* 39 (1952): 168–180.

58. "Medical News." *Journal of the American Medical Association* 219 (1972): 981. Copyright 1972, American Medical Association.

59. Minkoff, Eli C. "A Fossil Baboon from Angola, with a Note on *Australopithecus*." *Journal of Paleontology* 46 (1972): 836–844.

60. Mosteller, Frederick. *Fifty Challenging Problems in Probability with Solutions*. Reading, Mass.: Addison-Wesley, 1965, 73–77.

61. Munford, A. G. "A Note on the Uniformity Assumption in the Birthday Problem." *American Statistician* 31 (1977): 119.

62. *Newsweek,* March 6, 1978, 78.

63. *Newsweek,* May 5, 1980, 24–36.

64. Papoulis, Athanasios. *Probability, Random Variables, and Stochastic Processes.* New York: McGraw-Hill, 1965, 206–207.

65. Pareto, V. *Cours d'Economie Politique, Rouge et Cie.* Geneva: Bousquet and Busino, 1964.

66. Parzen, Emmanuel. *Modern Probability Theory and Its Applications.* New York: John Wiley, 1960.

67. Polya, G. "Über den zentral Grenzweitsatz der Wahrscheinlichkeitsrechnung und das Momentenproblem." *Mathematische Zeitschrift* 8 (1920): 171–181.

68. Polya, G. *Mathematical Discovery.* Vol. 1. New York: John Wiley, 1962, 68–73.

69. Premack, David. "Language in Chimpanzee?" *Science* 172 (1971): 808–827. Copyright 1971 by the American Association for the Advancement of Science.

70. Quetelet, L. A. J. *Lettres sur la Theorie des Probabilites, appliquee aux Sciences Morales et Politiques.* M. Hayez, Imprimeur de L'Academie Royal des Sciences, des Lettres et des Beaux-Arts de Belgique, Bruxelles, 1846, 400.

71. Reichler, Joseph L., ed. *The Baseball Encyclopedia.* 4th ed. New York: Macmillan, 1979, 1350.

72. Richardson, Lewis F. "The Distribution of Wars in Time." *Journal of the Royal Statistical Society* 107 (1944): 242–250.

73. Riordan, John. *An Introduction to Combinatorial Analysis.* New York: John Wiley, 1958.

74. Rochat, Roger W. "Cervical Cancer Screening: The Effect of Infrequently Occurring Disease on the Accuracy of Diagnosis." Presented to the Society for Epidemiological Research, Toronto, Canada, 1976.

75. Rohatgi, V. K. *An Introduction to Probability Theory and Mathematical Statistics.* New York: John Wiley, 1976, 81.

76. Rosenthal, David. *Genetic Theory and Abnormal Behavior.* New York: McGraw-Hill, 1970, 139.

77. Rudin, W. *Principles of Mathematical Analysis.* 2nd ed. New York: McGraw-Hill, 1964.

78. Rutherford, Sir Ernest, James Chadwick, and C. D. Ellis. *Radiations from Radioactive Substances.* London: Cambridge University Press, 1951, 172.

79. Savage, I. Richard. *Statistics: Uncertainty and Behavior.* Boston: Houghton-Mifflin Company, 1968.

80. Schell, Emil D. "Samuel Pepys, Isaac Newton, and Probability." *The American Statistician* 14 (1960): 27–30.

81. Sichel, Herbert S. "The Estimation of Parameters of a Negative Binomial Distribution with Special Reference to Psychological Data." *Psychometrika* 16 (1951): 107–127.

82. Srb, Adrian, Ray D. Owen, and Robert S. Edgar. *General Genetics.* 2nd ed. San Francisco: W. H. Freeman, 1965, 12–16.

83. *State Regulations for Protection Against Radiation.* Tennessee Department of Public Health, Division of Radiological Health, Nashville, 1978, 1200-2-6-.05(3)(C).

84. Strutt, John William (Baron Rayleigh). "On the Resultant of a Large Number of Vibrations of the Same Pitch and of Arbitrary Phase." *Philosophical Magazine* X (1880): 73–78.

85. *Tables of the Incomplete Γ-Function.* Edited by Karl Pearson. London: Cambridge University Press, 1922.

86. *Tennessean* (Nashville), Jan. 30, 1973.

87. Thatcher, A. R. "The Distribution of Earnings of Employees in Great Britain." *Journal of the Royal Statistical Society,* Series A. 131 (1968): 133–180.

88. Thorndike, Frances. "Applications of Poisson's Probability Summation." *Bell System Technical Journal* 5 (1926): 604–624.

89. Tucker, Alan. *Applied Combinatorics.* New York: John Wiley, 1980.

90. Tucker, Howard G. *A Graduate Course in Probability.* New York: Academic Press, 1967.

91. Vilenkin, N. Y. *Combinatorics.* New York: Academic Press, 1971, 24–26.

92. Vol'Kenschtein, Mikhail. *Molecules and Life.* New York: Plenum Press, 1973, 301–309.

93. Wadsworth, George P., and Joseph G. Bryan. *Applications of Probability and Random Variables.* 2nd ed. New York: McGraw-Hill, 1974, 276–278.

94. Walsh, John E. *Handbook of Nonparametric Statistics.* Princeton, N.J.: D. Van Nostrand, 1962, 54–69.

95. *Weather Atlas of the United States,* 1968. U.S. Department of Commerce, 61.

96. Whitworth, William Allen. *Choice and Chance.* New York: Hafner, 1965.

97. Wilks, Samuel S. *Mathematical Statistics.* New York: Wiley, 1962, 198–200.

98. Wiorkowski, John J. "A Curious Aspect of Knockout Tournaments of Size 2^m." *The American Statistician* 26 (1972): 28–30.

Appendix

CUMULATIVE AREAS UNDER THE STANDARD NORMAL DISTRIBUTION

Z	0	1	2	3	4	5	6	7	8	9
-3.	0.0013	0.0010	0.0007	0.0005	0.0003	0.0002	0.0002	0.0001	0.0001	0.0000
-2.9	0.0019	0.0018	0.0017	0.0017	0.0016	0.0016	0.0015	0.0015	0.0014	0.0014
-2.8	0.0026	0.0025	0.0024	0.0023	0.0023	0.0022	0.0021	0.0021	0.0020	0.0019
-2.7	0.0035	0.0034	0.0033	0.0032	0.0031	0.0030	0.0029	0.0028	0.0027	0.0026
-2.6	0.0047	0.0045	0.0044	0.0043	0.0041	0.0040	0.0039	0.0038	0.0037	0.0036
-2.5	0.0062	0.0060	0.0059	0.0057	0.0055	0.0054	0.0052	0.0051	0.0049	0.0048
-2.4	0.0082	0.0080	0.0078	0.0075	0.0073	0.0071	0.0069	0.0068	0.0066	0.0064
-2.3	0.0107	0.0104	0.0102	0.0099	0.0096	0.0094	0.0091	0.0089	0.0087	0.0084
-2.2	0.0139	0.0136	0.0132	0.0129	0.0126	0.0122	0.0119	0.0116	0.0113	0.0110
-2.1	0.0179	0.0174	0.0170	0.0166	0.0162	0.0158	0.0154	0.0150	0.0146	0.0143
-2.0	0.0228	0.0222	0.0217	0.0212	0.0207	0.0202	0.0197	0.0192	0.0188	0.0183
-1.9	0.0287	0.0281	0.0274	0.0268	0.0262	0.0256	0.0250	0.0244	0.0238	0.0233
-1.8	0.0359	0.0352	0.0344	0.0336	0.0329	0.0322	0.0314	0.0307	0.0300	0.0294
-1.7	0.0446	0.0436	0.0427	0.0418	0.0409	0.0401	0.0392	0.0384	0.0375	0.0367
-1.6	0.0548	0.0537	0.0526	0.0516	0.0505	0.0495	0.0485	0.0475	0.0465	0.0455
-1.5	0.0668	0.0655	0.0643	0.0630	0.0618	0.0606	0.0594	0.0582	0.0570	0.0559
-1.4	0.0808	0.0793	0.0778	0.0764	0.0749	0.0735	0.0722	0.0708	0.0694	0.0681
-1.3	0.0968	0.0951	0.0934	0.0918	0.0901	0.0885	0.0869	0.0853	0.0838	0.0823
-1.2	0.1151	0.1131	0.1112	0.1093	0.1075	0.1056	0.1038	0.1020	0.1003	0.0985
-1.1	0.1357	0.1335	0.1314	0.1292	0.1271	0.1251	0.1230	0.1210	0.1190	0.1170
-1.0	0.1587	0.1562	0.1539	0.1515	0.1492	0.1469	0.1446	0.1423	0.1401	0.1379
-0.9	0.1841	0.1814	0.1788	0.1762	0.1736	0.1711	0.1685	0.1660	0.1635	0.1611
-0.8	0.2119	0.2090	0.2061	0.2033	0.2005	0.1977	0.1949	0.1922	0.1894	0.1867
-0.7	0.2420	0.2389	0.2358	0.2327	0.2297	0.2266	0.2236	0.2206	0.2177	0.2148
-0.6	0.2743	0.2709	0.2676	0.2643	0.2611	0.2578	0.2546	0.2514	0.2483	0.2451
-0.5	0.3085	0.3050	0.3015	0.2981	0.2946	0.2912	0.2877	0.2843	0.2810	0.2776
-0.4	0.3446	0.3409	0.3372	0.3336	0.3300	0.3264	0.3228	0.3192	0.3156	0.3121
-0.3	0.3821	0.3783	0.3745	0.3707	0.3669	0.3632	0.3594	0.3557	0.3520	0.3483
-0.2	0.4207	0.4168	0.4129	0.4090	0.4052	0.4013	0.3974	0.3936	0.3897	0.3859
-0.1	0.4602	0.4562	0.4522	0.4483	0.4443	0.4404	0.4364	0.4325	0.4286	0.4247
-0.0	0.5000	0.4960	0.4920	0.4880	0.4840	0.4801	0.4761	0.4721	0.4681	0.4641

Z	0	1	2	3	4	5	6	7	8	9
0.0	0.5000	0.5040	0.5080	0.5120	0.5160	0.5199	0.5239	0.5279	0.5319	0.5359
0.1	0.5398	0.5438	0.5478	0.5517	0.5557	0.5596	0.5636	0.5675	0.5714	0.5753
0.2	0.5793	0.5832	0.5871	0.5910	0.5948	0.5987	0.6026	0.6064	0.6103	0.6141
0.3	0.6179	0.6217	0.6255	0.6293	0.6331	0.6368	0.6406	0.6443	0.6480	0.6517
0.4	0.6554	0.6591	0.6628	0.6664	0.6700	0.6736	0.6772	0.6808	0.6844	0.6879
0.5	0.6915	0.6950	0.6985	0.7019	0.7054	0.7088	0.7123	0.7157	0.7190	0.7224
0.6	0.7257	0.7291	0.7324	0.7357	0.7389	0.7422	0.7454	0.7486	0.7517	0.7549
0.7	0.7580	0.7611	0.7642	0.7673	0.7703	0.7734	0.7764	0.7794	0.7823	0.7852
0.8	0.7881	0.7910	0.7939	0.7967	0.7995	0.8023	0.8051	0.8078	0.8106	0.8133
0.9	0.8159	0.8186	0.8212	0.8238	0.8264	0.8289	0.8315	0.8340	0.8365	0.8389
1.0	0.8413	0.8438	0.8461	0.8485	0.8508	0.8531	0.8554	0.8577	0.8599	0.8621
1.1	0.8643	0.8665	0.8686	0.8708	0.8729	0.8749	0.8770	0.8790	0.8810	0.8830
1.2	0.8849	0.8869	0.8888	0.8907	0.8925	0.8944	0.8962	0.8980	0.8997	0.9015
1.3	0.9032	0.9049	0.9066	0.9082	0.9099	0.9115	0.9131	0.9147	0.9162	0.9177
1.4	0.9192	0.9207	0.9222	0.9236	0.9251	0.9265	0.9278	0.9292	0.9306	0.9319
1.5	0.9332	0.9345	0.9357	0.9370	0.9382	0.9394	0.9406	0.9418	0.9430	0.9441
1.6	0.9452	0.9463	0.9474	0.9484	0.9495	0.9505	0.9515	0.9525	0.9535	0.9545
1.7	0.9554	0.9564	0.9573	0.9582	0.9591	0.9599	0.9608	0.9616	0.9625	0.9633
1.8	0.9641	0.9648	0.9656	0.9664	0.9671	0.9678	0.9686	0.9693	0.9700	0.9706
1.9	0.9713	0.9719	0.9726	0.9732	0.9738	0.9744	0.9750	0.9756	0.9762	0.9767
2.0	0.9772	0.9778	0.9783	0.9788	0.9793	0.9798	0.9803	0.9808	0.9812	0.9817
2.1	0.9821	0.9826	0.9830	0.9834	0.9838	0.9842	0.9846	0.9850	0.9854	0.9857
2.2	0.9861	0.9864	0.9868	0.9871	0.9874	0.9878	0.9881	0.9884	0.9887	0.9890
2.3	0.9893	0.9896	0.9898	0.9901	0.9904	0.9906	0.9909	0.9911	0.9913	0.9916
2.4	0.9918	0.9920	0.9922	0.9925	0.9927	0.9929	0.9931	0.9932	0.9934	0.9936
2.5	0.9938	0.9940	0.9941	0.9943	0.9945	0.9946	0.9948	0.9949	0.9951	0.9952
2.6	0.9953	0.9955	0.9956	0.9957	0.9959	0.9960	0.9961	0.9962	0.9963	0.9964
2.7	0.9965	0.9966	0.9967	0.9968	0.9969	0.9970	0.9971	0.9972	0.9973	0.9974
2.8	0.9974	0.9975	0.9976	0.9977	0.9977	0.9978	0.9979	0.9979	0.9980	0.9981
2.9	0.9981	0.9982	0.9982	0.9983	0.9984	0.9984	0.9985	0.9985	0.9986	0.9986
3.	0.9987	0.9990	0.9993	0.9995	0.9997	0.9998	0.9998	0.9999	0.9999	1.0000

SOURCE: Reprinted with permission of Macmillan Publishing Company from *Statistical Theory* by B. W. Lindgren, pp. 391–393. Copyright © 1976 by B. W. Lindgren.

Answers to Selected Questions

CHAPTER 2

Section 2.2

1. $S = \{(17850, 17851), (17850, 17852), \ldots, (17854, 17855)\}$. Altogether, S contains 15 outcomes.

2. $(1, 3, 4), (1, 3, 5), (1, 3, 6), (2, 3, 4), (2, 3, 5), (2, 3, 6)$ **3.** $A = \{(b, c): b^2 < c\}$

5. (a) *Hint:* S contains eight outcomes. (b) $E = \{(H, H, H), (T, T, T)\}$. (c) If m coins are tossed, S will contain 2^m outcomes.

6. $A \otimes B = \{(3, 0), (3, 1), (5, 0), (5, 1), (7, 0), (7, 1)\}$ **7.** A cylinder of radius 1 and height 1.

8. Let A be the event "player makes at least three hits in a row." Then $A = \{(o, h, h, h),$ $(h, h, h, o), (h, h, h, h)\}$.   **10.** $A = (A_1 \cup A_2) \cap (A_3 \cup A_4)$

12. (a) $(A \cap B^C \cap C^C) \cup (A^C \cap B \cap C^C) \cup (A^C \cap B^C \cap C)$

14. $A \cap B$ contains 40 outcomes (there are 10 straight flushes in each of the four suits).

15. $(A \cup B)^C$ **17.** $B \subset A$

18. 15%. *Hint:* Think of a 2×2 table with the columns labeled "male" and "female" and the rows labeled "black" and "white." The information given is one row total, one column total, and one entry in the body of the table.   **19.** (a) $\cup_{i=1}^{k} A_i = \{x: 0 \le x < 1\}$

20. (a) $A \cap B^C \cap C$ refers to all females with tenure who are not math professors.

21. 1/3. *Hint:* Draw a Venn diagram. How many club members are either lawyers or liars or both?

24. (a) $A \cup B = \{U, V, X, Y\}$; (c) $A^C \cap B^C \cap C^C = \{Z\}$

27. The figures given are inconsistent. Let J, H, and N denote the numbers of fans who like Jennings, Haggard, and Nelson, respectively. Then $J \cup H \cup N$ *should* contain 96 people, but the information as given suggests that the three-way union includes 122 people, an impossibility.

Section 2.3

3. (a) $\frac{1}{12}$; (b) $\frac{11}{12}$; (c) $\frac{5}{12}$ 4. Since $p + p^2 = 1$, $p = (-1 + \sqrt{5})/2$.

6. $(A \cap B^C) \cup (A^C \cap B)$ is the event that exactly one of the two events, A and B, occurs.

9. The proof follows by comparing $1 - P(A^C) - P(B^C) = P(A) + P(B) - 1$ with the equality $P(A \cap B) = P(A) + P(B) - P(A \cup B)$ and noting that $P(A \cup B) \le 1$.

10. (a) $P(A^C \cup B^C) = 1 - P(A \cap B)$; (b) $P(A^C \cap (A \cup B)) = P(B) - P(A \cap B)$

Section 2.4

1. Mathematicians are like Frenchmen; Whatever you say to them, They translate into their own language/And forthwith it is something entirely different.—*Goethe*.

2. $P(\text{sum} = 7) = 0.1875$ if dice are loaded; $P(\text{sum} = 7) = 0.167$ if dice are fair.

3. $P(s)$ sums to 1 if $c = \frac{3}{4}$; $P(s \le 4) = P(2) + P(3) + P(4) = \frac{19}{27}$; $P(s \text{ is even}) = \sum_{k=1}^{\infty} (\frac{3}{4}) \cdot (\frac{2}{3})^{2k} = (\frac{3}{4}) \sum_{k=1}^{\infty} (\frac{4}{9})^k = \frac{3}{5}$.

5. $P(A \cup B) = P(A) + P(B) - P(A \cap B) = \frac{6}{12} + \frac{4}{12} - \frac{2}{12} = \frac{2}{3}$

6. $P(\text{odd man out}) = P(\text{exactly 1 head or exactly 1 tail}) = m/2^{m-1}$

7. If $P(i) = ki$, then $k = \frac{1}{21}$; $P(5 \text{ or } 6) = \frac{11}{21}$.   8. $12(\frac{1}{6})^3$

9. Expected frequency for $0 = 576 \cdot P(0) = 227.3$, etc.

10. *Hint:* $\sum_{k=0}^{\infty} \left(\frac{\lambda}{1 + \lambda}\right)^x = 1 / \left(1 - \frac{\lambda}{1 + \lambda}\right)$.

12. *Hint:* Are the three outcomes listed by D'Alembert equally likely?

Section 2.5

1. (a) $k = 3 \times 10^{-9}$; (c) *Hint:* Solve the equation $\int_0^{x_m} f(t)dt = 0.5$ for x_m.

2. 82%   3. $e^{-0.21} - e^{-0.42} \doteq 0.15$

4. $P(\text{batch is accepted}) = (1 - e^{-4})/(1 - e^{-9}) \doteq 0.982$

5. (a) $e^{-1} \doteq 0.37$; (b) Median life $= 693$ h   6. 0.56   8. $\sqrt{\pi}/4$

10. $\frac{3}{5}$ 11. $20 \cdot (0.215) \doteq 4.3$

12. $f(x, y) = \frac{1}{4\pi}$, for $x^2 + y^2 \le 4$, and 0, elsewhere; $P(A) = \frac{1}{2}$.

13. 0.487

14. 4.81. *Hint:* Draw the line $x + y = z_{80}$ over the rectangular sample space and get an expression for the area above the line in terms of z_{80}.

15. 0.6   17. $\frac{1}{20}$ 18. $\frac{3}{4}$

Section 2.6

1. No. If age is ignored, the three family types would not be equally likely—"(girl, boy)" would occur twice as often as either of the other two (Why?). The conditional probability, then, of both children being boys given that at least one is a boy would still be one third.

2. (a) $\frac{1}{13}$; (b) $\frac{1}{4}$; (c) $\frac{1}{10}$ 3. $2/m(m + 1)$

4. *Hint:* For any two events A and B, $P(A \cup B) \le 1$.

5. If $P(A|B) < P(A)$, then $P(A \cap B) = P(B|A)P(A) < P(A)P(B)$.

7. *Hint:* $P(B|A) + P(B^C|A) = 1$.   8. $\frac{3}{10}$ 9. 0.35   10. $\frac{7}{15}$

11. Consider the keys numbered 1 through n. A priori, each is equally likely to be the one capable of opening the door, so $P(K_i) = 1/n$, $i = 1, 2, \ldots, n$.

12. Since $P(x > s + t | x > t) = P(x > s)$, the exponential model assumes no wearout. (Over limited ranges of x, that is sometimes not a particularly unrealistic assumption.)

13. $P\{(w, w, w)\} = (\frac{2}{3}) \cdot (\frac{3}{4}) \cdot (\frac{4}{5})$, $P\{(b, w, w)\} = (\frac{1}{3}) \cdot (\frac{2}{4}) \cdot (\frac{3}{5})$, etc.

14. $\dfrac{P(A)}{P(A) + P(B)}$

16. 0.52   17. $\frac{5}{18}$

18. *Hint:* The sampling is random but are the six outcomes listed by Foster equally likely, as claimed?   **19.** 0.13   **20.** 0.74   **21.** 0.43   **22.** $\frac{2}{3}$
23. (a) 0.40; (b) 0.30; (c) 0.23   **24.** $\frac{44}{64}$ **26.** 0.46
27. Let B be the event "second card is an ace"; let A be the event "first card is an ace." Then $P(B) = P(B|A)P(A) + P(B|A^C)P(A^C) = \frac{4}{52}$. If 51 cards were turned over face down, what would P(fifty-second card is an ace) be?
28. *Hint:* This can be solved by using a continuous analog of Theorem 2.6.1. If B is an event dependent on the value of x, then $P(B) = \int_{-\infty}^{\infty} P(B|x)f(x)dx$, where $f(x)$ is the probability function for x. Let B be the event that the two points are less than a distance q apart.
30. $\frac{4}{7}$. Notice the similarity between Questions 2.6.27 and 2.6.30.
31. (a) P(two similar disorders) $= \frac{93}{160} > \frac{1}{2}$; (b) 91/132
32. *Hint:* Draw a Venn diagram showing two events A and B. How can $P(A \cap B^C)$ be reexpressed?
34. $\left(\frac{4}{9}\right) \cdot \left(\frac{4}{10}\right) \cdot \left(\frac{3}{11}\right) \cdot \left(\frac{5}{12}\right)$

Section 2.7

1. 0.015   **2.** 0.14   **3.** 0.84   **4.** 0.44   **6.** 0.48   **8.** 0.078   **10.** 0.54   **11.** 14
12. 0.25 if P(child is abused) $= \frac{1}{90}$; 0.029 if P(child is abused) $= \frac{1}{1000}$; 0.38 if P(child is abused) $= \frac{1}{50}$. Compare these figures with the pattern shown in Table 2.7.1.

Section 2.8

1. To make the intersection probabilities equal to the product of the row and column probabilities, the number of black females must equal 24. No, because the demographics of their staff may not reflect the demographics of their applicant pool.
2. No. P(fails both) $= 1 - P$(passes at least one) $= 0.37$. Why is it incorrect to say that P(fails both) $= (1 - 0.35)(1 - 0.40)$?
3. No. Is $P(A|B) = P(A)$?   **4.** 11   **5.** No
6. *Hint:* Start by using the identity proved in Example 2.2.14.
11. *Hint:* What does the union $A_1 \cup A_2 \cup \cdots \cup A_n$ represent?
15. 0.56. Probably not.   **16.** $4\left(\frac{1}{3}\right)^3\left(\frac{2}{3}\right) + \left(\frac{1}{3}\right)^4 = \frac{1}{9}$ **17.** 0.88   **18.** See Case Study 4.1.
19. $\frac{6}{36}$ **20.** $p^5 + 5p^4(1 - p) + 8p^3(1 - p)^2 + 2p^2(1 - p)^3$ **21.** $\frac{29}{90}$
22. P(different) $= 1 - P$(same) $= 1 - 0.375 = 0.625$. Independence. Yes, because people do not choose a spouse on the basis of blood type.
23. $\sum_{i=1}^{6}(i/21)^3 = \frac{441}{9261}$ **24.** Trip.   **25.** $1 - \left(\frac{4}{7}\right)^{rm}$ **26.** $\frac{1}{2}$
28. 0.43. *Hint:* The area of an inscribed equilateral triangle is $\left(\frac{3}{4}\right) \cdot r^2\sqrt{3}$.
29. $\frac{1}{36}$. *Hint:* If the X_i's are randomly chosen, what should be the probability associated with any particular ordering?
30. 0.57. *Hint:* Let A be the event "Amanda tells truth"; let B be the event "Bitsy tells truth." Let $Q = (A \cap B^C) \cup (A^C \cap B)$. Find $P(A|Q)$.
31. $(0.50)(0.8185)(0.50) = 0.20$

Section 2.9

1. 7   **2.** 0.075   **3.** 25   **4.** $1 - (0.95)^8 = 0.34$
5. $P(A \text{ wins}) = \frac{8}{14}$; $P(B \text{ wins}) = \frac{4}{14}$; $P(C \text{ wins}) = \frac{2}{14}$
6. $w/(w + r)$
7. After the fact, our estimate of the probability of the coin being fair is $1/(1 + 2^n)$.
8. 0.45
9. P(first "6" appears on kth roll) $= \left(\frac{5}{6}\right)^{k-1} \cdot \left(\frac{1}{6}\right)$; P(first "6" appears on even-numbered roll) $= \frac{5}{11}$
11. (a) $1 - (1 - p)^j$
12. 1. *Hint:* Let $p = P$(stock goes up 1 point on a given day) and let B be the event "client eventually goes bankrupt." Then $P(B) = 1 - p + p[P(B)]^2$.

CHAPTER 3

Section 3.2

1. $2 \cdot 2 \cdot 3 \cdot 2 = 24$ 2. $20 \cdot 9 \cdot 6 \cdot 20 = 21,600$ 3. 45; *aeu* and *cdx*
4. $4 \cdot 14 \cdot 6 \cdot 5 = 1680$ 5. $4 \cdot 3 \cdot 2 \cdot 1 = 24; 2 \cdot 2 = 4$
6. Duplication is inevitable because the total number of distinct initial sets is $26 \cdot 26 \cdot 26 = 17,576$, a figure smaller than the school's enrollment.
7. *Hint:* Consider two cases, numbers where the last digit is 0 and numbers where the last digit is a 2, 4, 6, or 8.
8. $4 \cdot 4 = 16$, if "nothing on" is considered a mode of operation; otherwise, 15.
10. $(8 \cdot 7 \cdot 6 \cdot 5)/365 \doteq 4.6$ 11. $6 \cdot 5 \cdot 4 = 120$ 13. $4! - 1 = 23$
14. $2 \cdot 4! \cdot 4! = 1152; 4 \cdot 3 \cdot 6! = 8640$ 15. $4! \cdot 5! = 2880$ 16. $5! = 120$
17. $(13!)^4; 13 \cdot (13)^4 \cdot (12!)^4$
18. $6!; (6!)^2$, assuming the teams have not yet been formed; if the teams *have* been formed, $6!$.
19. Write $(n + 1)! = (n + 1) \cdot n!$ and let $n = 0$.
20. 4, because $2^1 + 2^2 + 2^3 < 26$ but $2^1 + 2^2 + 2^3 + 2^4 > 26$.
21. $2^{20} = 1,048,576$ 22. 7^8 23. (a) $26^2 \cdot 10^4$; (b) $26^2 \cdot 10 \cdot 9 \cdot 8 \cdot 7$; (c) $26^2 \cdot (10^4 - 1)$
24. $10!/3! \, 2! \, 4! \, 1!$ 25. *FLORIDA*   26. $6!/3! \, 3!$ 27. $18!/15! \, 2! \, 1!$
28. $n!/r! \, w! \, b! \, g!$ 29. $(2n)!/n! \, (2!)^n = 1 \cdot 3 \cdot 5 \cdot 7 \cdots (2n - 1)$ 30. 5040
31. $8!; (8!)^2/4! \, 4!; (8!)^2$
32. (a) $11!/5! \, 4! \, 2!$; (b) No, because not all "5/4/2" records are equally likely—certain teams will be easier to beat than others.
33. $4!/(7!/3! \, (1!)^4) \cdot 11!/3! \, 2!(1!)^6 = 95,040$ 34. $6!/3! \, (1!)^3 + 6!/2! \, 2! \, (1!)^2 = 300$
35. $11 \cdot 10!/3! \, (1!)^7$
38. Consider $k!$ objects categorized into $(k - 1)!$ groups, each group of size k. By Theorem 3.2.3, the number of ways to arrange the $k!$ objects is $(k!)!/(k!)^{(k-1)!}$, and the latter must be an integer.
39. 12
40. $6!/2$. *Note:* The 2 appears because the necklace can be turned over, a reorientation that effectively eliminates half the circular permutations.
41. 2   42. 8   43. $6^4 \cdot 4^3/2 = 41,472$

Section 3.3

1. (a) $\binom{9}{2}\binom{4}{1} = 144$; (b) $9 \cdot 8 \cdot \binom{4}{1} = 288$ 2. $\binom{5}{2} = 10$ 4. $\binom{7}{2} = 21; \binom{2}{1}\binom{5}{1} = 10$
5. *Hint:* Use the formula for the sum of the first n integers.
6. $\binom{10}{6}\binom{15}{3} = 95,550$ 7. $\binom{5}{2} \cdot (4!)^2$
8. $\binom{11}{1} \cdot \binom{8}{3} = 3080$. *Note:* Think of each admissible path from, say, X to 0 as an ordered sequence of 11 steps. The number of different such paths is equal to the number of ways to choose where in the sequence the two steps "north" will come—namely, $\binom{11}{2}$.
10. $\binom{10}{5}/2$. Why do we need to divide by 2?
11. (a) $8!$; (b) $2 \cdot 4! \cdot 4!$; (c) $\binom{8}{4}$; (d) $\binom{8}{4} \cdot 4!$ 12. $\binom{6}{2} \cdot 26^2 \cdot 10^4$
13. 9, because $\binom{8}{4} < 120$ but $\binom{9}{4} > 120$.   14. $\binom{8}{4} \cdot \dfrac{7!}{4! \, 2! \, 1!}$
15. For the approximation to be adequate, x should be very small, in which case terms involving x^k can be ignored, even for relatively small k.
16. $\binom{4 + 3 - 1}{3} = 20$ 18. $\binom{10 + 4 - 1}{4} = 715$ 20. $\binom{3 + 20 - 1}{20} = 231$
22. $766,480$ 23. 2^8. No.   24. $\sum_{k=0}^{8} \binom{8}{k}k!$ 25. 24 26. $2^7 \cdot 5 = 640$

Section 3.4

2. $\binom{10}{5} = 252$. *Hint:* Superimpose Pascal's triangle on top the *ABRACADABRA* array. What do the entries in the triangle represent? Alternatively, every admissible path is an ordered sequence of five moves to the left and five moves to the right. By Theorem 3.2.3, the number of such sequences is $10!/5! \, 5!$, or 252.

3. $\left(\begin{array}{c}n\\k+1\end{array}\right) = \dfrac{n-k}{k+1}\cdot\left(\begin{array}{c}n\\k\end{array}\right)$

8. Consider the $n + 1$ equations, $\binom{0}{1} = \binom{1}{2} - \binom{0}{2}$, $\binom{1}{1} = \binom{2}{2} - \binom{1}{2}$, $\binom{2}{1} = \binom{3}{2} - \binom{2}{2}$, ..., $\binom{n}{1} = \binom{n+1}{2} - \binom{n}{2}$. Adding the left-hand sides gives $1 + 2 + \cdots + n$. All terms on the right-hand sides cancel when added except for $\binom{n+1}{2} = n(n + 1)/2$, the formula for the sum of the first n integers.

9. $(x + y)^n = \sum_{k=0}^{n} \binom{n}{k} x^k y^{n-k}$. Let $x = y = 1$.

10. Consider forming a sample from a set of n distinct objects. Each object can either be (1) part of the sample or (2) *not* part of the sample. By the corollary to Theorem 3.3.3, then, the number of different samples is 2^n. Equivalently, the total number of samples will necessarily equal the number of 0-object samples $(=\binom{n}{0})$, the number of 1-object samples $(=\binom{n}{1})$, and so on. It follows that $\binom{n}{0} + \binom{n}{1} + \cdots + \binom{n}{n} = 2^n$.

14. *Hint:* Write out the binomial expansion for $(x + y)^n$, differentiate twice, and set $x = y = 1$.

17. $n/2$ molecules in each chamber maximizes entropy.

Section 3.5

1. $1/8\cdot7\cdot6\cdot5 \doteq 0.0006$ **2.** $2(n!)^2/(2n)!$ **3.** $49\cdot48!\cdot4!/52! \doteq 0.00018$

4. $\binom{2}{1}^{50} / \binom{100}{50}$ **5.** $P(\text{sum} \le 5) = 10/216$

6. $13/\binom{15}{3} \doteq 0.028$. The relatively small probability of three infested trees being adjacent, just by chance, suggests that contagion might be a factor.

7. $2\cdot11\cdot13!/15! \doteq 0.10; \frac{1}{2}$

8. $7!/7^7$. The assumption being made is that all possible departure patterns are equally likely, which is probably not true, since residents living on lower floors would be less likely to ride the elevator than someone living on the eighth floor.

9. $2/2^7$. Answer is approximate, but reasonable if the numbers of each fur type available are large. An exact answer would require an application of the hypergeometric distribution (see Theorem 3.6.3).

10. 16 is the smallest n for which $1 - (0.9)^n > 0.8$.   **11.** (b) $3!/7!$

12. $\binom{5}{3}/(9 + \binom{5}{3}) = 10/19$ (Recall Bayes' rule.)

13. $\frac{1}{360}$. No, because wives are likely to be buried next to their husbands.

14. $\sum_{r=1}^{k} \binom{k}{r}\left(\dfrac{1}{n}\right)^r\left(\dfrac{r-1}{n}\right)^{k-r}$ **15.** $\left(\frac{2}{16}\right)\left(\frac{1}{15}\right)\left(\frac{1}{14}\right)\left(\frac{1}{13}\right)\left(\frac{3}{12}\right) \doteq 0.00001$

17. $\binom{k}{2}\cdot\dfrac{365\cdot364\cdots(365-k+2)}{365^k}$ **18.** $60/7!$ **19.** $6!/6^6$ **21.** $\binom{n}{2n-k}/2^{2n-k}$

22. $2^{10}/\binom{20}{10}$

23. The expression $\binom{12}{1}\binom{11}{1}\binom{10}{1}$ *orders* the three single cards, which would make the numerator inconsistent with the (unordered) denominator.

24. $10\cdot4/\binom{52}{5}; \left(10\cdot\binom{4}{1}^5 - 10\cdot4\right)/\binom{52}{5}$ **25.** $13\cdot\binom{4}{4}\cdot12\cdot\binom{4}{1}/\binom{52}{5}$

26. $3\cdot\binom{4}{1}\binom{4}{1}/\binom{47}{2} \doteq 0.044$ **27.** $\binom{12}{2}\binom{5}{3}\cdot4^3/\binom{52}{5}$ **28.** $\binom{32}{13}/\binom{52}{13}$

29. $\dfrac{13!}{3!\,6!\,4!\,0!}\cdot\dfrac{39!}{10!\,7!\,9!\,13!}/\dfrac{52!}{(13!)^4}$ **30.** $\left(\binom{2}{1}\binom{2}{1}\right)^4\binom{32}{4}/\binom{48}{12}$

31. *Hint:* Suppose 47 cards are dealt off the top of a deck, face down. Intuitively, what must be the probability of the bottom five cards being a full house? Why?

34. The U-shaped nature of the distribution is probably a bit of a surprise. Intuitively, we would expect the cumulative sums of most random sequences to be greater than zero half the time and less than zero half the time. What $p_{2j,\,2n}$ shows to the contrary is that extreme sequences—always positive or always negative—are the most likely.

Section 3.6

1. $P(\text{at least one six with 6 dice}) = 1 - \left(\frac{5}{6}\right)^6 = 0.66$; $P(\text{at least two sixes with 12 dice}) = 1 - \{\left(\frac{5}{6}\right)^{12} + 12\left(\frac{1}{6}\right)\left(\frac{5}{6}\right)^{11}\} = 0.62$; $P(\text{at least three sixes with 18 dice}) = 1 - \{\left(\frac{5}{6}\right)^{18} + 18\left(\frac{1}{6}\right)\cdot\left(\frac{5}{6}\right)^{17} + \binom{18}{2}\left(\frac{1}{6}\right)^2\left(\frac{5}{6}\right)^{16}\} = 0.60$

2. $\binom{4}{2}(\frac{1}{4})^2(\frac{3}{4})^2 = 0.21$ **3.** $\Sigma_{k=0}^{6} \binom{12}{k}(\frac{1}{2})^{12} = 0.61$

4. We are assuming that the success or failure of each helicopter is an independent event. In a battle context, independence may very well not hold.

5. $(\frac{99,999}{100,000})^{10,000}$ **6.** $P(Y = 3) = \binom{25}{3}(0.06)^3(0.94)^{22} = 0.13$

8. $\binom{4}{3}(\frac{1}{2})^3(\frac{1}{2})^1 = 0.25$ **10.** $P(Y = k) = \binom{n}{k}\left(\frac{b-a}{T}\right)^k\left(1 - \frac{b-a}{T}\right)^{n-k}$

11. P(better team wins "best two out of three") $= 0.575$; P(better team wins "best three out of five") $= 0.593$. Yes, the longer the series the greater the opportunity for the better team to prevail.

12. $P(Y = 2) = \binom{3}{2}(\frac{1}{27})^2(\frac{26}{27})^1 = 0.004$

13. $2(0.76)^{44}(0.24) + (160 - 44 + 1)(0.24)(0.76)^{44}(0.24) = 0.000041$

14. 7   **15.** 0.54   **16.** If Y is the number of boys, $P(Y = 1) + P(Y = 3) > P(Y = 2)$.

17. P(plane is shot down) $= 0.35$; P(boat is sunk) $= 0.40$

18. *Hint:* P(exactly n heads) $= \binom{2n}{n}(\frac{1}{2})^{2n} < e^{-\frac{1}{2}(1 + 1/2 + \cdots + 1/n)} < e^{-\frac{1}{2}\ln n} = 1/\sqrt{n}$.

19. *Hint:* If $m + n - 1$ additional games are played, meaning that either A or B would *have* to win, $P(A \text{ wins}) = \Sigma_{k=m}^{m+n-1} \binom{m+n-1}{k}p^k q^{m+n-1-k}$,

21. $\dfrac{10!}{5!\ 5!\ 0!\ 0!}\ (0.4)^5(0.3)^5(0.1)^0(0.2)^0 = 0.006$

22. $\dfrac{12!}{(2!)^6}\ (1/21)^2(2/21)^2(3/21)^2(4/21)^2(5/21)^2(6/21)^2 = 0.0005$

23. $\dfrac{9!}{2!\ 3!\ 3!\ 1!}\ (0.156)^2(0.344)^3(0.344)^3(0.156)^1 = 0.032$

24. $\dfrac{93!}{23!\ 50!\ 20!}\ (1/4)^{23}(1/2)^{50}(1/4)^{20}$

25. $\binom{54}{7}\binom{44}{0}\Big/\binom{98}{7} = 0.013$; $\binom{54}{6}\binom{44}{1}\Big/\binom{98}{7} = 0.082$; $\binom{54}{5}\binom{44}{2}\Big/\binom{98}{7} = 0.216$

26. 3   **28.** $\left\{\binom{4}{2}\binom{3}{2} + \binom{3}{2}\binom{4}{2} + \binom{2}{2}\binom{5}{2}\right\}\Big/\binom{9}{2}\binom{12}{2} = 0.019$

29. $\binom{10}{2}\binom{356}{1}\Big/\binom{366}{3}$ **30.** 9/20   **31.** $\binom{2}{1}\binom{1}{1}\binom{5}{2}\Big/\binom{8}{4}$

33. $\left\{\binom{2}{2} + \binom{6}{2} + \binom{2}{2}\right\}\Big/\binom{10}{2} = 0.38$

Section 3.7

1. Let $D(n)$ denote the number of derangements of n distinct objects. Then P(rearrangement has exactly r matches) $= \binom{n}{r} \cdot D(n - r)/n!$.

2. P(no matches) $=$ number of derangements$/n! \doteq e^{-1} = 0.37$

3. $1 - 14833/8! = 0.632$ (See Table 3.7.1.)

4. $\Sigma_{j=0}^{n} (-1)^j \binom{n}{j}(2n - j)!$ **5.** 14833

6. 0. *Hint:* Draw a Venn diagram showing four sets—Republicans, Democrats, liars, and cheats. What relationship does the Republican set have to the other three?

7. $10!\left[1 - \dfrac{1}{1!} + \dfrac{1}{2!} - \dfrac{1}{3!} + \cdots + \dfrac{1}{10!}\right] \doteq 10!\ e^{-1}$ **11.** 18

12. *Hint:* P(exactly r matches) $= [\binom{n}{r}/n!] \cdot$ number of derangements of $n - r$ objects.

CHAPTER 4

Section 4.2

2. $\frac{20}{50}$ **5.** $\frac{1}{4}$ **6.** $F_X(200) = 0.686$, $G_X(200) = 0.673$

7. $f_X(x) = \dfrac{\binom{4}{x}\binom{48}{5-x}}{\binom{52}{5}}$, $x = 0, 1, 2, 3, 4$ **8.** $f_Y(y) = \dfrac{\binom{5}{y}\binom{4}{3-y}}{\binom{9}{3}}$, $y = 0, 1, 2, 3$

9. $f_Y(y) = \binom{5}{y} (0.8)^y (0.2)^{5-y}$, $y = 0, 1, 2, 3, 4, 5$ **10.** $\dfrac{m(m+1)}{n(n+1)}$

11. $F_X(x) = \begin{cases} 0 & x < a \\ \dfrac{x+a}{b-a} & a \le x \le b \\ 1 & x > b \end{cases}$

13. (a) $F_X(t) = 1 - e^{-t}$; (b) $e^{-2} = 0.135$

14. (a) 0.9463; (b) $0.8770 - 0.6879 = 0.1891$; (c) $1 - 0.1093 = 0.8907$; (d) $z = 1.645$

15. $\frac{5}{16}$ **18.** (a) 0.648; (b) 0.005

19. $1 - F_X(7) = 0.001$. A small value for $1 - F_X(7)$ would tend to support the claim that the new treatment is better.

25. (a) ln 2; (b) 0.223; (c) 0.223; (d) $f_X(x) = 1/x$, $\quad 1 < x < e$ **26.** 0.971

27. $\binom{3}{x}\left(\frac{1}{4}\right)^x\left(\frac{3}{4}\right)^{3-x}$, $x = 0, 1, 2, 3$ **28.** (a) 0.135; (b) 0.118; (c) 0.517

30. (a) $1 - 4\left(x^2/2 + x/2 + 1/4\right)e^{-2x}$; (c) 0.938 **31.** $\binom{n}{r}(e^{-5\lambda})^r(1-e^{-5\lambda})^{n-r}$

32. $\alpha\, x_0^\alpha\, x^{-\alpha-1}$, $x \ge x_0$ **36.** $e^{-(2/3)\lambda t^{3/2}}$, $t > 0$

Section 4.3

1. (a) 0; (b) $\frac{2}{3}$ **2.** $\frac{90}{495}$

7. (a) $\dfrac{\binom{3}{x}\binom{2}{y}\binom{4}{3-x-y}}{\binom{9}{3}}$, $0 \le x \le 3$, $0 \le y \le 2$, $x + y \le 3$ **8.** $\frac{13}{50}$

9. $f_{X,Y}(x,y) = 1/4\pi$, $x^2 + y^2 \le 4$; otherwise, 0. **11.** (a) $\frac{3}{8}$; (b) $\frac{1}{2}$

12. $f_X(1) = \frac{1}{10}$, $f_X(2) = \frac{6}{10}$, $f_X(3) = \frac{3}{10}$ **13.** $f_X(x) = 1, 0 < x < 1; f_Y(y) = -\ln(y), 0 < y < 1$

14. $f_X(x) = 6x(1-x), 0 < x < 1; f_Y(y) = 3(1-y)^2, 0 < y < 1$

15. (a) $f_X(x) = \frac{1}{2}, 0 < x < 2; f_Y(y) = 1, 0 < y < 1$; (b) $f_X(x) = xe^{-x}, x > 0$; (c) $f_X(x) = 2e^{-2x}, x > 0$; (e) $f_X(x) = c(x + \frac{1}{2}), 0 < x < 1, c = 1$; (f) $f_X(x) = 2 - 2x, 0 < x < 1$; $f_Y(y) = 2y, 0 < y < 1$; (h) $f_X(x) = 2x, 0 < x < 1$

17. *Hint:* See Example 2.5.5.

18. (a) $\dfrac{6!}{2!\,2!\,2!} (\frac{1}{4})^2(\frac{1}{2})^2(\frac{1}{4})^2$; (b) $\frac{15}{64}$ **20.** (a) $c = 1$; (b) $\frac{1}{8}$ **21.** (a) $\frac{29}{156}$; (b) $\frac{1}{26}$

24. $\frac{1}{8}$ **25.** $\frac{95}{96}$ **26.** 0.264

27. $f_Z(z) = \begin{cases} z & 0 \le z \le 1 \\ 2-z & 1 \le z \le 2 \end{cases}$ **32.** $e^{-4.2} = 0.015$

34. (a) $f_{X,Y}(x, y) = (x + y), 0 < x < 1, 0 < y < 1$; (b) $f_{Y,Z}(y, z) = e^{-z}(\frac{1}{2} + y), z > 0, 0 < y < 1$; (c) $f_Z(z) = e^{-z}, z > 0$

35. $\binom{50}{10,15,15,10}(5/32)^{10}(11/32)^{15}(11/32)^{15}(5/32)^{10}$

36. (a) $f_{W,X}(w, x) = 4wx, 0 < w, x < 1$; (b) 3/8

Section 4.4

1. $f_{Y_1, Y_2, \ldots, Y_n}(y_1, \ldots y_n) = \lambda^{-n}e^{-(1/\lambda)\sum_{i=1}^n y_i}$ **3.** Yes.

4. $F_{X_1, X_2}(t_1, t_2) = \dfrac{t_1^2\, t_2^2}{16}$, $0 < t_1, t_2 < 2$

Section 4.5

1. $f_Y(-1) = \frac{3}{21}, f_Y(1) = \frac{12}{21}, f_Y(3) = \frac{6}{21}$ **3.** $f_Y(-1) = \frac{1}{3}, f_Y(5) = \frac{1}{3}, f_Y(14) = \frac{1}{3}$

4. $f_Y(y) = \frac{1}{3}, -7 < y < -4$ **5.** $f_Y(y) = e^{-(y-4)}, 4 < y$

6. $f_Y(y) = \left(\frac{-3}{4}\right)(y+3)(y+1), -3 < y < -1$

7. $Y = \dfrac{X - a}{b - a}$ **8.** $f_Y(y) = \dfrac{3}{16\,\pi^{3/2}}\sqrt{y},\ 0 < y < 4\pi$ **9.** $f_Y(y) = 1/2\sqrt{y},\ 0 < y < 1$

10. $f_Y(y) = e^{-y},\ y > 0$ **11.** $f_Y(y) = \sqrt{y}/18,\ 0 < y < 9$

12. $f_{X^3}(y) = 2(y^{-1/3} - 1),\ 0 < y < 1$ **13.** $f_{X^{-t}}(y) = \lambda e^{-\lambda y^{-1/t}} \cdot \dfrac{1}{t}\, y^{-t+1/t},\ y > 0$

16. (b) $f_Z(z) = e^{-(r+s)}\dfrac{(r+s)^z}{z!},\ z = 0, 1, 2, \ldots$ **17.** $f_Z(z) = \begin{cases} z^2 & 0 < z \le 1 \\ 2z - z^2 & 1 < z < 2 \end{cases}$

19. $f_Z(z) = (\tfrac{1}{2})z^2 e^{-z},\ z > 0$ **21.** $f_{X+Y}(z) = \tfrac{4}{3}\,[e^{-z} - e^{-4z}],\ z > 0$ **22.** $f_Z(z) = (\tfrac{1}{2})z^2 e^{-z},\ z > 0$

23. (a) $F_Z(z) = \begin{cases} z/2 & 0 < z < 1 \\ 1 - (1/2z) & 1 < z < \infty \end{cases}$; (b) $f_Z(z) = \begin{cases} 1/2 & 0 < z \le 1 \\ 1/2z^2 & 1 < z \end{cases}$

24. $F_Z(z) = \begin{cases} (1/2)z^2 & 0 \le z \le 1 \\ 1 - (1/2z^2) & 1 < z < \infty \end{cases}$ **25.** $f_Z(z) = 2 - 2z,\ 0 < z < 1$ **26.** $7/16$

Section 4.6

2. 0.015 **3.** $1/3$

4. $\dfrac{\dbinom{6}{y}\dbinom{4}{3 - x - y}}{\dbinom{10}{3 - x}},\ 0 \le x,\ 0 \le y,\ x + y \le 3$

9. (a) $(\tfrac{1}{4})\,(3 - x),\ 0 < x < 2$; (b) $(\tfrac{1}{2})\,[(6 - x - y)/(3 - x)],\ 2 < y < 4$

11. (a) 0.400; (b) 0; (c) $e^{-y} \cdot e^{x}$

12. $f_{Y|x}(y) = \dfrac{x + y}{x + 1/2},\ 0 < y < 1$ **14.** $(1/3)\,(2y + 2),\ 0 < y < 1$

15. (a) $f_X(x) = (2/5)\,[2x + (\tfrac{3}{2})],\ 0 < x < 1$; (b) $f_{Y|x}(y) = \dfrac{2x + 3y}{2x + 3/2},\ 0 < y < 1$; (c) $\tfrac{1}{2}$

16. $\tfrac{2}{3}$ **18.** $8x_1, x_2, x_3,\ 0 < x_1, x_2, x_3 < 1$

CHAPTER 5

Section 5.2

1. $\$100(\tfrac{1}{2}) + (-\$100 + \$200)(\tfrac{1}{2})^2 + (-\$100 - \$200 + \$400)(\tfrac{1}{2})^3 +$
$(-\$100 - \$200 - \$400)(\tfrac{1}{2})^3 = 0$

2. $E(S) = \displaystyle\sum_{s=0}^{\infty} s \cdot f_S(s) = (0.82)\sum_{s=1}^{\infty} \dfrac{e^{-0.82}(0.82)^{s-1}}{(s - 1)!} = 0.82$

3. $E(\text{length}) = 4(\tfrac{1}{8}) + 5(\tfrac{1}{4}) + 6(\tfrac{5}{16}) + 7(\tfrac{5}{16}) = 5.8$ games, if the teams are evenly matched.

4. 4.5

5. (a) *Hint:* Add the complements of $P(X \ge k)$ for several values of k and use the fact that $\Sigma_{k=0}^{\infty} P(X = k) = 1$.

6. 1.66 **7.** 2.5 **8.** $\$10.95$ **9.** (a) $\dfrac{c}{2 - c}$

10. Expected number of drawings $= \tfrac{1}{2} + \tfrac{1}{3} + \tfrac{1}{4} + \cdots$, a series that diverges.

12. *Hint:* Recall the condition that the sum defining $E(X)$ converge absolutely.

13. 2 **14.** 2.6

15. *Hint:* Factor r/p out of the expression for $E(X)$ and show that the sum remaining equals 1.

17. *Hint:* When will N be greater than $n - 1$, given the sequence $x_0, x_1, \ldots, x_{n-1}$?

18. (a) $\tfrac{3}{2}$; (b) $\tfrac{1}{4}$; (c) 1; (d) 1; (e) 1

20. *Hint:* $E(X) = \dfrac{\alpha}{2\pi}\left[\ln\left(1 + \dfrac{(x-\mu)^2}{\alpha^2}\right)\Big|_{-\infty}^{\infty}\right] + \mu$

Section 5.3

1. 25.5; 14.7   **2.** By symmetry, 9.6.   **3.** $\sum_{i=1}^{n} a_i = 1$ **4.** μ **5.** $n\left(\dfrac{p}{4} - \dfrac{1}{8}\right)$

6. 2   **8.** 112   **9.** $\dfrac{r(n+1)}{2}$; no.   **10.** 36   **11.** 35

12. 7.3. *Hint:* Let R be the event "red has not appeared in t throws"; let B be "black has not appeared in t throws," and G, "green has not appeared in t throws." Then $P(R \cup B \cup G) = P(\text{success is not achieved in } t \text{ throws}) = f(t) = (\frac{2}{3})^t + (\frac{5}{6})^t + (\frac{1}{2})^t - (\frac{1}{2})^t - (\frac{1}{3})^t - (\frac{1}{6})^t$. Also, $E(\text{number of tosses}) = \sum_{t=0}^{\infty} f(t)$. (Recall Question 5.2.5.)

13. *Hint:* See Case Study 5.1.

14. By Theorem 5.3.2, $E(X^3) = \int_0^1 x^3 \cdot 2(1-x)\,dx = \frac{1}{10}$; by finding the pdf for Y, $E(Y) = \int_0^1 y \cdot (\frac{2}{3}y^{-2/3} - \frac{2}{3}y^{-1/3})\,dy = \frac{1}{10}$.

15. \$50,000   **16.** $\frac{1}{4}$ **17.** 2

18. *Hint:* What is the probability of getting, for instance, a 6 on the first draw and a 4 on the second, assuming $n \geq 6$?

20. 12.25

Section 5.4

1. $\frac{12}{25}$ **2.** 1.36   **3.** $\frac{5}{6}$ **4.** 0.75   **5.** $\frac{1}{12}$ **6.** 0.58   **7.** $\frac{3}{80}$

9. 1.07. *Hint:* Y is binomial with $p = \int_0^{0.60} 6x(1-x)\,dx = 0.648$.   **10.** Choose $\dfrac{b+a}{2}$.

11. q/p^2

13. $E\left(\dfrac{X-\mu}{\sigma}\right) = \dfrac{1}{\sigma}[E(X-\mu)] = \dfrac{1}{\sigma}(\mu - \mu) = 0$

15. $\text{Var}(Y) = (1000)^2\text{Var}(X) = (1000)^2(\frac{1}{3})$ **16.** $\sigma_L = \{50(\frac{1}{32})^2 + 49(\frac{1}{16})^2\}^{1/2} = 0.49$ in.

17. 10.06   **18.** $E(T) = \dfrac{n(n+1)p}{2}$ **19.** $\text{Var}(\overline{X}) = (1/n)^2\sum_{i=1}^{n}\text{Var}(X_i) = \sigma^2/n$ **21.** 0.16

25. $\frac{-8}{484}$

26. $\dfrac{k(n^2-1)}{12}\left(1 - \dfrac{k-1}{n-1}\right)$. *Note:* If the numbers on the chips drawn were independent—that is, if the samples were taken *with* replacement—the variance would be $\dfrac{k(n^2-1)}{12}$.   **27.** 0.7 ft

Section 5.5

1. $E(X^r) = \dfrac{b^{r+1} - a^{r+1}}{(r+1)(b-a)}$ **2.** $6\lambda^3$ **5.** $E(X^{2k}) = n^k \cdot \dfrac{\Gamma(k + \frac{1}{2})\Gamma(n/2 - k)}{\Gamma(\frac{1}{2})\Gamma(n/2)}$, $2k < n$

7. 2   **9.** (a) -0.6; (b) -0.6; (c) -1.2 **13.** $\dfrac{e^{2t} - 2e^t + 1}{t^2}$

15. $\psi_X(t) = M_X(\ln t)$, because $t^X = e^{X \ln t}$.

17. $E(X^2) = np + n(n-1)p^2$. *Hint:* $M_X^{(2)}(t) = n(q + pe^t)^{n-1}pe^t + pe^t n(n-1)(q + pe^t)^{n-2}pe^t$.

18. *Hint:* Write $e^{tx} \cdot \dfrac{e^{-\lambda}\lambda^x}{x!} = e^{-\lambda} \cdot \dfrac{(\lambda e^t)^x}{x!}$ and use the series definition of e^x.   **20.** $e^{-\mu t}M_X(t)$

21. *Hint:* $E(X^r) = 0$ for r odd.

22. No, $M_Z(t) = (\lambda/(\lambda - t))^2$, which is not the form of the moment-generating function for an exponential random variable.

23. Yes, because $M_{X+Y}(t) = e^{-2\lambda + 2\lambda e^t} = e^{-\mu + \mu e^t}$ for $\mu = 2\lambda$.

Section 5.6

1. 0.88   **3.** 250
4. (a) $P(|X - \mu| \geq \sigma) = 0.47$; using the Bienaymé-Chebyshev inequality, $P(|X - \mu| \geq \sigma) < 1$.
6. *Hint:* Use the fact that $\text{Var}(\overline{X}) = \sigma^2/n$.

Section 5.7

1. $\dfrac{3x + 2}{6x + 3}$ **2.** $\frac{7}{6}$ **5.** $\dfrac{n + 3}{4}$ **7.** 21 days

CHAPTER 6

Section 6.2

1. (a) 0.865; (b) binomial, $n = 20{,}000$, $p = \frac{1}{10{,}000}$ **2.** 0.099   **3.** 0.095   **4.** 0.602
5. 6.9×10^{-12} **8.** 0.090   **9.** f_Z is Poisson with $\lambda = 10$ **10.** 0.154
11. 0.264. Yes, the assumption is reasonable.   **12.** 0.908   **13.** (a) 0.195; (b) 0.430

15.

Hits	Expected number
0	227.3
1	211.4
2	98.3
3	30.5
4	7.1
5^+	1.3

16. 0.472   **20.** 0.036   **21.** (a) 0.242; (b) 0.465

Section 6.3

1. (a) 0.0994; (b) 0.1210   **4.** 0.0465 Hypnosis seems to improve guessing ability.
5. 0.0013   **6.** Probability $= P(Z \geq 13.3) \doteq 0$. Conclusion: People postpone dying until after
their birthday.   **7.** 0.8944   **8.** 0.1802, using the continuity correction.
9. (a) $\displaystyle\sum_{k=75}^{80} \binom{100}{k}(0.7)^k(0.3)^{100-k}$; (b) 0.1840   **11.** 0.7745   **13.** 216   **14.** (a) 0.067; (b) 0.159
15. No. Probability of mileage exceeding 25,000 is 0.8413.   **16.** 0.16
17. (a) 8.71; (b) Variability   **18.** 24.8
19. Probability of molar length ≥ 9 mm $= 0.04$. It is unlikely the skull belongs to genus *Papio*.
20. (a) cutoff between A and $B = 83$; (b) cutoff between B and $C = 75$
21. (a) 0.0062; (b) 0.3372; (c) 0.5934   **22.** (a) 0.1357; (b) 0.1922
23. (a) 583.8; (b) *Hint:* use the formula for the standard deviation of a binomial pdf.   **24.** 1.28
26. 0.064

Section 6.4

3. 2   **5.** (a) 0.531; (b) 10   **6.** 125/1296

Section 6.5

1. 0.647   **5.** (a) $\binom{9}{4}(0.3)^5(0.7)^5$; (b) $(0.3)^5 + \binom{5}{4}(0.3)^5(0.7) + \binom{6}{4}(0.3)^5(0.7)^2$
7. (a) 0.988; (b) 17

Section 6.6

4. $\lambda^{-m}\dfrac{\Gamma(m + r)}{\Gamma(r)}$ **6.** $r = \frac{4}{7}$; $\lambda = \frac{2}{7}$
9. The pdf of Y is gamma with $r = 3$ and $\lambda = 0.001$.

CHAPTER 7

Section 7.2

3. *Hint:* First use the infinite analog of Question 2.5.11 to show that there is some $\epsilon > 0$ so that $|X - Y| > \epsilon$ on a set of positive probability.
5. *Hint:* Split up the Y_n as in the proof of Theorem 7.2.3.

Section 7.3

1. *Hint:* Compare $\sin(n\pi s)$ with $\sin(2n\pi s)$.   **2.** ϕ
4. $A^* = \{s \in S \,|\, s$ is in infinitely many of the $A_n\}$ **5.** $A_* = \{0\}$, $A^* = [-1, 1)$
6. *Hint:* Use Corollary 3 to Theorem 5.6.1 and Question 7.3.6.
8. *Hint:* For this particular sequence, $\bigcup_{k=n}^{\infty} C_{k,\epsilon}^c = C_{n,\epsilon}^c$

Section 7.4

2. Probability $= 0.49$
3. The Bienyamé-Chebyshev inequality does not give useful information in this case.
4. No.   **5.** No.

Section 7.5

2. (a) *Hint:* Perform a change of variables and use the gamma function. (b) $\int_0^\infty (1/2)e^{ty-\sqrt{y}}\, dy$ is not finite, since $ty-\sqrt{y} > 0$ for large y. (c) The terms of the series are $> (k + 1)^k t^k$, so they do not converge to 0.

Index